COMPREHENSIVE CHEMICAL KINETICS

COMPREHENSIVE

Section 1. THE PRACTICE AND THEORY OF KINETICS

Volume 1 The Practice of Kinetics

Volume 2 The Theory of Kinetics

Volume 3 The Formation and Decay of Excited Species

Section 2. HOMOGENEOUS DECOMPOSITION AND ISOMERISATION REACTIONS

Volume 4 Decomposition of Inorganic and Organometallic Compounds

Volume 5 Decomposition and Isomerisation of Organic Compounds

Section 3. INORGANIC REACTIONS

Volume 6 Reactions of Non-metallic Inorganic Compounds

Volume 7 Reactions of Metallic Salts and Complexes, and Organometallic Compounds

Section 4. ORGANIC REACTIONS (6 volumes)

Volume 9 Addition and Elimination Reactions of Aliphatic Compounds

Volume 10 Ester Formation and Hydrolysis and Related Reactions

Volume 13 Reactions of Aromatic Compounds

Section 5. POLYMERISATION REACTIONS (2 volumes)

Section 6. OXIDATION AND COMBUSTION REACTIONS (2 volumes)

Section 7. SELECTED ELEMENTARY REACTIONS (2 volumes)

Additional Sections

HETEROGENEOUS REACTIONS

SOLID STATE REACTIONS

KINETICS AND TECHNOLOGICAL PROCESSES

CHEMICAL KINETICS

EDITED BY

C. H. BAMFORD

M.A., Ph.D., Sc.D. (Cantab.), F.R.I.C., F.R.S.

Campbell-Brown Professor of Industrial Chemistry,

University of Liverpool

AND

C. F. H. TIPPER

Ph.D. (Bristol), D.Sc. (Edinburgh)

Senior Lecturer in Physical Chemistry,

University of Liverpool

VOLUME 4

DECOMPOSITION OF INORGANIC AND ORGANOMETALLIC COMPOUNDS

ELSEVIER PUBLISHING COMPANY

AMSTERDAM - LONDON - NEW YORK

1972

ELSEVIER PUBLISHING COMPANY
335 JAN VAN GALENSTRAAT
P. O. BOX 211, AMSTERDAM, THE NETHERLANDS

AMERICAN ELSEVIER PUBLISHING COMPANY, INC.
52 VANDERBILT AVENUE
NEW YORK, NEW YORK 10017

LIBRARY OF CONGRESS CARD NUMBER 73-151741
ISBN 0-444-40936-X

WITH 23 ILLUSTRATIONS AND 54 TABLES

PRINTED IN THE NETHERLANDS

COMPREHENSIVE CHEMICAL KINETICS

ADVISORY BOARD

Professor S. W. BENSON
Professor SIR FREDERICK DAINTON
Professor G. GEE
the late Professor P. GOLDFINGER
Professor G. S. HAMMOND
Professor W. JOST
Professor G. B. KISTIAKOWSKY
Professor V. N. KONDRATIEV
Professor K. J. LAIDLER
Professor M. MAGAT
Professor SIR HARRY MELVILLE
Professor G. NATTA
Professor R. G. W. NORRISH
Professor S. OKAMURA
Professor SIR ERIC RIDEAL
Professor N. N. SEMENOV
Professor Z. G. SZABÓ
Professor O. WICHTERLE

Contributors to Volume 4

D. A. ARMSTRONG — Department of Chemistry, Faculty of Arts and Science, University of Calgary, Calgary, Alberta, Canada

R. J. CVETANOVIC — Division of Chemistry, National Research Council of Canada, Ottawa, Canada

A. HAAS — Lehrstuhl für Anorganische Chemie II, Ruhr-Universität Bochum, 463 Bochum-Querenburg, Germany

J. L. HOLMES — Department of Chemistry, University of Ottawa, Ottawa, Canada

K. H. HOMANN — Lehrstuhl für Physikalische Chemie II, Eduard Zintl Institut, Technische Hochschule, Darmstadt, Germany

K. F. PRESTON — Division of Chemistry, National Research Council of Canada, Ottawa, Canada

S. J. W. PRICE — Department of Chemistry, University of Windsor, Windsor, Ontario, Canada

Preface

The rates of chemical processes and their variation with conditions have been studied for many years, usually for the purpose of determining reaction mechanisms. Thus, the subject of chemical kinetics is a very extensive and important part of chemistry as a whole, and has acquired an enormous literature. Despite the number of books and reviews, in many cases it is by no means easy to find the required information on specific reactions or types of reaction or on more general topics in the field. It is the purpose of this series to provide a background reference work, which will enable such information to be obtained either directly, or from the original papers or reviews quoted.

The aim is to cover, in a reasonably critical way, the practice and theory of kinetics and the kinetics of inorganic and organic reactions in gaseous and condensed phases and at interfaces (excluding biochemical and electrochemical kinetics, however, unless very relevant) in more or less detail. The series will be divided into sections covering a relatively wide field; a section will consist of one or more volumes, each containing a number of articles written by experts in the various topics. Mechanisms will be thoroughly discussed and relevant non-kinetic data will be mentioned in this context. The methods of approach to the various topics will, of necessity, vary somewhat depending on the subject and the author(s) concerned.

It is obviously impossible to classify chemical reactions in a completely logical manner, and the editors have in general based their classification on types of chemical element, compound or reaction rather than on mechanisms, since views on the latter are subject to change. Some duplication is inevitable, but it is felt that this can be a help rather than a hindrance.

Section 2 deals with reactions involving only one molecular reactant, i.e. decompositions, isomerisations and associated physical processes. Where appropriate, results from studies of such reactions in the gas phase and condensed phases and induced photochemically and by high energy radiation, as well as thermally, are considered. The effects of additives, *e.g.* inert gases, free radical scavengers, and of surfaces are, of course, included for many systems, but fully heterogeneous reactions, decompositions of solids such as salts or decomposition flames are discussed in later sections. Rate parameters of elementary processes involved, as well as of overall reactions, are given if available.

In Volume 4 the decompositions of inorganic and metal organic compounds are discussed (except for homonuclear diatomic molecules, considered in a later section). Chapter 1 covers hydrides (and deuterides) of oxygen, sulphur, nitrogen, boron, etc, Chapter 2 deals with oxides, sulphides and derivatives, Chapter 3 with halogens, halides and related molecules and finally in Chapter 4 carbonyls, alkyls, aryls and other compounds of metals and also elements such as silicon, phosphorus, arsenic and antimony are considered.

The Editors wish to express their sincere appreciation for the continued support and advice from the members of the Advisory Board.

Liverpool
October, 1971

C. H. BAMFORD
C. F. H. TIPPER

Contents

Chapter 1

Kinetics of the Homogeneous Decomposition of Hydrides

K. H. HOMANN AND A. HAAS

1. Introduction

In this review we try to give a critical survey of the kinetics of the decomposition of compounds consisting only of atoms of hydrogen or deuterium and another element, exclusive of hydrocarbons and hydrogen halides. This rather arbitrary selection of substances has to be understood in view of the contents of the whole monograph, the chapters of which are devoted to the decomposition reactions of oxides, halides, hydrides, etc. This section deals mainly with reactions involving only one molecular reactant, the emphasis being thus on the thermally induced unimolecular reactions in the gas phase. In nearly all cases the unimolecular reaction is an initiating step for a more or less complex decomposition mechanism. Consecutive steps will often involve reactions of atoms and free radicals with the parent molecule or, more frequently, with intermediate species. Our selection of specific reactions involved in thermal decomposition is certainly somewhat arbitrary. We have followed the policy of reviewing the unimolecular reaction and discussing the consecutive reactions of the parent molecule with active species only if the rates of these reactions are known and may contribute significantly to the consumption of the original molecule.

Some reactions of the type $H+\text{hydride} \rightarrow \text{hydride radical}+H_2$ have been studied, mainly at lower temperatures, with H atoms generated by an external source. There might be appreciable errors in extrapolation of these rate coefficients to temperatures where thermal decomposition takes place. In many cases only a lower or upper limit of the rate of consecutive reactions can be given, especially if the decomposition takes place at temperatures appreciably above 1000 °K. We will not discuss reaction mechanisms in detail which lead to untested rate phenomena nor those which are based upon product analysis without a well-defined time history. It is true, however, that no decomposition of a hydride consisting of more than two atoms has a mechanism which is fully understood and which can be completely described in terms of the kinetics of the elementary reactions.

A great many papers have been published on the photochemically induced decomposition of hydrogen compounds. Naturally, these experiments give no information about the kinetics of the first unimolecular reaction step. Wherever there is information about the kinetics of secondary reactions with the reactant

References pp. 42–45

molecule, the results of photochemical investigation will be included. Quantum yields, which only serve to test a mechanism without rendering kinetic data, will be disregarded. Although experimental results of unimolecular decomposition kinetics will frequently be reported there is no discussion of the theories of unimolecular reactions, since this has been given in Volume 2, Chapter 3. (For an excellent summary of unimolecular reactions of small molecules together with a satisfactory theoretical treatment, the reader is also referred to the paper by Troe and Wagner[1].)

The compounds of the elements with hydrogen are grouped into three fairly distinct classes, the volatile, the salt-like and the interstitial hydrides. The first class comprises hydrides of the elements from the third to the seventh group of the periodic table. The second class is formed by the hydrides of the alkali and alkaline earth metals and others including those of the rare earth metals. In the last group are the hydrogen compounds of some transition elements as well as those of uranium. In some cases the classification in either of the last two groups is not clear. However, we are not concerned with representatives of the salt-like and the interstitial hydrides, since none of them decompose under homogeneous conditions. Gaseous hydrogen is evolved when these compounds are heated in the solid or liquid phase. For example, copper hydride (CuH) is unstable at room temperature. The decomposition occurs *via* a gas–solid reaction without any formation of gaseous CuH during the pyrolysis[2]. Thus, we need only deal with the gas phase decomposition kinetics of some hydrides of the first class. These hydrides have covalent bonds and are formed from the following elements

Group:	III	IV	V	VI	VII
	B	(C)	N	O	F
	Al	Si	P	S	Cl
	Ga	Ge	As	Se	Br
	(In)	Sn	Sb	Te	I
		Pb	Bi	Po	

(Hydrocarbons and hydrogen halides are omitted since they will be dealt with elsewhere.) The chemical properties of most of these hydrides are rather well known, but this cannot be asserted for their decomposition kinetics. Some of them are very stable (H_2O, HF, NH_3) while others decompose easily at room temperature (TeH_2, PbH_4). A study of the homogeneous decomposition has only been undertaken for those elements inside the frame in the Table. The pyrolyses of the others have either been found to proceed heterogeneously or the kinetics is unknown.

Many of the elements in the table form hydrides with more than one heavy atom per molecule, for example H_2O, H_2O_2, or NH_3, N_2H_4, N_3H, etc. Accordingly, certain hydrides need not decompose into the elements but can form lower hydrides as products of pyrolysis. Since the different hydrides of one element have different thermal stabilities, the temperature of the reaction may be important in determining the reaction stoichiometry. Thus, the mechanism and kinetics may be a function of temperature. The tendency to form higher and polymeric hydrides before decomposing into the elements is most pronounced in the upper left part of the table including B, Al, Ga, C, Si, Ge; in contrast to this, higher hydrides of elements in the upper right corner of the table usually decompose homogeneously to form lower hydrides, *e.g.* $H_2O_2 \rightarrow H_2O$ $N_2H_4 \rightarrow NH_3$ which are more stable at the temperature of decomposition. Of the heavier elements (Sn, Pb, Sb, Bi, Se, Te, and Po), only hydrides having one central atom are known. Accordingly, the final reaction products are the elements. In describing the decomposition kinetics of the hydrides we shall proceed from right to left according to the table. If available, data on the deuterides will be cited.

2. Water and heavy water

Because of its high thermal stability compared to that of other hydrides, water does not decompose extensively below 2000 °K. Thus, at one atmosphere and 2500 °K it is only dissociated to the extent of 9 %. Accordingly, it is impossible to study the homogeneous decomposition by classical methods. It is only with the shock tube technique that the rates of pyrolysis of water and heavy water have been measured.

The overall stoichiometric equation for this decomposition leading to equilibrium depends on the temperature. A considerable amount of the final products are H, OH, and O. Bauer *et al.*[3] were the first to report an investigation of the water dissociation by the shock-tube method. The temperature range for this study was 2400–3200 °K. They followed the reaction by measuring the UV absorption of the hydroxyl radical produced during the decomposition. The apparent activation energy for the parameter $(1/[H_2O])(d[OH]/dt)$ of about 50 kcal.mole^{-1} seemed to indicate that the reaction

$$H_2O+M \rightarrow OH+H+M, \ \Delta H_0^0 = 118.0 \text{ kcal.mole}^{-1}* \quad (1)$$

was not the rate-determining step in the formation of OH. The authors assumed that a complicated mechanism of secondary reactions, involving hydrogen peroxide and HO_2, led to this result. Unless this mechanism and the rates of the reactions involved are known, the variation of hydroxyl concentration cannot be correlated with that of consumption of water.

The development of very fast infrared detectors in recent times make it pos-

* All reaction enthalpies are calculated according to JANAF Tables[7].

References pp. 42–45

sible to follow the emission of the water molecule at 2.8 μ during the reaction behind a shock front. Since the intensity of emission is proportional to the concentration of water under the conditions of decay, this method is more likely to give reliable results than the OH absorption method. Along these lines, Olschewski *et al.*[4] studied the unimolecular decomposition of water at temperatures between 2700 and 6000 °K. They used mixtures of water vapour and argon (H_2O, 0.02–0.2 %; Ar, 1×10^{-3}–6×10^{-2} mole.l^{-1}) and measured the emission of H_2O by means of a liquid nitrogen cooled indium-antimonide infrared detector. This system was calibrated using the known initial concentration of water directly behind the shock front. The rate coefficient for the unimolecular reaction in its second order region is given by

$$k_1 = -[\mathrm{Ar}]^{-1}[\mathrm{H_2O}]^{-1}\frac{\mathrm{d}[\mathrm{H_2O}]}{\mathrm{d}t}$$

according to reaction (1) with M = Ar. The temperature dependence of the right hand side of this expression as measured by Olschewski *et al.* is shown in Fig. 1. It is only above *ca.* 4500 °K that the rate of reaction (1) is sufficiently higher than that of the next consecutive reaction

$$\mathrm{H+H_2O \rightarrow OH+H_2}, \; \Delta H_0^0 = 14.75 \text{ kcal.mole}^{-1} \qquad (2)$$

so that the coefficient is k_1. Below 4500 °K one obtains a value of $2k_1$, since reaction (2) then immediately follows the unimolecular dissociation, so that two molecules of water are decomposed each time reaction (1) takes place. This is demonstrated by the occurrence of two parallel straight lines in Fig. 1. The transition region between the measured coefficients k_1 and $2k_1$ can be shifted by variation of the initial water concentration.

In a region of temperature where the consecutive reactions are much faster than the unimolecular step, an equilibrium between the species H_2O, OH, H_2, H and O_2 will soon be established and the disappearance of water is then governed by the complicated mechanism including (2) and the reactions

$$2\,\mathrm{OH} \rightleftharpoons \mathrm{H_2O+O}$$

$$\mathrm{O+OH} \rightleftharpoons \mathrm{O_2+H}$$

$$\mathrm{OH+H} \rightleftharpoons \mathrm{H_2+O}$$

Close to the shock front, however, reaction (1) will be followed immediately only by (2) so that for the initial rate of disappearance of water a coefficient $2k_1$ is measured. This is no longer true some distance behind the shock front, however.

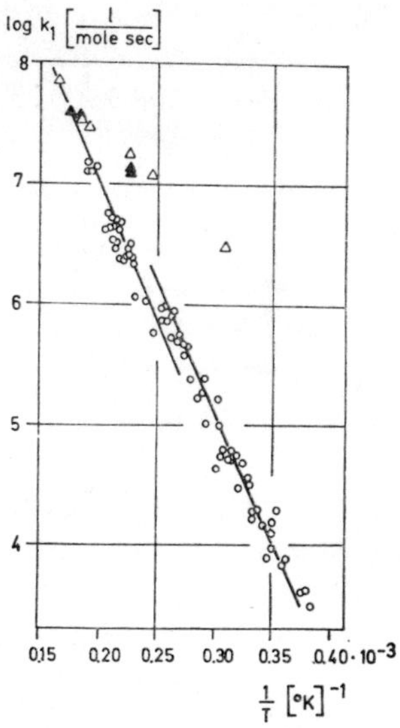

Fig. 1. Rate coefficients for the low-pressure region of the unimolecular decomposition of water Circles represent measurements by IR emission (2.8 μ); lower curve at higher temperatures, k_1; upper curve at lower temperatures, $2k_1$. Triangles represent measurements by UV absorption (3100 A), evaluated according to a rate law $c = c_{\infty}\{1-\exp(k_1[\mathrm{Ar}]t)\}$; $\triangle$: Ar $= 0.5\text{–}1\times10^{-2}$ mole.l^{-1}; $\blacktriangle$: Ar $= 2\text{–}3\times10^{-2}$ mole.l^{-1}. (From Olschewski *et al.*[4])

From Fig. 1 the value of k_1 can be given approximately by

$$k_1 = 5.0\times10^{11} \exp\left(\frac{-105{,}000}{RT}\right) \text{l.mole}^{-1}.\text{sec}^{-1}$$

for Ar = M. Another form of this expression is

$$k_1 = 1.7\times10^{18} T^{-\frac{1}{2}} \exp\left(\frac{-117{,}600}{RT}\right) \text{l.mole}^{-1}.\text{sec}^{-1}$$

Furthermore, Olschewski *et al.*[4] showed that results from IR emission and OH absorption are comparable only above 5000 °K, where the formation of OH is governed by reaction (1).

The decomposition of heavy water (D_2O) has been studied by the same authors at 5000 °K and an argon concentration of about 10^{-2} mole.l^{-1}. Preliminary measurements did not show any difference from the behaviour of normal water within the limits of experimental error.

A remark would be in order at this point concerning the calculation of the dissociation rate coefficient by means of the rate coefficient for recombination of H and OH and the equilibrium constant $K = [H_2O]/[H][OH]$. Getzinger[5] determined the rate coefficient for the recombination reaction H+OH+Ar →

H_2O+Ar using shock heated dilute hydrogen–oxygen mixture in argon. He obtained a value of $k_{rec} = 5.4 \pm 2.7 \times 10^9$ [$l^2.mole^{-2}.sec^{-1}$] without a systematic variation with temperature in the range 1400–1900 °K. In a similar experiment, Schott and Bird[6] obtained $4 \pm 2 \times 10^9$ for this coefficient. They also noted no marked dependence on temperature in the same range. Using the relation $k_{rec} = k_{diss} \times K$ (k_{diss} is obtained by extrapolation of the data of Olschewski *et al.*[4] and K is based on the JANAF Tables[7]), one calculates k_{rec} at 1400 °K to be 4.6×10^9 and at 1900 °K to be 1.2×10^9 $l^2.mole^{-2}.sec^{-1}$. This can be regarded as excellent agreement with the experimental results within their limits of error.

There is not as much agreement with the recombination rate coefficients that have been obtained from studies of hydrogen–oxygen–nitrogen flames. Padley and Sugden[8] deduced a rate coefficient for the termolecular reaction $H+OH+N_2 \rightarrow H_2O+N_2$ in atmospheric pressure flames at temperatures between 2085 and 2400 °K of about 1.2×10^{10} $l^2.mole^{-2}.sec^{-1}$. McAndrew and Wheeler[9] estimate (also from recombination rates in flames at 2080 °K) a value of 2×10^{10} $l^2.mole^{-2}.sec^{-1}$. The calculated rate coefficient ($k_{diss} \times K$) at an average temperature of 2200 °K is 7×10^8 $l^2.mole^{-2}.sec^{-1}$. The fact that argon was used in the dissociation experiments while in the flame work N_2 was the third body, could account only for a difference of a factor of about three in k_{rec}; this can be judged from the relative efficiencies of N_2 and Ar in other termolecular reactions. There is still a factor of ten to be accounted for.

3. Hydrogen peroxide (deuterium peroxide)

The thermal decomposition of hydrogen peroxide in the temperature range involved occurs according to the overall reaction

$$H_2O_2 = H_2O+\tfrac{1}{2}O_2, \ \Delta H_0^0 = -26.1 \text{ kcal.mole}^{-1}$$

It was recognized very early that the homogeneous reaction could hardly be studied at temperatures below about 400 °C, since heterogeneous decomposition is much faster than the gas-phase reaction at these temperatures. This fact has stimulated a search for treatments or coatings of the surface of the reaction vessel to provide a low and reproducible catalytic activity. Satisfactory results were obtained by rinsing pyrex or silica with 40 % hydrofluoric acid[10] and by coating the vessel surface with boric acid[11]. With these treatments a transition from a heterogeneous to a homogeneous reaction was observed between 400 and 450 °C.

The gas-phase reaction has been studied using static reaction systems[12,13], flow reactors[10,11,14,15] and, more recently, using the shock-tube technique[16,17]. The decomposition was followed in static experiments both measuring the amount of hydrogen peroxide decomposed and by observing the pressure increase.

To allow for the heterogeneous reaction, its rate was determined at lower temperatures (< 400 °C) and then extrapolated to the temperature of investigation of the homogeneous reaction. Forst[13] reported that the surface reaction on pyrex treated with sulphuric acid is almost completely inhibited by the addition of a large amount of helium. This specific influence of inert gases has yet to be re-examined by other investigators. The study of the reaction behind a shock wave avoids such heterogeneous decomposition.

The majority of investigators found the reaction to be first order in hydrogen peroxide. Accordingly, the decay of H_2O_2 can be described formally by the rate law

$$\frac{d[H_2O_2]}{dt} = -k_{homo}[H_2O_2]$$

The homogeneous rate coefficient k_{homo} is found to vary linearly with total pressure indicating that it is proportional to the rate coefficient of a unimolecular reaction in its second-order range. It is generally accepted that the initiating decomposition step is

$$H_2O_2+M \rightarrow 2\,OH+M,\ \Delta H_0^0 = 49.6 \text{ kcal.mole}^{-1} \tag{1}$$

This is followed immediately by the fast reaction

$$OH+H_2O_2 \rightarrow H_2O+HO_2,\ \Delta H_0^0 = -29.3\pm2 \text{ kcal.mole}^{-1} \tag{2}$$

with the HO_2 being consumed by the reactions

$$HO_2+HO_2 \rightarrow H_2O_2+O_2 \tag{3}$$

and

$$HO_2+OH \rightarrow H_2O+O_2 \tag{4}$$

Assuming rapid attainment of a quasi stationary state and that $k_2[H_2O_2] \gg k_4[HO_2]$, the relation between k_{homo} and k_1 is

$$k_{homo} = 2\,k_1[M]$$

A chain mechanism for the pyrolysis of H_2O_2 has been proposed by Satterfield and Stein[14] to account for the $\frac{3}{2}$ order observed by them. They include the reaction $HO_2+H_2O_2 \rightarrow H_2O+OH+O_2$ in the above scheme. However, other investigators have failed to observe the $\frac{3}{2}$ order reaction. Furthermore, the usual tests, addition of inhibitors etc., do not indicate the occurrence of a chain reaction.

References pp. 42–45

Recently, Meyer *et al.*[16,17] have studied the H_2O_2 pyrolysis by the shock-tube method. They measured the decay of H_2O_2 by following its absorption at wavelengths between 2300 and 2900 A in the temperature range 875–1425 °K. The mixtures used were about 1 % H_2O_2 in Ar at total densities between 3.8×10^{-2} and 2.2×10^{-1} mole.l^{-1}. As with the static experiments, a first-order rate law was found, with k_{homo} being proportional to the total density. The rate coefficient of the unimolecular reaction in its second-order range at temperatures above 1000 °K can be expressed by

$$k_1 = 4 \times 10^{12} \exp\left(\frac{-40{,}500}{RT}\right) \text{ l.mole}^{-1}\text{.sec}^{-1}$$

Although the concentrations of H_2O_2 in the shock-tube experiments were much lower than those used in earlier experiments, it appears probable that k_{homo} was equal to $2\,k_1[M]$. Since a measurable deviation from first-order kinetics does not occur, the rate of reaction (2) must be greater or less than that of reaction (1) by about a factor of five or more. This would mean that, if $2\,k_1$ was measured at a temperature of 1400 °K

$$k_2 \geqq 8.5 \times 10^9 \text{ l.mole}^{-1}\text{.sec}^{-1}$$

This calculation is for 1 % H_2O_2 in Ar with $[OH] \leqq 0.1[H_2O_2]$. If, on the other hand, only k_1 was measured, a similar estimation, now assuming that $[OH] \approx [H_2O_2]$, gives

$$k_2 \leqq 7 \times 10^7 \text{ l.mole}^{-1}\text{.sec}^{-1}$$

as an upper limit. Meyer *et al.* observed an absorption by the reacting gases in the region below 2500 A which might be attributed to the HO_2 radical. The comparatively rapid rise of this absorption and the fact that the absorption of OH at 3064 A was beyond the limit of detectability supports the assumption that twice k_1 has been measured. From flash photolysis of H_2O_2 vapour, Greiner[18] estimates an upper limit for k_2 of $(3.5 \pm 1.1) \times 10^7$ l.mole^{-1}.sec^{-1} at room temperature.

Baldwin *et al.*[19] determined the ratio $k_2/k_{(OH+H_2)}$ to be 7.1 from studies of the slow H_2/O_2 reaction in aged boric acid-coated vessels at 773 °K. From an investigation of the hydrogen peroxide decomposition in the presence of hydrogen Baldwin and Bratton[15] obtained for the same ratio at 713 °K a value of about 6. A reasonable value for $k_{(OH+H_2)}$ at 773 °K appears to be 8×10^8 l.mole^{-1}.sec^{-1}, averaging from measurements of Fenimore and Jones[20], Dixon-Lewis and Williams[21] and Kaufmann and Del Greco[22]. This would give $k_2 \approx 5 \times 10^9$ at 773 °K so that k_2 at 1400 °K can surely assumed to be greater than 10^{10} l.mole^{-1}.

TABLE 1

EXPERIMENTAL CONDITIONS AND RESULTS OF DIFFERENT WORKERS ON THE HOMOGENEOUS DECOMPOSITION OF HYDROGEN PEROXIDE

Method	*Reaction vessel*			*Temperature of transition hetero to homo (°K)*	*Pressure range (torr)*		*Temperature range (°K)*	*Inert gases*	*Reaction order in H_2O_2*	*Kinetic parameters*	*Ref.*
	Surface treatment	*Diam. (cm)*	*Length (cm)*		*Total*	*H_2O_2*					
Flow system	Boric acid	0.6 2.5	32–105 4.3	Not determined	760	1.5	743–813	N_2, O_2	1	$E \approx 50$ kcal.mole^{-1}	11
Flow system	Pyrex rinsed with 4 N H_2SO_4	1	92.5	675–725	760	15.2	490–765	H_2O	1.5	$E \approx 55$ kcal.mole^{-1}	14
Static system	Vycor, pyrex Hot H_2SO_4 then conc. H_2O_2		Flask 2000 cm^3	~ 675			575–875		1 (Pressure increase measured)	$k_{homo} = 10^{13} \exp\left(\frac{48,000}{RT}\right)$ sec^{-1}	12
Static system	Vycor, pyrex Hot H_2SO_4 then conc. H_2O_2		Spherical flash 2000 cm^3		10–100	6–22	705–742	He, O_2 H_2O	1 (Pressure increase measured)	$k_1(M = H_2O) = 10^{(16 \pm 0.9)} \exp\left\{-\frac{48.1 \pm 2.9 \times 10^3}{RT}\right\}$ $k_1(M = He) = 1.6 \times 10^{(10+0.9)} \exp\left(-\frac{42,500}{RT}\right)$ l.mole^{-1}.sec^{-1}	13
Flow system	Pyrey, silica rinsed with 40 % HF	3.8	20	~695	1.6–7.0; 760	0.27–3.1	515–750 (760 torr) 842–932 (low pressure)	He, O_2, N_2, CO_2, H_2O	1	$k_1(M = H_2O_2) = 5 \times 10^{15} \exp\left(-\frac{48,000}{RT}\right)$ l.mole^{-1}.sec^{-1}	10
Flow system	Boric acid (aged)	0.6 1.2 2.4	20	~715	25–760	0.1–1	715–835	N_2, O_2 H_2O	1	$k_1(M = N_2) = 8.5 \times 10^{13} \exp\left(-\frac{45,500}{RT}\right)$ l.mole^{-1}.sec^{-1}	15
Shock tube					~3–18 atm	1 % of total	875–1425	Ar	1	$k_1(M = Ar) = 4 \times 10^{12} \exp\left(-\frac{40,500}{RT}\right)$ l.mole^{-1}.sec^{-1} for $T = > 1000°$ K	16, 17

TABLE 2

RESULTS OBTAINED BY DIFFERENT WORKERS FOR THE EFFICIENCIES OF M RELATIVE TO N_2 AS UNITY

Molecule M	$H_2O_2+M \rightarrow 2\,OH+M$		
	Forst[13]	*Hoare et al.*[10]	*Baldwin and Bratton*[15]
H_2O_2	5.4	5.9	6.6
H_2O	4.0	4.3	6.0
O_2	(0.78)	0.71	0.78
Ar			0.67
He	0.57	0.53	
CO_2		1.24	

sec^{-1}. This is in agreement with the lower limit for k_2 if the decay of H_2O_2 is governed by 2 k_1.

Table 1 summarizes some details of experiments and results of different investigators of the homogeneous decomposition of H_2O_2. Since k_1 is the rate coefficient in the second-order regime of the unimolecular reaction, it is dependent on the nature of M. Table 2 (after Baldwin and Bratton[15]) gives the values of k_1 for different added gases relative to k_1 with M as nitrogen. For the table it is assumed that, in every case, $2k_1$ has been measured experimentally.

In Fig. 2 an Arrhenius plot is given for $k_1(M = Ar)$ in the temperature range 720–1425 °K, combining the results of all authors who observed first-order kinetics for the decomposition of H_2O_2. The line appears to be slightly curved corresponding to an activation energy of 45.5 kcal.mole^{-1} below 900 °K and approaching the

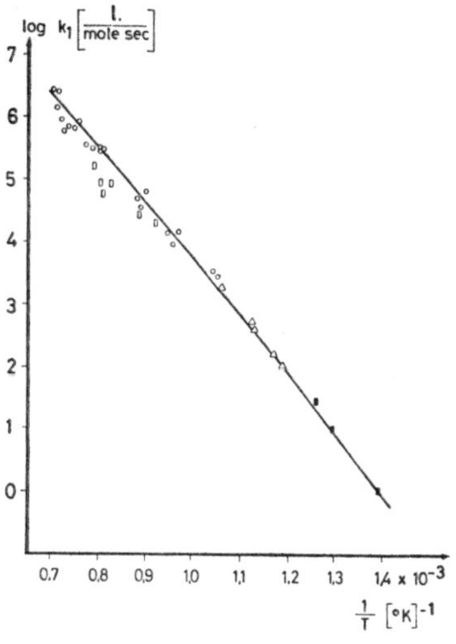

Fig. 2. Rate coefficients for the low-pressure region of the unimolecular decomposition of hydrogen peroxide. △, Ref. 10; ■, ref. 15; ○, refs. 16, 17 (total density 10^{-2} mole.l^{-1}); □, refs. 16, 17 (total density 10^{-1} mole.l^{-1}).

value of about 40.5 kcal.mole^{-1} above 1050 °K. However, this curvature might still be within the limits of experimental error. The values for k_1 represented by the symbol □ have been obtained from experiments at total densities of 10^{-1} mole.l^{-1}. They are systematically low in this plot and indicate a transition to the high-pressure region of the unimolecular reaction.

The pyrolysis of D_2O_2 has so far only been studied by Giguère and Liu[12], who found no difference in the reaction rate compared to that of H_2O_2 within the limits of experimental error.

4. Hydrogen sulphide

Very little data on the homogeneous decomposition of hydrogen sulphide are available in the literature up to 1967. Darwent and Roberts[23] published some results on the pyrolysis of H_2S in a quartz vessel (10 cm long, 5 cm in diameter). The experiments were carried out with pure H_2S at a pressure between 30 and 300 torr and temperatures ranging from 770 to 970 °K. The reaction was followed by measuring the formation of hydrogen. Below 900 °K hydrogen was produced mainly heterogeneously, while at higher temperatures the reaction rate became fairly independent of the type of surface and its area. This suggests but does not establish the occurrence of a homogeneous reaction. The formation of hydrogen was found to be second order with respect to the H_2S concentration throughout the pressure range studied. The dependence on temperature of the reaction rate above 900 °K gives an apparent activation energy of about 50 kcal.mole^{-1}. This cannot be reconciled with the unimolecular reaction

$$H_2S+M \rightarrow HS+H+M, \ \Delta H_0^0 = 87.65 \text{ kcal.mole}^{-1}$$

as a rate-determining step for hydrogen formation, and hardly with the reaction

$$H_2S+M \rightarrow H_2+S+M, \ \Delta H_0^0 = 69.2 \text{ kcal.mole}^{-1}.$$

A chain reaction involving S and H atoms and SH radicals may be operative. The authors discuss a bimolecular step $2\,H_2S \rightarrow 2\,H_2+S_2$ but it is hard to see how this one-reaction mechanism would take place.

Shock-tube experiments on the decomposition of hydrogen sulphide have been performed but were unsuccessful because traces of oxygen and other oxidizers could not be removed from the reactant[24]. No data are available on the homogeneous decomposition of hydrogen polysulphides, nor have the kinetics of pyrolysis of selenium and tellurium hydrides been studied.

5. Ammonia and deuterated ammonia

5.1 INITIATION REACTIONS

The homogeneous pyrolysis of ammonia can only be studied at temperatures above 2000 °K. This relatively high temperature is required because the NH_2–H bond dissociation energy is near 100 kcal.mole^{-1}. Furthermore, the heterogeneous decomposition studied on various surfaces is much faster than the homogeneous reaction in the temperature region that can be covered by classical methods. The rate of the heterogeneous reaction strongly depends on the kind of surface, but none has been found which is so inert to ammonia that the homogeneous decomposition can be studied in a conventional vessel. Bodenstein and Kranendieck[25] studied the pyrolysis in silica vessels and found exclusively heterogeneous reaction; the same result was obtained by Hinshelwood and Burke[26], who extended these studies to a temperature of 1050 °C.

Data on the rate of the homogeneous reaction have been obtained by following the decay of ammonia behind shock waves. The stoichiometry of the ammonia decomposition is

$$NH_3 = \tfrac{1}{2}N_2 + \tfrac{3}{2}H_2,\ \Delta H_0^0 = 9.36 \text{ kcal.mole}^{-1}$$

According to the dissociation equilibrium $H_2 \rightleftharpoons 2$ H, part of the hydrogen will be present as atoms after the decomposition at temperatures of some thousand degrees K.

Shock tube experiments by Jacobs[27] have shown that it is essential to purify the ammonia and the diluent from oxygen or other oxidizing components, otherwise oxidation would seriously interfere with decomposition. Jacobs followed the decay of ammonia through its infrared emission at 3 μ in the temperature range 2100–3000 °K. He argued that an assumed reaction order of $\frac{3}{2}$ in ammonia and of $\frac{1}{2}$ in the inert gas would best fit the observed concentration–time records, *i.e.*

$$\frac{d[NH_3]}{dt} = -k_{homo}[NH_3]^{\frac{3}{2}}[Ar]^{\frac{1}{2}}$$

From an Arrhenius plot for k_{homo}, an expression

$$k_{homo} = 2.5 \times 10^{13} \exp\left(\frac{-77{,}700}{RT}\right) \text{ l.mole}^{-1}.\text{sec}^{-1}$$

results. The $\frac{3}{2}$ order makes it clear that the mechanism of the decomposition is complicated, involving consecutive reactions. This might be expected for studies such as these, carried out with high ammonia concentrations (8 % NH_3 in Ar).

Due to the heat of reaction there was an uncertainty in the temperature of up to 150° so that the results obtained with high concentrations of ammonia might be subject to some criticism. The $\frac{1}{2}$ order with respect to the diluent can hardly be reconciled with a reasonable reaction mechanism and the author himself does not put much weight on this result.

While Jacobs followed the decay of ammonia by its IR emission, Michel and Wagner[28] used the absorption of the molecule at 2300–2500 A. The advantage of observing the concentration of NH_3 in a spectral region where the absorption coefficient is high lies in the ability to follow the reaction using initial concentrations of less than 0.1 % in argon. Under these conditions, a first-order reaction is obtained, the partial density of NH_3 being varied between 1.3×10^{-5} and 150×10^{-5} mole.l^{-1}. The first-order rate coefficient is proportional to the argon density which covered a range of 1.2×10^{-2}–10×10^{-2} mole.l^{-1}. The fact that the reaction rate varies linearly with the total density of an inert gas is evidence for the measured rate coefficient being closely connected to that of a unimolecular reaction in its second order range. The second-order rate coefficient

$$k = -\frac{1}{[NH_3]}\frac{1}{[Ar]}\frac{d[NH_3]}{dt}$$

could be expressed in the temperature region 2100–2900 °K according to the Arrhenius equation

$$k = 4.4 \times 10^{12} \exp\left\{\frac{-(79.5 \pm 2.5) \times 10^3}{RT}\right\} \text{l.mole}^{-1}.\text{sec}^{-1}$$

The studies of Michel and Wagner, however, give no information about the mechanism of the reaction, so that the measured rate coefficient, k, cannot be correlated safely with that of an initiating unimolecular step such as

$$NH_3 + M \rightarrow NH_2 + H + M, \Delta H_0^0 = 102 \pm 4 \text{ kcal.mole}^{-1} \tag{1}$$

If it is assumed that the unimolecular decomposition of NH_3 can be associated with an activation energy of about 80 kcal.mole^{-1}, then there is an obvious divergency with the results obtained from theoretical calculations, if (1) is supposed to be the observed reaction[1,29]. Independently of the absolute value of the dissociation energy, the calculation gives an activation energy of the unimolecular reaction in its second-order region of about 11 kcal.mole^{-1} less than the dissociation energy[30].

A second possible initiating reaction is

$$NH_3 + M \rightarrow NH + H_2 + M, \Delta H_0^0 \approx 88 \text{ kcal.mole}^{-1} \tag{2}$$

References pp. 42–45

The activation energy as determined by Michel and Wagner would tend to favour reaction (2) if it corresponds to a unimolecular step. The appearance of the NH radical immediately after the beginning of the reaction behind the shock front lends support to the supposition that reaction (2) might be involved in the decomposition[30,31]. There are reasons, however, which suggest that the energy barrier for this reaction path is higher than the heat of reaction[30].

A comparative study of the decomposition of NH_3 and ND_3 was reported by Henrici[30] using an experimental technique similar to that of Michel and Wagner[28]. The concentration of NH_3 (ND_3) was measured by means of its absorption at 2300 A and that of NH (ND) at a wavelength of 3360 A. The temperature range for the study was from 2100 to 2900 °K. The results do not reveal, within the limits of experimental error, any difference in the reaction rates upon deuteration. No differences in the shape of time–concentration records (NH *versus* ND radicals) nor in the rate relationships with respect to the parent molecules were observed. An isotopic effect of more than a factor of two should have been recognized. It would seem possible to obtain additional information on the mechanism of the first step by isotope experiments in the high-pressure region of the unimolecular reaction.

An experimental activation energy which seems to be too low from a theoretical point of view may be caused by an early acceleration due to a chain reaction. However, it is not clear what mechanism would be operative in this case. The possibility that the intermediate formation of hydrazine plays a role cannot be fully excluded.

In principle, there are two ways of suppressing the influence of secondary reactions: (*a*) to extend the measurements to higher temperatures where the unimolecular reaction is much faster than its consecutive reactions, or (*b*) to use lower concentrations of ammonia in the temperature region below *ca.* 2500 °K. Henrici[30] re-examined the ammonia decomposition at temperatures between 2200 and 2400 °K using mixtures with as low as 0.03–0.07 % NH_3 in Ar at total gas densities of about 3×10^{-2} mole.l^{-1}. He observed an induction period for the NH_3 decay immediately behind the shock front. At longer times the reaction accelerated to rates corresponding to those obtained by Michel and Wagner[28] who used higher initial NH_3 concentrations. If the rate coefficients from the decay during the induction period are plotted together with the high-temperature rate coefficients using an Arrhenius equation (Fig. 3), an activation energy of 91 kcal. mole^{-1} results, which agrees excellently with that predicted theoretically for reaction (1). The uncertainty in this value may well be ± 5 kcal.mole^{-1}, but it now seems highly probable that the activation energy of the unimolecular reaction is larger than 86 kcal.mole^{-1}. This result favours decomposition by reaction (1), but does not quite rule out reaction (2).

Figure 3 shows that the unimolecular decay is most probably measured at temperatures above 2900 °K. However, above 3300 °K the reaction rate becomes

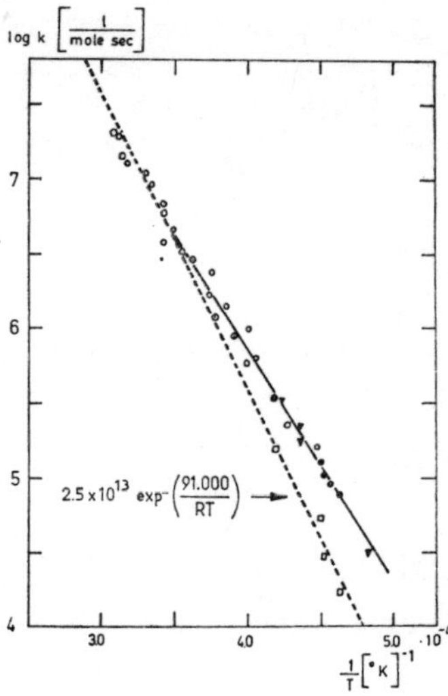

Fig. 3. Rate coefficients of the unimolecular decay of ammonia in the low-pressure region ▼ 0.4–0.5 × 10^{-2} mole.l^{-1} Ar, 0.4–0.6 % NH_3; ○ 0.9–1.2 × 10^{-2} mole.l^{-1} Ar, 0.4–0.6 % NH_3; ⊙ 2.8–3.8 × 10^{-2} mole.l^{-1} Ar, 0.6 % NH_3; □ 2.8–3.8 × 10^{-2} mole.l^{-1} Ar, 0.07 % NH_3 (during induction period); ● 2.8–3.8 × 10^{-2} mole.l^{-1} Ar, 0.07 % NH_3 (after induction period). (From Henrici[30].)

too high to be determined accurately; the temperature range of 400° is too small to give a precise activation energy. At temperatures below 2800 °K the reaction is soon accelerated by secondary reactions the rates of which are dependent on the initial NH_3 concentration. A transition in the mechanism of decay at a temperature of about 2800 °K has been reported by Avery and Bradley[32] from the measurements of NH emission during the reaction. The rate of ammonia consumption is sensitive to oxygen impurities. This effect might be the cause for the scatter of the rate coefficients after the induction period at lower temperatures.

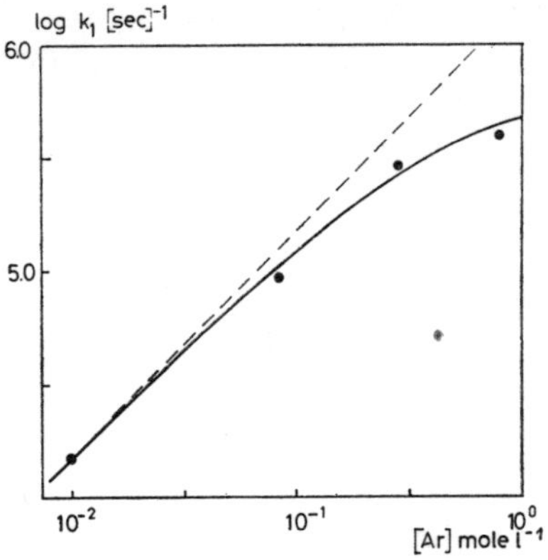

Fig. 4. Dependence on density of the first-order rate coefficient of the unimolecular decomposition of ammonia at 2700 °K. (From Henrici[30].)

From Fig. 3, the following Arrhenius expression for k_1 results

$$k_1 = (2.5 \pm 1) \times 10^{13} \exp - \left\{ \frac{(91 \pm 5) \times 10^3}{RT} \right\} \text{l.mole}^{-1}.\text{sec}^{-1}$$

Information about the enthalpy change of the unimolecular reaction can most reliably be obtained from the temperature dependence of the rate in the high-pressure regime. Until now it has not been possible to reach this region by experiment. Figure 4 shows a part of the fall-off curve of the first-order rate coefficient with decreasing total density. According to this, the reaction is in its second-order range at densities up to 2×10^{-2} mole.l^{-1}.

5.2 CONSECUTIVE REACTIONS

If it is assumed that

$$NH_3 + M \rightarrow NH_2 + H + M$$

is the initiation step in the thermal decomposition of ammonia, then the reaction

$$NH_3 + H \rightarrow NH_2 + H_2, \ \Delta H_0^0 \approx 0 \text{ kcal.mole}^{-1} \tag{3}$$

appears to be one of the most important consecutive reactions at temperatures between 2000 and 2900 °K. Reaction (3) is not observable at room temperature under conditions where H atoms can be generated by an electric discharge, *i.e.* at pressure of a few torr[33]. At 473 °K, k_3 is still below 10^4 l.mole^{-1}.sec^{-1} [34]. Assuming that Henrici has measured k_1 at temperatures above 2900 °K, *i.e.* if reaction (1) is then at least five times faster than reaction (3), and that he measured $2\,k_1$ or a higher multiple of k_1 at temperatures around 2300 °K using higher NH_3 concentrations, some limits of k_3 may be estimated. With the further assumption that a maximum value for the pre-exponential factor of k_3 is 10^{11}, an Arrhenius value for k_3 would lie between $5 \times 10^9 \exp(-12{,}500/RT)$ and $10^{11} \exp(-24{,}500/RT)$ l.mole^{-1}.sec^{-1}. Comparing this with the rate coefficient of a related reaction such as $H + CH_4 \rightarrow CH_3 + H_2$ ($E \approx$ 9–12 kcal.mole^{-1})[35,36], an activation energy closer to the lower value would be expected.

The reaction

$$NH_2 + NH_2 \rightarrow NH_3 + NH, \ \Delta H_0^0 \approx -12 \pm 8 \text{ kcal.mole}^{-1} \tag{4}$$

is apparently rather slow at temperatures between 300 and 1000 °K. This can be inferred from the fact that an absorption spectrum of NH_2 but not of NH is obtained after the flash photolysis of NH_3 or N_2H_4 under non-explosive conditions[37]. If in the latter case explosion occurs and the temperature rises adiabatically to about 1900 °K, NH is observed.

It is not known whether a sequence of reactions such as

$$NH_2 + NH_3 \rightarrow N_2H_4 + H \qquad (5)$$

$$N_2H_4 \rightarrow 2\,NH_2 \qquad (6)$$

is possible under the conditions of ammonia decomposition. They might give rise to a chain reaction.

Bradley *et al.*[38] have proposed that NH is formed by the sequence

$$NH_2 + NH_3 \rightarrow N_2H_3 + H_2 \qquad (7)$$

$$N_2H_3 \rightarrow NH + NH_2 \qquad (8)$$

They assumed that reaction (7) is rate-determining and estimated an activation energy of about 27 kcal.mole^{-1}.

6. Hydrazine

The vivid interest in hydrazine as a powerful propellant has stimulated many investigations both of its thermal decomposition and of its oxidation. Although hydrazine decomposes much more readily than ammonia, the study of its homogeneous decomposition by classical means using a static system is complicated considerably by wall catalysis. Thus, other experimental techniques have had to be applied, *e.g.* decomposition flames, flash photolysis, studies of explosion characteristics and the shock-tube technique.

6.1 STOICHIOMETRY

The most exothermic decomposition reaction of hydrazine would be

$$N_2H_4 = \tfrac{4}{3}\,NH_3 + \tfrac{1}{3}\,N_2,\ \Delta H_0^{298} = -37.47\ \text{kcal.mole}^{-1} \qquad (a)$$

This has so far only been observed for the heterogeneous decomposition in glass or silica vessels at temperatures between 250 and 400 °C [39]. However, the formation of small quantities of hydrogen in these experiments indicates that (a) is accompanied by a pyrolysis corresponding to

$$N_2H_4 = NH_3 + \tfrac{1}{2}N_2 + \tfrac{1}{2}H_2,\ \Delta H_0^{298} = -33{,}80\ \text{kcal.mole}^{-1} \qquad (b)$$

References pp. 42–45

This is the dominant overall reaction for the decomposition on platinum or tungsten at 200 and 380 °C, respectively[40]. Most workers on hydrazine decomposition flames[41-44], in which the reactions are homogeneous, report a stoichiometric equation similar to (b) for final flame temperatures up to 1900 °K. Measurements of MacLean and Wagner[45] on decomposition flames and of Husain and Norrish[37] on the flash photolysis of hydrazine indicate the contribution of the overall reaction

$$N_2H_4 = N_2 + 2\,H_2,\ \Delta H_0^{298} = -22.75 \text{ kcal.mole}^{-1} \qquad \text{(c)}$$

Michel and Wagner[28] have shown that the homogeneous decomposition can be described by a combination of (b) and (c) with (c) becoming more dominant with increasing temperature. Their results were obtained from shock-tube studies in which the concentration of ammonia could be measured after the decomposition of the hydrazine. If the ratio of the contribution of (b) and (c) is expressed in terms of a stoichiometric parameter, $\nu = [NH_3]/[N_2H_4]_0 - [N_2H_4]$, *viz.* the ammonia formed divided by the amount of hydrazine decomposed, a value $\nu = 1$ corresponds to (b) while (c) is represented by $\nu = 0$. The decrease of ν from unity is illustrated in Fig. 5. N_2H_4 decomposing at temperatures higher than 2000 °K

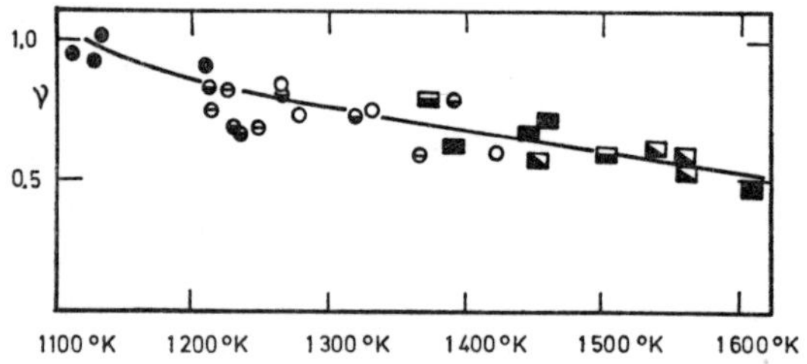

Fig. 5. Temperature dependence of the stoichiometry coefficient ν for the homogeneous decomposition of hydrazine from shock tube data. ●, $\rho_{N_2H_4} = 11.6$, $\rho_\tau = 0.7$; ◒, $\rho_{N_2H_4} = 5.0$, $\rho_\tau = 1.2$; ⊖, $\rho_{N_2H_4} = 2.2$, $\rho_\tau = 2.5$; ○, $\rho_{N_2H_4} = 1.2$, $\rho_\tau = 1.2$; ■, $\rho_{N_2H_4} = 11$, $\rho_\tau = 2.5$; ◪, $\rho_{N_2H_4} = 11$, $\rho_\tau = 5.7$, ⬓, $\rho_{N_2H_4} = 7.4$; $\rho_\tau = 7.5$. $\rho_{N_2H_4}$ = partial density of N_2H_4 in 10^{-5} mole.l^{-1}; ρ_τ = total density at reaction conditions in 10^{-2} mole.l^{-1}. (From Michel and Wagner[28].)

forms less than 10% NH_3 ($\nu < 0.1$). There is no systematic dependence of ν on the partial density of hydrazine or on the total gas density. A stoichiometry corresponding to $\nu = 1.01$–1.04 in the temperature range 750–1000 °K was reported by Palmer *et al.*[46]. Eberstein and Glassman[47] found a ν of about 0.9 in the temperature regime 750–1000 °K. Both figures are good evidence for the fact that ν does not appreciably exceed unity for the homogeneous decomposition at lower temperatures. Decomposition after reaction (a) is likely to be possible only through catalytic action.

6.2 INITIATION REACTION

Most of our knowledge about the kinetics of the homogeneous decomposition has come from shock-tube experiments. These have been performed in several laboratories under a variety of experimental conditions. However, their results are contradictory in some respects especially with regard to activation energy and on the question of the importance of chain reactions. In some cases the experimental conditions are such that consecutive reactions have to be taken into account or at least cannot be safely excluded. Until recently, one reason for the difficulty of reconciling the results of different investigators was that, if they were interpreted in terms of the unimolecular reaction[48]

$$N_2H_4 + M \rightarrow 2\,NH_2 + M,\ \Delta H_0^{298} = 58 \pm 9 \text{ kcal.mole}^{-1} \quad (1)$$

this reaction appeared to be somewhere between its first- and second-order regime. Some dependence of the first-order rate coefficient on total density was observed. The experimental accuracy, however, was not sufficiently high and the range of total densities studied was too small to obtain a fall-off curve and calculate the limiting high-pressure rate coefficient. This led to corresponding uncertainties in the calculated values for the dissociation energy of the NH_2–NH_2 bond, for which Foner and Hudson[48] had obtained 58 ± 9 kcal.mole^{-1} by mass spectrometric methods. A reliable value for this bond dissociation energy can be obtained from the apparent activation energy of the unimolecular reaction (1) in its first-order region. It therefore appears reasonable to start a discussion of the N_2H_4 decomposition with some new results on the kinetics in the high-pressure region of this reaction.

Olschewski *et al.*[49] have investigated the unimolecular decomposition of hydrazine in its first-order region using a high-pressure shock tube. They report data on the kinetics of the reaction at total densities between 10^{-1} and 2 mole.l^{-1}, corresponding to a pressure range of 12–250 atm at the temperatures studied. The first-order rate coefficient becomes independent of total pressure at about 100 atm. The range of temperatures covered was 1070–1420 °K and the concentration of N_2H_4 in Ar was varied over the relatively wide range of 0.1–0.003 %. The concentration of N_2H_4 was followed spectrophotometrically by its absorption at 2300 A. Figure 6 shows an Arrhenius plot for the limiting high-pressure rate coefficient at various N_2H_4 concentrations. Changing the concentration of N_2H_4 in a region higher than 5×10^{-5} mole.l^{-1} at temperatures < 1300 °K had the effect of altering the observed rate coefficient; that is, due to secondary reactions the rate is not strictly first-order in N_2H_4. Only at about 2×10^{-5} mole.l^{-1} was the rate coefficient independent of hydrazine concentration. It is most likely that under these conditions the true rate of the unimolecular reaction (1) has been obtained. At worst, a value of two times k_1 has been measured because of a fast

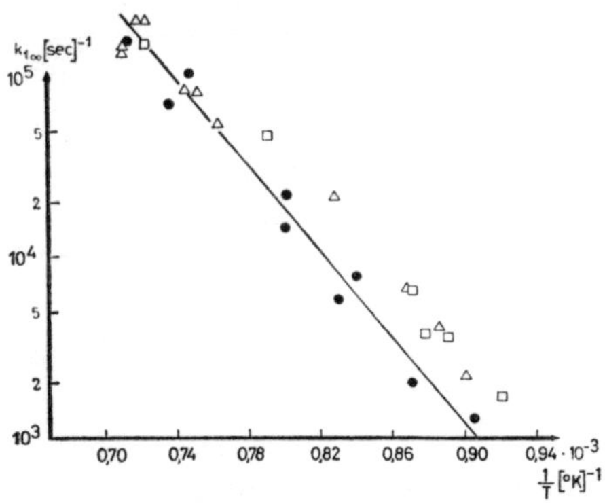

Fig. 6. High-pressure rate coefficients of the unimolecular decomposition of hydrazine ●, 1.5–5 × 10^{-5} mole.l^{-1} N_2H_4; △, 9–14 × 10^{-5} mole.l^{-1} N_2H_4; □, 18–28 × 10^{-5} mole.l^{-1} N_2H_4. (From Olschewski *et al.*[51].)

consecutive reaction. This matter will be discussed later.

From the experiments with the lowest N_2H_4 concentrations, an Arrhenius expression for the high-pressure rate coefficient

$$k_1 = 5.6 \times 10^{13} \exp \left\{ -\frac{(55 \pm 2) \times 10^3}{RT} \right\} \text{sec}^{-1}$$

is obtained for the temperature range 1070–1420 °K. The difference between the activation energy of the reaction in its high-pressure region and the bond dissociation energy is assumed to be not larger than $\pm RT$. For the temperature range studied by Olschewski *et al.*[49] this would correspond to $RT \approx 2.5$ kcal. mole^{-1}. The authors, however, do not think that a treatment according to the HKRRM Theory (see Volume 2, Chapter 3) could be profitably applied in the case of hydrazine since not enough is known about the properties of the activated complex. They assume instead that $E_0 \approx$ activation energy $= 55 \pm 2$ kcal.mole^{-1}. This agrees within the limits of experimental error with the dissociation energy obtained by Foner and Hudson[48].

From recent experiments by Olschewski *et al.*[49] a low-pressure activation energy of about 41 kcal.mole^{-1} was obtained. This is in excellent agreement with theoretical calculation[49], assuming a dissociation energy of 55 kcal.mole^{-1}. Figure 7 shows the pressure dependence of the rate coefficient k_1 as a plot of log k_1 *versus* log {[Ar] mole.l^{-1}}. The solid lines have been calculated according to theories of unimolecular reaction taking into account the characteristic temperatures of the N_2H_4 oscillators. The curve has been fitted to the experimental high-pressure and low-pressure rate. The values represented by the symbol ● are taken from the work of Olschewski *et al.*[49].

In addition, the figure includes the data of other investigators who also studied the reaction in its pressure-dependent region, together with the early results by

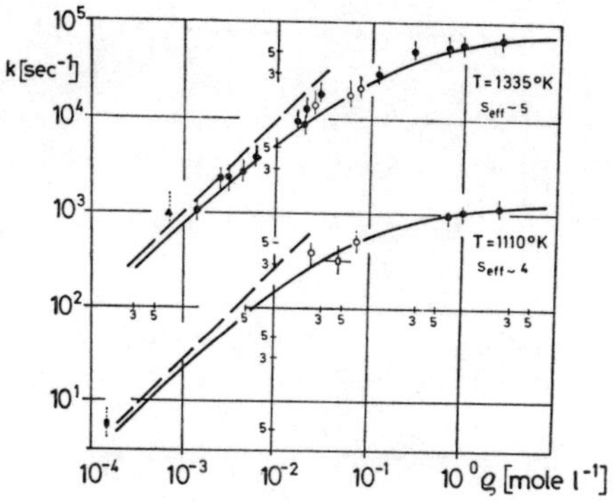

Fig. 7. Dependence on density of the first-order rate coefficient of the unimolecular hydrazine decomposition at 1110° and 1335° K. ●, Ref. 51; ○, ref. 28; –□–, ref. 46; ▲, ref. 52; ■, ref. 50. S_{eff} = no. of effective oscillators. (From Olschewski *et al.*[51].)

Szwarc[50] obtained at very low pressure. The symbols O refer to the work of Michel and Wagner[28], who studied the homogeneous decomposition by means of a shock-tube method similar to that of Olschewski *et al.*[51]. The reaction temperatures were between 1100° and 1600 °K, with reaction mixtures containing 0.5–0.03 % N_2H_4 in Ar at total densities between 2.5 and 7.5×10^{-2} mole.l^{-1}. The reaction was not first order in hydrazine under all conditions. At temperatures above 1260 °K and partial densities of $N_2H_4 < 20 \times 10^{-5}$ mole.l^{-1} Michel and Wagner observed an initial first-order decay (the first half-life of the reaction), then there appeared to be a stage of varying reaction order which became lower with decreasing initial N_2H_4 concentration. Rate coefficients from the initial decay are plotted in an Arrhenius manner in Fig. 8 for two different total densities. The two plots correspond to the equations

$$k_1 = 6.3 \times 10^{12} \exp\left(-\frac{52{,}200}{RT}\right) \text{ sec}^{-1}$$

$$at = 7.5 \times 10^{-2} \text{ mole.l}^{-1}$$

and

$$k_1 = 10^{12} \exp\left(-\frac{47{,}500}{RT}\right) \text{ sec}^{-1}$$

$$at = 2.5 \times 10^{-5} \text{ mole.l}^{-1}.$$

The low-pressure activation energy can be calculated using HKRRM-Theory and by setting the dissociation energy E_0 equal to the high-pressure activation energy. In this way a figure of about 41 kcal.mole^{-1} results[49]. Apparent activation energies of 52.2 and 47.5 kcal.mole^{-1} as measured in the fall-off region fit well into this scheme. Strong deviations from a first order rate law have been observed

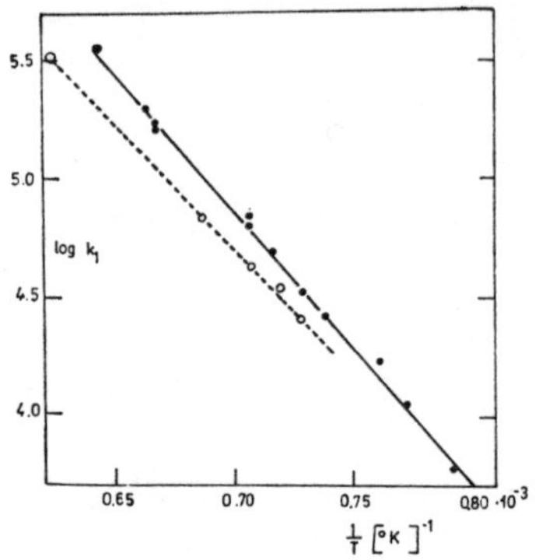

Fig. 8. First order rate coefficients of N_2H_4 pyrolysis at temperatures from 1260 to 1600 °K. ●, total density 7.5×10^{-2} mole.l^{-1}; ○, 2.5×10^{-2} mole.l^{-1}. (From Michel and Wagner[28].)

by Michel and Wagner[28] at partial densities of N_2H_4 above 20×10^{-5} mole.l^{-1} and temperatures between 1100 and 1200 °K (total densities of about 8×10^{-2} mole.l^{-1}). Under these conditions a complicated reaction occurred, involving an induction period. This behaviour suggests a chain mechanism operating at higher N_2H_4 concentrations and lower temperatures. However, some consecutive reactions become relatively slow as one approaches higher temperatures and lower N_2H_4 densities.

Figure 7 includes a value of k_1 (□) obtained by Palmer *et al.*[46] who decomposed N_2H_4 also diluted with argon by the so-called single-pulse shock-tube method. The temperature range covered by these experiments was 970–1120 °K. Although the data for the rate coefficients of N_2H_4 disappearance show considerable scatter, the authors obtain activation energies between 45 and 50 kcal.mole^{-1} for the pyrolysis depending somewhat upon the rate law they used for evaluating the experimental results. This range of activation energies is not unreasonable for the fall-off region in view of the recent results of Olschewski[49]. On account of the relatively high initial concentration of N_2H_4 ($10–140 \times 10^{-5}$ mole.l^{-1}) and the low temperatures used by Palmer *et al.*, however, some doubts remain as to whether they really measured the rate coefficient of the unimolecular reaction.

Figure 7 demonstrates clearly that the second-order region of the reaction is only reached at total densities below 10^{-3} mole.l^{-1}. This region of pressure corresponding to less than a tenth of an atmosphere cannot be easily studied in shock-tube experiments with spectrophotometric recording since the hydrazine absorption becomes too small. Diesen[52] has applied mass spectrometric analysis to a shock-heated gas mixture under these conditions. The reacting gas behind a reflected shock wave was sampled through a small hole in the end plate of the shock tube and analysed by a high-speed time-of-flight mass spectrometer. The importance of this work is that it clarified the concentration–time relationships between the N_2H_4, the products NH_3, N_2 and H_2 and the intermediate NH_2.

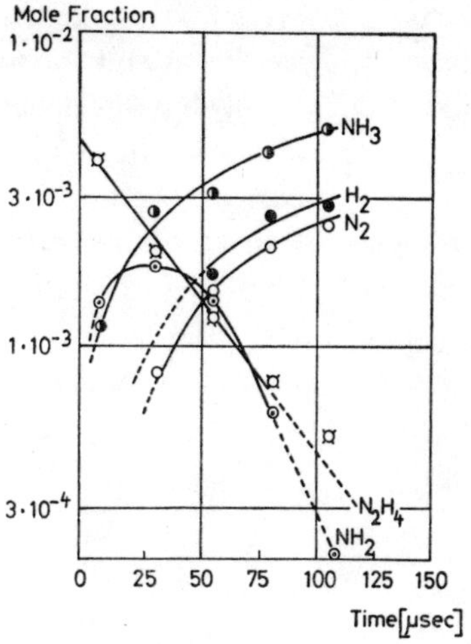

Fig. 9. Concentration–time curves for the N_2H_4 decomposition behind a shock wave at 2060 °K and 0.13 atm. (0.5 % N_2H_4 in argon.) (From Diesen[52].)

This was the only radical to be detected with certainty. The formation of NH_2 preceeds in time the appearance of NH_3 followed by the production of H_2 and N_2 (Fig. 9). The abundance of the amino radical relative to hydrazine increases with temperature and with decreasing N_2H_4 concentration. At temperatures below 1600 °K its concentration becomes too low to be detected by the mass spectrometer. From the rate of disappearance of N_2H_4, first-order rate coefficients were determined, which were approximately proportional to the pressure, as would be expected for the second-order region of the reaction. In order to incorporate Diesen's results into Fig. 7, his rate coefficients had to be extrapolated from 1430 °K to 1335 °K. The uncertainty of the value is partly due to this extrapolation. The rate coefficient at the lowest total density originates from the early work of Szwarc[50], who was the first to discriminate experimentally between the homogeneous and the heterogeneous decomposition of hydrazine. He used a flow system with toluene as a carrier gas (0.7–5.0 % N_2H_4 in toluene) at total densities of about 1.5×10^{-4} mole.l^{-1}. The temperatures ranged from 894 to 1057 °K. The NH_2 radicals produced by the unimolecular step (1) were removed in the presence of excess toluene by the process

$$C_6H_5CH_3 + NH_2 \rightarrow C_6H_5CH_2 + NH_3$$

A subsequent reaction is the recombination of the benzyl radicals to form dibenzyl, so that the rate of formation of dibenzyl measures the rate of reaction (1). Szwarc evaluated his results in terms of first-order kinetics for the overall process, since he could not find a systematic variation of his first-order rate coefficients with total pressure (varied from 5 to 15 torr). However, in view of the fact that only one tenth to one third of the reaction he observed was homogeneous, a proportionality between the apparent first-order rate coefficient and the total density, which was only varied by a factor of three, would hardly have been detected. The temperature dependence of the rate coefficient, which he quotes to be $k_1 = 4\times10^{12}\exp(-60{,}000/RT)\ \mathrm{sec}^{-1}$, is certainly not correct; the low-pressure activation energy should be somewhere in the region of 40 kcal.mole^{-1} (*vide infra*). The value of k_1 in Fig. 7 results from an extrapolation of Szwarc's results to 1110 °K. It should be correct within an order of magnitude. However, there is still an uncertainty as to the collision efficiency of toluene relative to that of argon. The value in Fig. 7 may, therefore, be too high.

6.3 CONSECUTIVE REACTIONS

Investigators of the unimolecular decomposition of hydrazine have taken great care to minimise the influence of secondary reactions upon their results. Thus, very little is known about reactions, for example, of the amino radical, NH_2. Many investigators of hydrazine decomposition flames assume a chain reaction to be operative. From the temperature dependence of the normal flame velocity, an overall activation energy of about 35 kcal.mole^{-1} is found[43]. In view of the low-pressure activation energy for the unimolecular reaction of 60 kcal.mole^{-1} (Szwarc's value), formerly assumed, a comparatively long chain seemed to be responsible for the low overall activation energy in the flame. The more recent figure[49] of about 40 kcal.mole^{-1} shows that this assumption must be at least partly revised. A chain reaction is probably operative under flame conditions but the chain length should not be large.

A reaction mechanism involving

$$N_2H_4(+M) \rightarrow 2\,NH_2(+M) \qquad (1)$$

as the primary decomposition step and reaction

$$NH_2+N_2H_4 \rightarrow N_2H_3+NH_3 \qquad (2)$$

as the first consecutive step appears to be generally accepted for a temperature range up to 1600 °K. Reaction (2) has not yet been investigated separately. Michel and Wagner[28] calculated an Arrhenius expression for k_2. *viz.*

$$k_2 \approx 3\times10^{10}\exp\left(-\frac{17{,}000}{RT}\right)\ \mathrm{l.mole^{-1}.sec^{-1}}$$

This implies that they measured k_1 and not an integral multiple of it and that chain termination steps are not significant in the first reaction stage which they assume to be governed by reactions (1)–(4) (see (3) and (4) below). At higher temperatures the disproportionation reaction

$$2\,NH_2 \rightarrow NH+NH_3$$

which has already been discussed (decomposition of ammonia, p. 16) may play a role. Although the imino radical could not be detected by absorption spectroscopy during decomposition of diluted hydrazine below 1600 °K in the shock-tube experiments[31], Diesen[52] believes he has evidence from mass spectrometric analysis for its appearance at temperatures above *ca.* 1800 °K. The absorption of NH has only been determined in the explosive flash photolysis[37] and in a decomposition flame at temperatures of 1900 °K[45]. The reactions of the NH radical and their rates are unknown. For the formation of hydrogen atoms, which most probably are chain carriers, the reaction

$$N_2H_3(+M) \rightarrow N_2H_2+H(+M) \tag{3}$$

is proposed[28], followed by the attack on hydrazine by H atoms, *via*

$$H+N_2H_4 \rightarrow NH_2+NH_3,\ \Delta H_0^{298} \approx 46\ \text{kcal.mole}^{-1} \tag{4}$$

or

$$H+N_2H_4 \rightarrow N_2H_3+H_2 \tag{5}$$

The reaction of hydrogen atoms with hydrazine has only been studied at temperatures between 25 and 200 °C[53, 54]. At these temperatures the reaction proceeds *via* (5). This is confirmed by the observation that in the reaction $D+N_2H_4$ no NH_2D could be detected; this rules out reaction (4) at low temperatures. For the rate coefficient of reaction (5) Schiavello and Volpi[54] quoted the Arrhenius expression

$$k_5 = 3.5\times10^8 \exp\left(-\frac{2000}{RT}\right)\ \text{l.mole}^{-1}.\text{sec}^{-1}$$

Birse and Melville[55] measured the disappearance of N_2H_4 in the presence of hydrogen atoms generated by mercury-sensitized photolysis of H_2[55]. They could not discriminate between reactions (4) and (5) and assumed that the H atoms were removed by reaction (4). In view of the more recent experiments with D atoms it is, however, more likely that they determined the rate of reaction (5). They report a rate coefficient

$$k_5 = 10^{10} \exp\left(-\frac{7000}{RT}\right) \text{ l.mole}^{-1}.\text{sec}^{-1}$$

which is considerably different from that of Schiavello and Volpi.

Dimide (N_2H_2) has been detected and determined in hydrazine decomposition flames but its method of formation and further reactions are unknown[56]. Recently an interesting experiment on the UV photolysis of hydrazine has been reported by Stief and De Carlo[57], who irradiated a mixture of $^{14}N_2H_4$ and $^{15}N_2H_4$ with the Kr resonance lines at 1165 and 1236 A and with the continuum of a hydrogen lamp (> 1850 A). Among the reaction products only very little $^{14}N^{15}N$ was found, so that, in contrast to pyrolysis, most of the N_2 was formed without any fission of the N–N bond.

7. Other hydrides of Group V elements

The slow, thermal decomposition of hydrazoic acid in a static system has been studied by Meyer and Schumacher[58]. It turned out to be completely governed by heterogeneous catalysis. There are no studies on the kinetics of the homogeneous decomposition of this substance save for the investigation of its decomposition flame[59]. From the variation of flame properties with pressure it can be deduced that second-order reactions control the over-all rate. The unimolecular reaction

$$HN_3 \rightarrow HN + N_2$$

has been proposed as a rate-determining step in the slow decomposition or in auto-ignition processes[60], but no data on its rate and activation energy are available. The thermal decomposition of hydrazoic acid has been used for the generation of NH radicals[61], but there are no quantitative data on the reaction rates. Besides NH, N_3 also could be detected in the flash photolysis of HN_3[62].

The kinetics of the thermally induced homogeneous decomposition of phosphine (PH_3) have not yet been studied. The species PH_2, PH and P_2 are formed on flash photolysis of PH_3 and could be identified by their absorption spectra[63]. There are proposals as to the mechanism of the consecutive process after the photochemical primary step, but nothing is known about the kinetic parameters of these reactions. With arsine and antimony hydride only the heterogeneous decomposition has been studied[64,65].

8. Silanes

Silanes have molecular structures analogous to those of saturated hydrocarbons. The decomposition reactions, however, are not similar to those of the alkanes. This is in great part due to the fact that there are no unsaturated volatile silanes. Silanes are more reactive than the hydrocarbon analogues and all are decomposed at red heat to silicon and hydrogen[66]. At somewhat lower temperatures cracking reactions occur, the products of which are hydrogen, lower and higher silanes and polymeric silicon–hydrogen compounds. The stability of the silanes decreases with increasing molecular weight.

8.1 MONOSILANE AND PERDEUTERO-MONOSILANE

The most recent work on monosilane decomposition kinetics was published by Purnell and Walsh[67]. The pyrolysis was studied in a static system using reaction vessels the inner walls of which were coated with silicon. This was done in order to provide always a uniform surface at the beginning of the reaction. The other experimental conditions are summarized in Table 3. The reaction was followed by measuring changes of the total pressure and by determining the concentration–time relations using gas chromatography. Figure 10 shows the variations of reactant and product concentrations with time. From these and from the pressure change during the reaction, two stages of decomposition could be distinguished. (1) Up to about 3 % decomposition (first stage) the total pressure remained constant, and the amount of hydrogen as well as of disilane and of trisilane increased steadily. (2) The second stage was characterized by a pressure increase. The concentrations of disilane and trisilane levelled off and remained constant during the further pyrolysis while that of hydrogen continued to increase. During this stage, which has been observed up to 20 % decomposition, solid products deposited on the walls of the vessel.

The kinetics were evaluated using the initial slopes of the hydrogen and disilane concentration–time curves. From Fig. 10 it is obvious that the formation of hydrogen and disilane is accelerating from the very beginning of the reaction.

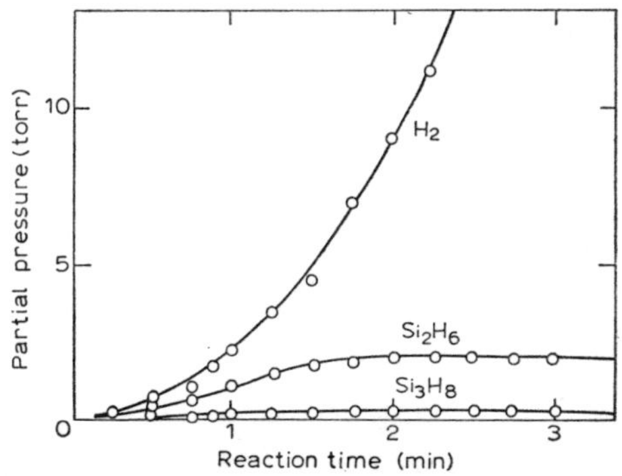

Fig. 10. Product–time curves for the initial stages of pyrolysis of 77 torr monosilane at 430 °C. (3 min = 20 % decomposition.) (From Purnell and Walsh[67].)

From the measurements published in the paper it cannot be inferred that the concentration–time curves can safely be extrapolated to zero time. The authors do not communicate details as to how well the beginning of the reaction could be defined. It should be pointed out that reaction kinetics derived by these means are subject to the uncertainties inherent in the extrapolation method. This holds particularly for a rapidly accelerated reaction. Furthermore, the possibility that the decomposition was initiated heterogeneously could not be excluded with certainty. These objections have to be considered when regarding the following kinetic results.

The stoichiometry of the first stage of pyrolysis can be described by the equation

$$2.39\ SiH_4 = Si_2H_6 + 0.13\ Si_3H_8 + 1.26\ H_2$$

Both hydrogen and disilane are formed initially according to a rate law involving an order of 1.5 ($\pm 5\ \%$) with respect to the monosilane. The energies of activation for the formation of hydrogen and disilane were identical within the limits of experimental error. This suggests that the production of both products was governed by the same rate-determining initial step. The initial rate of product formation can be described by the Arrhenius-type expression

$$\frac{d[Si_2H_6]}{dt} = \frac{1}{1.26}\frac{d[H_2]}{dt}$$

$$= (1.5 \pm 0.6) \times 10^9 \exp\left\{\frac{(55.9 \pm 5) \times 10^3}{RT}\right\} [SiH_4]^{\frac{3}{2}}\ \text{mole.l}^{-1}.\text{sec}^{-1}$$

Apparently the concentration of Si_3H_8 could not be measured with sufficient accuracy to determine the rate law for its initial formation.

In the second stage of decomposition the formation of hydrogen was found to be first order in the SiH_4 pressure. An appreciable amount of hydrogen then originated from the involatile products at the wall. The solid silicon hydride when first formed had the empirical formula SiH_2, but later it lost hydrogen rapidly. The enhancement of the rate of Si_2H_6 formation by the addition of gases such as CO_2, C_2H_6, SF_6, which are chemically inert to SiH_4 might be evidence for the participation of a unimolecular reaction in its pressure dependent region as a rate-determining step.

Thus, two following simplified reaction mechanisms appear possible, *viz.*

(*A*) $$SiH_4 \rightarrow SiH_2 + H_2 \quad (1)$$

$$SiH_2 + SiH_4 \rightleftharpoons Si_2H_6 \quad (2)$$

$$Si_2H_6 + SiH_2 \rightleftharpoons Si_3H_8 \quad (3)$$

(*B*) $$SiH_4 \rightarrow SiH_3 + H \quad (4)$$

$$H + SiH_4 \rightleftharpoons SiH_3 + H_2 \quad (5)$$

$$SiH_3 + SiH_4 \rightleftharpoons Si_2H_6 + H \quad (6)$$

$$2\,SiH_3 \rightarrow Si_2H_6 \quad (7)$$

If mechanism (*A*) is operative, the experimental results require that reaction (1) be a unimolecular reaction near its pressure-dependent region and that it be rate-determining for the formation of both H_2 and Si_2H_6. Mechanism (*B*) is analogous to a Rice–Herzfeld radical chain reaction[68], involving H atoms and SiH_3 radicals as chain carriers. A steady-state treatment, neglecting back reactions, then gives

$$\frac{d[H_2]}{dt} = \frac{d[Si_2H_6]}{dt} = k_6 \left(\frac{k_4}{k_7}\right)^{\frac{1}{2}} [SiH_4]^{\frac{3}{2}}$$

The predicted values of overall activation energies for the two mechanisms are $E_A = E_1$ and $R_B = E_6 + \frac{1}{2} E_4$.

Since E_1, E_4 and E_6 are not known, only thermochemical considerations concerning Si–H and Si–Si bonds are available to form a basis for choosing between mechanisms (*A*) and (*B*). Niki and Mains[69] and also Gunning *et al.*[70] proposed that the enthalpy change for the step

$$SiH_4 \rightarrow SiH_2 + 2\,H$$

is less than 112.7 kcal.mole^{-1}. Accepting the electron impact value[71] of D^0_{298} (SiH_3–H) = 94±3 kcal.mole^{-1} leads to the conclusion that SiH_3 is very unstable with respect to SiH_2. Accordingly, the reaction

$$2\,SiH_3 \rightarrow SiH_4 + SiH_2$$

would be exothermic by at least 73 kcal.mole^{-1}. This suggests SiH_2 as the dominant intermediate species, thus favouring mechanism (*A*). This conclusion is supported by the fact that the divalent state of Group IV elements becomes more stable relative to the tetravalent state with increasing atomic number. A further test for the absence of SiH_3 and H atoms could be the study of the pyrolysis products from a mixture of SiH_4 and D_2, as has been done for the monogermane decomposition (p. 34).

The work of Purnell and Walsh[67] will now be compared with previous studies of this decomposition. Hogness *et al.*[72] were the first to undertake a systematic investigation of the kinetics of the SiH_4 pyrolysis. They followed the course of the reaction by measuring the increase in total pressure and assumed a stoichiometry corresponding to

$$SiH_4 = Si_{solid} + 2\,H_2$$

TABLE 3

EXPERIMENTAL CONDITIONS AND RESULTS FOR STUDIES OF SiH_4 AND SiD_4 DECOMPOSITION

Method	*Vessel surface*	*Volume (cm^3)*	*Pressure range (torr)*	*Temperature range (°C)*	*Reaction order*	*Rates coefficient k (sec^{-1})*	*Remarks, ref.*
Static	Glass coated with silicon	250	8.5–55.7	380–490	1	$2\times10^{13}\exp\left\{-\frac{(51.7\pm2)\times10^3}{RT}\right\}$	Ref. 72
Static	Duran glass coated with Si	250	70–450	375–429	1	$1.4\times10^{13}\exp\left(-\frac{51{,}900}{RT}\right)$	For the later stage of reaction after apparent "induction period" Ref.72
Static	Molybdenum glass coated with Si	200	20.8–54.3	420–461	1	$1.9\times10^{14}\exp\left(-\frac{56{,}100}{RT}\right)$	Ref. 75
Static	Pyrex coated with Si	$A/V = 1.5$	35–230	375–430	1.5	$1.5\times10^{(15\pm0.16)}\exp\left\{-\frac{(55.9\pm0.5)\times10^3}{RT}\right\}$ $l^{\frac{1}{2}}.mole^{-\frac{1}{2}}.sec^{-1}$	For the initial stage of reaction (up to 3 % decomposition of SiH_4). Ref. 67
Static	Pyrex coated with Si	$A/V = 1.5$	5–150	388 and 430	1	$1.6\times10^{(10\pm1.0)}\exp\left(-\frac{51{,}200}{RT}\right)$	For the later stage of reaction (10–20 % decomposition of SiH_4). Ref. 67
Static	Duran glass coated with Si	250	70–450	375–429	1	$3.2\times10^{13}\exp\left(-\frac{53{,}800}{RT}\right)$	Pyrolysis of SiD_4. Ref. 74

They observed an "induction period" followed by a pressure increase, which for some time was first order with respect to the initial SiH_4 concentration. Clearly, the "induction period" can be identified with the first stage of the pyrolysis reported by Purnell and Walsh[67], when there is a reaction but without pressure rise. The first-order reaction is the second stage of the decomposition after solid material has been deposited on the wall of the reaction vessel. The retardation of the pyrolysis in a later stage of the reaction is due to inhibition of the decomposition by hydrogen. The mechanism of hydrogen inhibition is not known. It is not possible, for example, to generate any silanes by the action of hydrogen on a silicon film at 430 °C[73].

Further studies on the monosilane pyrolysis under conditions very similar to those of Hogness *et al.*[72] and using the same experimental technique were reported by Stokland[74] and by Devyatykh *et al.*[75]. Both groups observed the "induction period" followed by a first order reaction. Their results are identical to those of Hogness *et al.* within the limits of error to be expected for this type of kinetic experiment. Stokland detected the formation of higher silanes (Si_2H_6, Si_3H_8) and noticed that the amount of decomposed SiH_4 was always higher than would be expected from the increase in pressure. However, he disregarded this and preferred to evaluate the kinetics according to the pressure rise. He assumed that the experimental errors in analysis and in the quenching of the reaction for sampling were such that a further refinement was not justified.

The rate of reaction is reported to be independent of the nature of the surface[75]. This means only that the surface is always covered by the same reaction products. However, Hogness *et al.*[72], and also Purnell and Walsh[67], have observed that the pyrolysis is faster in a glass vessel having an initially clean surface which later is covered with reaction products. This is a clear indication of heterogeneous reaction on glass. After product deposition the rate is apparently insensitive to variations of the surface volume ratio, both for the reaction during the "induction period" and for the later stage of the decomposition. Obviously this does not mean that the later stage is homogeneous. It might suggest, but does not prove, that the initial stage of the reaction, when no solids are deposited, is mainly homogeneous in silicon-coated vessels.

Stokland[74] found that SiD_4 decomposed more slowly than SiH_4. There seemed to be a systematic increase in the ratio of the rate coefficients, k_D/k_H (for the second stage of the pyrolysis), from about 0.54 at 648 °K to *ca.* 0.7 at 764 °K. The experimental conditions used by all investigators of monosilane pyrolysis, and their results, are summarized in Table 3.

References pp. 42–45

8.2 DISILANE

Emeléus and Reid[76] were the first to report a kinetic investigation of the pyrolysis of disilane. The decomposition was studied in cylindrical soft-glass bulbs (3 cm i.d.) which were coated with silicon. The temperature range was 314–360 °C and the initial pressures of disilane varied from 130 to 380 torr. The reaction was followed by the rise in total pressure although there were considerable differences between the fraction of Si_2H_6 decomposed and the fractional pressure increase. At 40 % conversion, for example, the fractional pressure change is larger than the fraction of conversion by 4.5 %. It might be inferred that these discrepancies are still larger at the beginning of the reaction. Thus, it is doubtful that all stages of the decomposition can be represented by the quoted stoichiometric relation (*cf.* below)

$$Si_2H_6 = 1.1\ SiH_4 + 0.46\ H_2 + 0.9\ Si$$

It was found that the pressure rise was preceded by an "induction period". The decomposition then accelerated to a stage in which it was approximately first order. Later on it was retarded, probably due to inhibition by the hydrogen formed. From the variation of the first-order rate coefficient

T (°K)	587	605	622	633
k (sec^{-1})	10^{-4}	4.2×10^{-4}	1.1×10^{-3}	2.8×10^{-3}

an apparent activation energy of 51.3 kcal.mole^{-1} was obtained. Addition of hydrogen eliminated the induction period completely and increased the relative

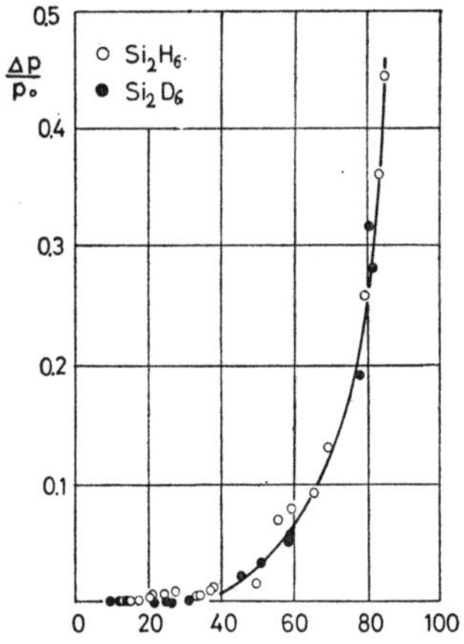

Fig. 11. The relative pressure changes during the decomposition of Si_2H_6 and Si_2D_6 as function of percentage decomposition. (From Stokland[74].)

amount of SiH_4 formed. The reaction was retarded by packing the vessel.

Stokland[74] reinvestigated the pyrolysis, also using a static system, and found that the relationship between pressure change and amount of decomposition was far more complicated than that found by Emeléus and Reid[76]. The percent decomposition is certainly not proportional to the pressure increase (Fig. 11). As in the pyrolysis of monosilane there is a first stage of reaction during which the pressure remains constant. Both tri- and tetrasilane are among the products, and in some experiments even higher silanes were found to be present. Since higher silanes are less stable, this shows that secondary reactions commence after only a small amount of decomposition has occurred. Furthermore, after deposition of the polymeric silicon hydrides, additional heterogeneity is present.

In spite of the complexity of the pyrolysis, a first-order rate law was found to fit the experimental results at least for the greater part of the reaction time. Stokland's first-order rate coefficients were consistently higher than those of Emeléus and Reid[76], this can be understood in the light of the relation between decomposition and pressure change. They could be represented formally by an Arrhenius expression for the overall coefficient

$$k_H = 5.8 \times 10^{14} \exp\left(-\frac{48{,}900}{RT}\right) \text{ sec}^{-1}$$

for Si_2H_6, and

$$k_D = 1.8 \times 10^{14} \exp\left(-\frac{47{,}900}{RT}\right) \text{ sec}^{-1}$$

for Si_2D_6. However, the degree of homogeneity is not known with any certainty.

8.3 TRISILANE

The pyrolysis of trisilane has been studied by Emeléus and Reid[76] in the same way as that of disilane. The temperature range was 303–360 °C while the initial pressure of trisilane was varied between 73 and 148 torr. Again, an apparent "induction period" before the pressure rise was observed. This "induction period" was shorter than for disilane under the same conditions. Stokland[74] showed that a great amount of solid products deposited soon after the beginning of the reaction, and that the consumption of the reactant was not proportional to pressure rise. Since Emeléus and Reid used the pressure rise to follow the decomposition, the kinetics derived from their observations are certainly not correct; they determined an overall activation energy of about 43 kcal.mole^{-1}. The mechanism of pyrolysis seems very complicated. Monosilane, disilane, tetrasilane and large amounts of solid material are formed[74]. Heterogeneous reactions seem to play

References pp. 42–45

a major role as soon as deposits appear. Only during the "induction period" is there some evidence for a homogeneous reaction, and then only if it can be assumed that the mechanism is similar to that of monosilane decomposition.

9. Germanes

9.1 MONOGERMANE

The first investigation on the thermal decomposition of GeH_4 was carried out by Hogness and Johnson[77]. They found rates proportional to the one third power of the germane pressure at higher temperatures, with an inhibition by hydrogen at lower temperatures. They evaluated their results in terms of a fully heterogeneous reaction involving Langmuir absorption.

Tamaru *et al.*[78] restudied this reaction by a static method using a cylindrical pyrex vessel (volume, 67 cm^3; inside diameter, 2.7 cm) with an initial pressure of undiluted germane between 50 and 400 torr. The temperature range was 278–330°C.

The reaction was followed by observing the total pressure of the system, assuming the overall reaction

$$GeH_4 = Ge_s + 2\,H_2$$

Their pressure records at the beginning of the reaction show that no "induction period" similar to that observed during the pyrolysis of monosilane occurs. This suggests that the overall reaction

$$2\,GeH_4 = Ge_2H_6 + H_2$$

is not relevant in the germane pyrolysis. The pressure records can be interpreted in terms of simultaneous heterogeneous and homogeneous reactions according to the rate law

$$-\frac{dp_{GeH_4}}{dt} = k_{het} + k_{homo}\,p_{GeH_4} \tag{1}$$

Within the pressure range studied the authors found no dependence of k_{homo} on the total pressure. In particular, they did not find the retardation of the reaction by hydrogen reported by Hogness and Johnson[77]. The temperature dependence of k_{homo}, as a rate coefficient for the homogeneous overall reaction, in the relatively small temperature range 551–603 °K gives the approximate expression

$$k_{homo} = 2.7\times10^{15} \exp\left\{-\frac{(51.4\pm3)\times10^3}{RT}\right\} \text{sec}^{-1}$$

Tamaru *et al.* proposed that the homogeneous decomposition involves the initial step

$$GeH_4 \rightarrow GeH_2 + H_2$$

This is inferred from the fact that addition of D_2 gave no HD during the pyrolysis. It should have been found if hydrogen atoms were present due to the reaction

$$GeH_4 \rightarrow GeH_3 + H$$

The subsequent fate of GeH_2 is not known. The authors do not exclude the possibility that the further reaction of GeH_2 occurs at the wall. The activation energy for the overall reaction at the wall, which is always covered by freshly formed Ge, comes out to ~ 42 kcal.mole^{-1}.

Devyatykh and Frolov[79] have recently investigated the decomposition under nearly the same conditions and using the same technique as Tamaru *et al.*[78]. They also assumed that the complete process consisted of a homogeneous reaction, first order in GeH_4, and a heterogeneous reaction of zero order. Their experimental results were also described by equation (1). The temperature dependence of their homogeneous rate coefficient in the range 588–652 °K gave the expression

$$k_{\text{homo}} = 3.7 \times 10^{15} \exp\left(-\frac{53{,}400}{RT}\right) \text{sec}^{-1}$$

The activation energy of the simultaneously occurring wall reaction was determined to be 37.5 kcal.mole^{-1}. There might, however, be considerable uncertainties in the calculated homogeneous rate coefficients, since the heterogeneous part of the reaction rate strongly depends on the amount of germanium deposited on the glass wall.

9.2 DIGERMANE

Emeléus and Jellinek[80] studied the decomposition of Ge_2H_6 in the same way as for Si_2H_6. They reported a stoichiometry corresponding to

$$Ge_2H_6 = 1.18\ GeH_4 + 0.49\ H_2 + 0.82\ GeH_{0.3}$$

However, this was determined when the reaction had gone almost to completion. It does not mean that this is the stoichiometry during the initial stage of the reaction. The rise in total pressure was used to follow the reaction and, as with Si_2H_6, an apparent "induction period" was observed. The pressure rise then accelerated to a stage where it was approximately first order. For the temperature range 195–222 °C a formal overall rate coefficient (calculated assuming the above stoichiometry)

$$k = 4 \times 10^{11} \exp\left(-\frac{33{,}700}{RT}\right) \text{sec}^{-1}$$

was given. This is again subject to uncertainties in the stoichiometry during pyrolysis. Furthermore, the reaction was retarded by packing the vessel, which is evidence for an unknown heterogeneous contribution.

10. Boron hydrides

Ever since the pioneering work of Stock[66] it has been clear that boron hydrides can easily undergo thermally induced decomposition and condensation reactions. These reactions result in the formation of new hydrides having more or fewer boron atoms. Both volatile hydrides and solid non-volatile boron–hydrogen compounds are produced until, finally, at temperatures above ~ 600 °C, elemental boron is formed. All boron–hydrogen compounds are thermodynamically unstable with respect to the elements at room temperature. However, their rates of decomposition differ considerably. Generally, one may say that boranes containing BH_2 groups are less stable than those without such groups. The relative rates of decomposition of some simple boranes with respect to the very unstable B_4H_{10} are B_5H_{11} (0.2), B_2H_6(0.02), B_5H_9 (0.002) and $B_{10}H_{14}$ (< 0.002)[81]. The BH_2 hydrides begin to decompose at room temperature. The condensation reactions forming higher hydrides and hydrogen generally are inhibited by hydrogen. This means they are, in a certain sense, "reversible". However, true equilibrium is not established in most of the cases. The whole process of boron hydride decomposition can be described in general terms as condensation–polymerisation, but in no case are the details fully understood. Since solid products are formed the process must necessarily come to a stage in which heterogeneous reactions play a role. However, this does not exclude the possibility that heterogeneous reactions are involved in the pyrolysis before involatile material is deposited.

As is usually the case in the study of complicated reactions that involve a great many different species, more attention has been given to the analysis of reaction products and intermediates than to the problems of the investigation of the kinetics of possible elementary reaction steps. Analytical studies of the systems have been advanced by the development of techniques such as gas chromatography for the analysis of multicomponent systems and mass spectrometry for the detection of free radicals and other highly unstable species. Furthermore, since most of the higher boranes are obtained by pyrolysis of diborane there is a more than usual interest in understanding the reaction mechanism of this complicated process. A major contribution to the diborane problem has come from the studies of isotope exchange between different borane molecules or between hydrogen and boranes[82-84].

Since we are primarily concerned with kinetics, we do not intend to discuss all the proposed mechanisms or the numerous boron hydrides formed during the pyrolyses. The literature available on this field up to 1964 has been reviewed by Adams[81] and by Lipscomb[85]. Rapid progress is being made in studies of the kinetics of boron hydride decomposition so that this discussion must be regarded as preliminary.

10.1 DIBORANE

Clarke and Pease[86] reported a general study of the kinetics of diborane pyrolysis using a static system (silica and pyrex vessels), the temperature range being 85–163 °C. The reaction was followed by determining the amount of unreacted diborane and of hydrogen formed. By pyrolysing B_2H_6–H_2 mixtures the authors showed that hydrogen inhibits the decomposition to some extent. Experiments using different initial pressures (50, 100, 200 torr) enabled the empirical rate law

$$-\frac{dx}{dt} = k' p_0^{\frac{1}{2}} x^{\frac{5}{2}} \tag{a}$$

where p_0 is the initial pressure and x the fraction of diborane unreacted, to be obtained. This formulation of a rate law is somewhat unfamiliar so that it isnot obvious that it corresponds to a $\frac{3}{2}$ reaction order in B_2H_6 at the beginning of the decomposition. With increasing extents of pyrolysis the order seems to fall, but this can be accounted for by inhibition due to the hydrogen produced. If one makes certain assumptions, the rate expression becomes compatible with the simplified mechanism

$$B_2H_6 \underset{k_{-1}}{\overset{k_1}{\rightleftharpoons}} 2\,BH_3 \tag{1}$$

$$BH_3 + B_2H_6 \underset{k_3}{\overset{k_2}{\rightleftharpoons}} \text{“intermediate hydrides”}^* + H_2 \tag{2, 3}$$

$$\text{“intermediate hydrides”} + B_2H_6 \xrightarrow{k_4} \text{higher hydrides} \tag{4}$$

Assuming equilibrium between B_2H_6 and BH_3 and a steady-state concentration for the “intermediate hydrides”, this gives a rate expression

* “Intermediate hydrides” is used here instead of B_3H_7 by the authors.

$$-\frac{d[B_2H_6]}{dt} = \frac{2k_2 \frac{k_4}{k_3} k_1^{\frac{1}{2}}[B_2H_6]^{\frac{3}{2}}}{\frac{k_4}{k_3}[B_2H_6]+[H_2]} \qquad \text{(b)}$$

In the initial stage of the reaction, where the concentration of hydrogen is negligible, this reduces to the $\frac{3}{2}$-order rate law. Later in the decomposition, however, it is equivalent to the empirical rate law only if the expression

$$[B_2H_6]+\frac{k_3}{k_4}[H_2]$$

remains equal to the initial concentration of B_2H_6. Thus, k_3/k_4 might be calculated from the amount of H_2 formed, but the accuracy of the measurements does not seem to warrant this refinement.

Thus, one shortcoming of the mechanism is that it does not allow for the experimentally determined amount of hydrogen produced, which was between 1 and 2 moles per mole of B_2H_6 reacted. The experimental rate coefficient k' varied from 8.6×10^{-5} at 85 °C to 5.8×10^{-2} $l^{\frac{1}{2}}$.mole$^{-\frac{1}{2}}$.sec^{-1} at 163.5 °C, corresponding to an apparent activation energy of 26 kcal.mole^{-1}. On the assumption that k' is approximately 2 k_2 $K_1^{\frac{1}{2}}$ in the initial stages of the reaction the experimental activation energy could be interpreted as $E_2+\frac{1}{2}\Delta H_1$. Using a value of about 35 kcal.mole^{-1} for $D_0(BH_3-BH_3)$, as estimated by Burg and Fu[87], this gives $E_2 = 8.5$ kcal.mole^{-1}. These conclusions, however, must be regarded with some reserve, since the proposed mechanism does not establish the correct stoichiometry for hydrogen formation, nor can the homogeneity of the reactions involved be safely assumed.

At about the same time, the pyrolysis of diborane was studied by Bragg *et al.*[88] in the temperature range 90–130 °C. These workers again used a static system (reaction vessel volume 212 cm^3) and followed the conversion both by measurement of pressure increase and by determination of the amount of hydrogen formed. The system was also examined by mass spectrometric analysis. The empirical rate law was found to be

$$-\frac{d[B_2H_6]}{dt} = k'[B_2H_6]^{\frac{3}{2}}$$

which has the same form as that derived from the initial rate measurements of Clarke and Pease[86]. Also, at a temperature of 110 °C k' was 7.8×10^{-4} $l^{\frac{1}{2}}$.mole$^{-\frac{1}{2}}$.sec^{-1}, which agrees quite well with the value of 8.7×10^{-4} $l^{\frac{1}{2}}$.mole$^{-\frac{1}{2}}$sec^{-1} obtained by interpolation from the results of Clarke and Pease. The experimental activation energy of 25.5 kcal.mole^{-1} also is nearly the same as that reported by Clarke and Pease.

The mass spectra showed B_5H_{11} to be an intermediate and B_5H_9 a relatively stable product, and thus Bragg *et al.* proposed the mechanism

$$B_2H_6 \rightleftharpoons 2\,BH_3 \quad (1)$$

$$BH_3+B_2H_6 \rightarrow \text{intermediate hydrides} \quad (2)$$

$$\text{intermediate hydrides}+B_2H_6 \rightarrow B_5H_{11}+2\,H_2 \quad (5)$$

$$B_5H_{11} \rightarrow B_5H_9+H_2 \quad (6)$$

$$B_5H_{11} \rightarrow \text{higher hydrides}+B_2H_6 \quad (7)$$

They emphasized that the above was a plausible but not necessarily the correct mechanism. It does not account for the inhibition by hydrogen. The occurrence of B_5H_{11} and later of B_5H_9 seems to be well established and has been confirmed by Borer *et al.*[89] using gas chromatography. There is, however, no definite information on the identity of intermediate boron hydrides, which must necessarily be formed in the production of B_5H_{11} from B_2H_6. Hill *et al.*[90] observed a pressure rise in the initial stage of the decomposition when the condensable products such as B_5H_{11} and higher boranes were removed in a cold trap kept at −78 °C. The increasing pressure could be evidence for a stoichiometry such as

$$3\,B_2H_6 = 2\,B_3H_7+2\,H_2$$

$$2\,B_2H_6 = B_4H_8+2\,H_2$$

Later the total pressure decreased in accordance with the overall conversion

$$5\,B_2H_6 = 2\,B_5H_{11(\text{condensed})}+4\,H_2$$

Fehlner and Koski[91] found from H–D exchange measurements that tetraborane is a precursor of B_5H_{11} but not of B_5H_9. This throws some doubt on the role of reaction (41) as a simple step in this mechanism. The way in which B_5H_{11} reacts to form higher hydrides is not clear. In an independent investigation Bragg *et al.*[88] found that B_2H_6 was formed from pyrolysis of B_5H_{11} at about 2.5 times the rate of B_5H_9 formation. This ratio is independent of initial pressure and temperature. With all these studies using static systems there is some doubt as to the complete homogeneity of the reactions involved. There has been no crucial test to decide if rapid reactions do not take place preferentially at the walls.

The experimental data of Bragg *et al.*[88] on diborane decomposition, like those of previous investigators, have been interpreted in terms of a mechanism involving monoborane (BH_3) as a reactive intermediate. Fehlner[92] proposed that a different mechanism based on the initial dissociation reaction

$$B_2H_6 \rightarrow BH_2+BH_4$$

References pp. 42–45

should also be considered. This suggestion was based on the mass spectrometric detection of BH_2, together with BH_3, during B_2H_6 pyrolysis at low pressure in a quartz reactor (5×10^{-4} torr at 300 °C)[93]. However, Stafford *et al.*[94], who performed similar experiments using a stirred reactor constructed of stainless steel, could detect BH_3 but not BH_2. In neither case was there any evidence for BH_4. It is most probable that under the conditions of pyrolysis (very low pressure and high temperature) the decomposition and formation of certain intermediates is completely governed by heterogeneous reactions.

There is no doubt that the mechanisms proposed by Clarke and Pease[86] and by Bragg *et al.*[88] are too simple to explain the change towards first-order kinetics that has been observed in shock-tube experiments at higher temperatures[95]. There is as yet no satisfactory explanation for the large kinetic isotope effect for the production of hydrogen in the pyrolysis of diborane. Enrione and Schaefer[96] observed that the rate of hydrogen formation is five times greater than that of deuterium from B_2D_6, in spite of the fact that BD_3 should be present in larger equilibrium concentration than BH_3 under the same conditions. This leads us to the conclusion that the mechanism of this complicated decomposition will remain conjectural until more stoichiometric and rate data become available.

10.2 TETRABORANE (10)

Little kinetic data are available on the decomposition of higher boron hydrides, though a study of the conversion of B_4H_{10} to B_5H_{11} through the process

$$2\,B_4H_{10} + B_2H_6 \rightleftharpoons 2\,B_5H_{11} + 2\,H_2$$

is reported by Dupont and Schaefer[97]. The reaction was studied in pyrex bulbs in the temperature range 72–93 °C. The rate expression found experimentally, *viz.*

$$-\frac{d[B_4H_{10}]}{dt} = k[B_4H_{10}]$$

was interpreted by the authors as being due to the mechanism

$$B_4H_{10} \rightarrow B_4H_8 + H_2 \qquad \text{(slow)}$$

$$B_4H_8 + B_2H_6 \rightarrow B_5H_{11} + BH_3$$

An Arrhenius expression, $k = 10^{12} \exp(-24{,}300/\boldsymbol{RT})\ \text{sec}^{-1}$, was quoted for the coefficient of the rate-determining step. It is doubtful whether this step is a true unimolecular reaction. The effect of pressures has not been studied. The range of temperatures used appears to be too narrow to discuss the temperature dependence in terms of a pre-exponential factor and an activation energy. Furthermore, the reaction does not seem to have been tested for a heterogeneous contribution.

Pearson and Edwards[98] report that the thermal conversion of tetraborane without diborane being present is also first order, but is about one order of magnitude slower. The experimental energies of activation for both reactions, however, have been found to be almost the same.

Mass spectrometric investigation of the pyrolysis of tetraborane by Stafford *et al.*[99] gave evidence for the intermediate B_4H_8. As in similar work under their experimental conditions the influence of heterogeneous catalysis might be dominant.

10.3 DECABORANE (14)

Beachell and Haugh[100] investigated the pyrolysis of decaborane, using a static system with pyrex vessels, in the temperature range 170–238 °C. Decaborane was consumed by a nearly first-order process (n = 1.1–1.2); however, the reaction was retarded during the latter stages by the presence of hydrogen. No products or intermediate more volatile than $B_{10}H_{14}$ were observed. From the variation with temperature of the first-order rate coëfficient an experimental activation energy for the overall reaction of 41 kcal.mole^{-1} was estimated. These results on the decaborane pyrolysis were supported by later work of Owen[101]. He also used a static system, the temperature range being 210–250 °C and the decaborane pressures between 60 and 600 torr. In this study non-volatile solid hydrides and hydrogen were again the final products. As before, the decomposition was first order and was slightly retarded by hydrogen. In the first stage of the reaction approximately one mole of hydrogen was formed per one mole of decaborane reacted. The activation energy was 41.6$\pm$0.5 kcal.mole^{-1}, in agreement with that reported by Beachell and Haugh[100]. However, again there is no evidence for the homogeneity of the reaction. Further, it was clearly demonstrated that in a later stage of the pyrolysis most of hydrogen was evolved from decomposition of solid products.

In most of the other investigations of boron hydride pyrolysis no kinetic data are reported; usually only the reaction products and in some cases also active intermediates are detailed. For these studies the reader is referred to existing literature reviews[81,85].

11. Conclusion

This survey has shown that the homogeneous decomposition of hydrides of the elements becomes more complex when moving from right to left in the main groups of the Periodic Table. One obvious reason for this increasing complexity is the greater number of atoms per molecule concomitant with the increased valency of the central element. Furthermore, the tendency to form element–element bonds is more pronounced in the third and fourth main groups, which gives rise to the appearance of a homologous series of the simple hydrides. In view of this variety of different hydrogen compounds it is difficult to select common features in the characteristics of decomposition. While it is quite well established that the first step in pyrolysis of water and ammonia is the rupture of a single element–hydrogen bond, there is good evidence that the decay of monosilane and monogermane begins by the splitting-off of a hydrogen molecule. In the decomposition of higher hydrides the weaker element–element bond is preferentially cleaved. This seems to be valid for thermal pyrolysis, while the situation is not so clear-cut in photolysis.

The temperatures of measurable decay range from room temperature for tetraborane to several thousand degrees for water. While the reactions at high temperature are real dissociations to smaller particles, the decomposition occurring at lower temperatures are generally condensations with the simultaneous formation of hydrogen. In the latter case the possibilities for the initial and subsequent elementary steps increase enormously. In these cases it is extremely difficult to obtain the rate of the initial step and even harder to describe the overall reaction in terms of the kinetics of the single steps.

In some cases where the gas-phase decomposition has been studied by classical methods using static systems the question of the degree of homogeneity of the reactions is unsolved.

Although the main purpose of this chapter is to provide a critical review of the field, we hope that it will also stimulate future work.

ACKNOWLEDGMENTS

The authors express their gratitude to Prof. Dr. Dres. h.c. W. Jost and Prof. Dr. O. Glemser for encouragement to this work. We further thank Prof. Dr. H. Gg. Wagner and Dr. J. Troe for valuable discussions and Dr. W. Solomon for reading and refining the manuscript.

REFERENCES

1 J. Troe and H. Gg. Wagner, *Ber. Bunsenges. Phys. Chem.*, 71 (1967) 937.
2 R. Foerthmann and A. Schneider, *Naturwiss.*, 53 (1966) 500.
3 S. H. Bauer, G. L. Schott and R. E. Duff, *J. Chem. Phys.*, 28 (1958) 1089.
4 H. A. Olschewski, J. Troe and H. Gg. Wagner, *Z. Physik. Chem. N.F.* 47 (1965) 383; 11*th Intern. Symp. on Combustion*, The Combustion Institute, Pittsburgh, 1967, p. 155.
5 R. W. Getzinger, 11*th Intern. Symp. on Combustion*, The Combustion Institute, Pittsburgh, 1967, p. 117.

6 G. L. Schott and P. F. Bird, *J. Chem. Phys.*, 41 (1964) 2869.
7 *JANAF Thermochem. Tables*, Dow Chemical Company, Midland, Michigan, 1965.
8 P. J. Padley and T. M. Sugden, *Proc. Roy. Soc. (London), Ser. A*, 248 (1958) 248.
9 R. McAndrew and R. Wheeler, *J. Phys. Chem.*, 66 (1962) 229.
10 D. E. Hoare, J. B. Protheroe and A. D. Walsh, *Trans. Faraday Soc.*, 55 (1959) 548.
11 C. K. McLane, *J. Chem. Phys.*, 17 (1949) 379.
12 P. A. Giguère and I. D. Liu, *Can. J. Chem.*, 35 (1957) 283.
13 W. Forst, *Can. J. Chem.*, 36 (1958) 1308.
14 C. N. Satterfield and T. W. Stein, *J. Phys. Chem.*, 61 (1957) 537.
15 R. R. Baldwin and D. Bratton, *8th Intern. Symp. on Combustion*, Williams and Wilkins, Baltimore, 1962, p. 110.
16 E. Meyer, *Diplomarbeit*, Univ. Göttingen, 1967.
17 E. Meyer, J. Troe and H. Gg. Wagner, *12th Intern. Symp. on Combustion*, The Combustion Institute, Pittsburgh, 1969, p. 345.
18 N. R. Greiner, *J. Chem. Phys.*, 45 (1966) 99.
19 R. R. Baldwin, P. Doran and L. Mayor, *8th Intern. Symp. on Combustion*, Williams and Wilkins, Baltimore, 1962, p. 103.
20 C. P. Fenimore and G. W. Jones, *J. Phys. Chem.*, 62 (1958) 693.
21 G. Dixon-Lewis and A. Williams, *9th Intern. Symp. on Combustion*, Academic Press, New York, 1963, p. 659.
22 F. Kaufman and F. P. Del Greco, *9th Intern. Symp. on Combustion*, Academic Press, New York, 1963, p. 576.
23 B. de B. Darwent and R. Roberts, *Proc. Roy. Soc. (London), Ser. A*, 216 (1953) 344.
24 J. Troe, private communication.
25 M. Bodenstein and G. Kranendieck, *Nernst-Festschr.*, 1912, p. 99.
26 C. N. Hinshelwood and R. T. Burke, *J. Chem. Soc.*, 127 (1925) 1105.
27 T. A. Jacobs, *J. Phys. Chem.*, 67 (1963) 665.
28 K. W. Michel and H. Gg. Wagner, *10th Intern. Symp. on Combustion*. The Combustion Institute, Pittsburgh, 1965, p. 353.
29 J. Troe and H. Gg. Wagner, Paper presented at the *28th AGARD Propulsion and Energetics Meeting on Selected Topics in Aerothermochemistry, Oslo, May* 1966.
30 H. Henrici, *Dissertation*, Univ. Göttingen, 1966.
31 W. Jost, K. W. Michel and H. Gg. Wagner, The NH Radical in the Pyrolysis of Ammonia and Hydrazine, *Tech. Status Rep., Contract No. AF* 61 (514)–1142, August 1961.
32 H. E. Avery and J. N. Bradley, *Trans. Faraday Soc.*, 60 (1964) 850.
33 G. E. Moore, K. H. Shuler, S. Silvermann and R. Hermann, *J. Phys. Chem.*, 60 (1956) 813.
34 E. A. Albers, K. H. Hoyermann, J. Wolfrum and H. Gg. Wagner, *12th Intern. Symp. on Combustion*, The Combustion Institute, Pittsburgh, 1969, p. 313.
35 T. G. Majury and E. W. R. Steacie, *Discussions Faraday Soc.*, 14 (1953) 45.
36 K. Schofield, *Planetary Space Sci.*, 15 (1967) 643.
37 D. Husain and R. G. W. Norrish, *Proc. Roy. Soc. (London), Ser. A*, 273 (1963) 145.
38 J. N. Bradley, R. N. Butlin and D. Lewis, *Trans. Faraday Soc.*, 63 (1967) 2962.
39 J. C. Elgin and H. S. Taylor, *J. Am. Chem. Soc.*, 51 (1929) 2059.
40 P. J. Askey, *J. Am. Chem. Soc.*, 52 (1930) 970.
41 R. C. Murray and A. R. Hall, *Trans. Faraday Soc.*, 47 (1951) 743.
42 G. K. Adams and G. W. Stocks, *4th Intern. Symp. on Combustion*, Williams and Wilkins, Baltimore, 1953, p. 239.
43 P. Gray, I. C. Lee, H. A. Leach and D. C. Taylor, *6th Intern. Symp. on Combustion*, Reinhold, New York, 1956, p. 255.
44 P. Gray and I. C. Lee, *7th Intern. Symp. on Combustion*, Butterworths, London, 1959, p. 61.
45 D. I. MacLean and H. Gg. Wagner, *11th Intern. Symp. on Combustion*, The Combustion Institute, Pittsburgh, 1967, p. 871.
46 E. T. McHale, B. F. Knox and H. B. Palmer, *10th Intern. Symp. on Combustion*, The Combustion Institute, Pittsburgh, 1965, p. 341.
47 I. J. Eberstein and I. Glassman, *10th Intern. Symp. on Combustion*, The Combustion Institute, Pittsburgh, 1965, p. 365.

48 S. N. FONER AND R. L. HUDSON, *J. Chem. Phys.*, 29 (1958) 442.
49 H. A. OLSCHEWSKI, J. TROE AND H. GG. WAGNER, The thermal decomposition of N_2H_4, *Admin. Rep.* 6, AF 61 (052)–946, Jan. 1968.
50 M. SZWARC, *Proc. Roy. Soc.* (*London*), *Ser. A*, 198 (1949) 267.
51 H. A. OLSCHEWSKI, J. TROE AND H. GG. WAGNER, *12th Intern. Symp. on Combustion*, The Combustion Institute, Pittsburgh, 1969, p. 345.
52 R. W. DIESEN, *J. Chem. Phys.*, 39 (1963) 2121.
53 J. K. DIXON, *J. Am. Chem. Soc.*, 54 (1932) 4262.
54 M. SCHIAVELLO AND G. G. VOLPI, *J. Chem. Phys.*, 37 (1962) 1510.
55 E. A. B. BIRSE AND H. W. MELVILLE, *Proc. Roy. Soc. (London)*, *Ser. A*, 175 (1940) 164.
56 K. H. HOMANN, D. I. MACLEAN AND H. Gg. WAGNER, *Naturwiss.*, 52 (1965) 12.
57 L. J. STIEF AND V. J. DECARLO, *J. Chem. Phys.*, 44 (1966) 4638.
58 R. MEYER AND H. J. SCHUMACHER, *Z. Phys. Chem.*, A170 (1934) 33.
59 P. LAFITTE, I. HAJAL AND J. COMBOURIEU, *10th Intern. Symp. on Combustion*, The Combustion Institute, Pittsburgh, 1965, p. 79.
60 P. GRAY AND T. C. WADDINGTON, *Nature*, 179 (1957) 576.
61 F. C. RICE AND M. FREAMO, *J. Am. Chem. Soc.*, 73 (1951) 5529.
62 B. A. THRUSH, *Proc. Roy. Soc.* (*London*), *Ser. A*, 235 (1956) 143.
63 D. KLEY AND K. H. WELGE, *Z. Naturforsch.*, 20a (1965) 124.
64 V. M. KEDYARKIN AND A. D. ZORIN, *Tr. Khim. y Khim. Tekhnol.*, (1965) 161.
65 K. TAMARU, *J. Phys. Chem.*, 59 (1955) 777, 1084;
P. S. SHANTOROVICH AND B. V. PAVLOV, *Zh. Fiz. Khim.*, 30 (1956) 811.
66 A. STOCK, *Hydrides of Boron and Silicon*, Cornell University Press, Ithaca, 1933.
67 J. H. PURNELL AND R. WALSH, *Proc. Roy. Soc.* (*London*), *Ser. A*, 293 (1966) 543.
68 F. O. RICE AND K. F. HERZFELD, *J. Am. Chem. Soc.*, 56 (1934) 284.
69 H. NIKI AND G. J. MAINS, *J. Phys. Chem.*, 68 (1964) 304.
70 H. E. GUNNING, O. P. STRAUSZ, M. A. NAY AND G. N. C. WOODALL, *J. Am. Chem. Soc.*, 87 (1965) 179.
71 W. C. STEELE, L. D. NICHOLS AND F. G. A. STONE, *J. Am. Chem. Soc.*, 84 (1962) 4441;
W. C. STEELE AND F. G. A. STONE, *J. Am. Chem. Soc.*, 84 (1962) 3599.
72 T. R. HOGNESS, T. L. WILSON AND W. C. JOHNSON, *J. Am. Chem. Soc.*, 58 (1936) 108.
73 J. H. PURNELL AND R. WALSH, *Proc. Roy. Soc.* (*London*), *Ser. A*, 293 (1966) 543.
74 K. STOKLAND, *Kgl. Norske Videnskabs Skrifter*, (3) (1948–49) 1; *Trans. Faraday Soc.*, 44 (1948) 545.
75 G. G. DEVYATYKH, V. M. KEDYARKIN AND A. D. ZORIN, *Russ. J. Inorg. Chem.*, 10 (1965) 833.
76 H. J. EMELÉUS AND C. REID, *J. Chem. Soc.*, (1939) 1021.
77 T. R. HOGNESS AND W. C. JOHNSON, *J. Am. Chem. Soc.*, 54 (1932) 3583.
78 K. TAMARU, J. BOUDART AND H. TAYLOR, *J. Phys. Chem.*, 59 (1955) 801.
79 G. G. DEVYATYKH AND J. A. FROLOV, *Russ. J. Inorg. Chem.*, 11 (1966) 385.
80 H. J. EMELÉUS AND H. H. G. JELLINEK, *Trans. Faraday Soc.*, 40 (1944) 93.
81 R. M. ADAMS, The boranes or boron hydrides in *Boron, Metallo–Boron Compounds and Boranes*, R. M. ADAMS, (Ed.), Interscience, New York 1964, p. 574.
82 J. J. KAUFMANN AND W. S. KOSKI, *J. Am. Chem. Soc.*, 78 (1956) 5774;
W. S. KOSKI, *Advances in Chemistry Series*, *No.* 32, The American Chem. Soc., 1961, p. 78.
83 M. HILLMAN, D. J. MANGOLD AND J. H. NORMAN, *Advances in Chemistry Series*, *No.* 32, The American Chem. Soc., 1961, p. 151.
84 W. S. KOSKI, *Proc. Symp. on Exchange Reactions*, Upton, N.Y., 1961.
85 W. N. LIPSCOMB, *Boron Hydrides*, Benjamin, New York, 1963, p. 175.
86 R. P. CLARKE AND R. N. PEASE, *J. Am. Chem. Soc.*, 73 (1951) 2132.
87 A. B. BURG AND Y. C. FU, *J. Am. Chem. Soc.*, 88 (1966) 1147.
88 J. K. BRAGG, L. V. MCCARTY AND F. J. NORTON, *J. Am. Chem. Soc.*, 73 (1951) 2134.
89 K. BORER, A. B. LITTLEWOOD AND C. S. G. PHILLIPS, *J. Inorg. Nucl. Chem.*, 15 (1960) 316.
90 J. R. MORREY, A. B. JOHNSON, Y. C. FU AND G. R. HILL, *Advances in Chemistry Series*, *No.* 32, The American Chem. Soc., 1961, p. 157.
91 T. P. FEHLNER AND W. S. KOSKI, *J. Am. Chem. Soc.*, 86 (1964) 1012.

92 T. P. Fehlner, *J. Am. Chem. Soc.*, 87 (1965) 4200.
93 T. P. Fehlner and W. S. Koski, *J. Am. Chem. Soc.*, 86 (1964) 2733.
94 A. B. Baylis, G. A. Pressley, Jr. and F. E. Stafford, *J. Am. Chem. Soc.*, 88 (1966) 2428.
95 G. B. Skinner and A. D. Snyder, paper presented at the meeting of the AIAA, Dec. 1963.
96 R. E. Enrione and R. Schaeffer, *J. Inorg. Nucl. Chem.*, 18 (1961) 103.
97 J. A. Dupont and R. Schaeffer, *J. Inorg. Nucl. Chem.*, 15 (1960) 310.
98 R. K. Pearson and L. J. Edwards, Abstract of paper presented at the 132*nd Meeting of the American Chemical Society, New York*, 1957.
99 A. B. Baylis, G. A. Pressly, Jr., M. E. Gordon and F. E. Stafford, *J. Am. Chem. Soc.*, 88 (1966) 929.
100 H. C. Beachell and F. J. Haugh, *J. Am. Chem. Soc.*, 80 (1958) 2939.
101 A. J. Owen, *J. Chem. Soc.*, (1961) 5438.

Chapter 2

The Decomposition of Inorganic Oxides and Sulphides

K. F. PRESTON AND R. J. CVETANOVIĆ

1. Introduction

The kinetics of decomposition of certain inorganic oxides and sulphides and a few related compounds are reviewed in this chapter. Discussion is limited to the gas and liquid phase and to the reactions of neutral species. Accordingly, reactions in ionizing solvents have been excluded. The decompositions of the following compounds are considered: C_3O_2, CO, CO_2, CS_2, COS, CSe_2, COSe, N_2O, NO, N_2O_3, $NO_2(N_2O_4)$, N_2O_5, HNO_2, HNO_3, O_3, SO_2, SO_3, F_2O, F_2O_2, Cl_2O, ClO_2, Cl_2O_6, Cl_2O_7 and $HClO_4$. A survey of the relevant literature was completed in September 1969, although we have also taken into consideration a number of more recent articles which have been brought to our attention.

Reactions and rate coefficients are numbered consecutively throughout each main section, the reaction number appearing as a subscript in the symbol for the corresponding rate coefficient. For a reaction

$$aA+bB+\ldots \rightarrow \tag{z}$$

the rate coefficient k_z is defined in accordance with the rate law

$$-\frac{d[A]}{dt} = ak_z[A]^a[B]^b$$

unless an alternative definition is adopted in the text. k_{-z} is the rate coefficient for the reverse step and $K_z = k_z/k_{-z}$ is the equilibrium constant for the reaction. Superscripts ^{0} and $^\infty$ attached to the rate coefficients indicate values of a "unimolecular" rate coefficient relevant to the regions of pressure where second-order and first-order behaviour, respectively, are observed.

Thermochemical data were taken from JANAF Thermochemical Tables[1] unless stated otherwise in the text. Spectral characteristics were obtained from Herzberg's book[2].

References pp. 132–141

2. Carbon oxides and sulphides

The decompositions of C_3O_2, CO, CO_2, CS_2, COS, CSe_2 and COSe are dealt with in this section. Apart from carbon suboxide, this is a group of stable, unreactive compounds. Considerable emphasis has been placed on the investigation of the photolytic decompositions of some of these compounds which are thought to provide useful sources of atoms (C, O, S and Se) and free radicals (C_2O). The photochemistry of carbon dioxide has particular relevance to the chemistry of planetary atmospheres, although to date the mechanism of CO_2 photolysis remains obscure.

2.1 CARBON SUBOXIDE

2.1.1 Thermal decomposition of C_3O_2

Palmer *et al.*[3] have studied the pyrolysis of C_3O_2 at temperatures in the range 900–1100 °K by following the rate of carbon deposition from a He stream containing 0.1–0.5 mole % C_3O_2. The reaction was first order in C_3O_2 and was inhibited by the addition of CO; a substance other than C_3O_2 or CO was responsible for carbon deposition at the wall. Reaction (1) and its reverse

$$C_3O_2 \underset{-1}{\overset{1}{\rightleftarrows}} C_2O+CO \qquad (1, -1)$$

were assumed to be rate controlling and C_2O was believed responsible for carbon formation. A zero activation energy was derived for reaction (−1) and the small value of k_{-1} (Table 1) was taken as an indication of a spin-forbidden reaction. Such is the case, of course, if C_2O is in its ground, triplet state. The pre-exponential factor derived from k_1 on the assumption that the unimolecular reaction was in its first-order region (although this was not established) was also unusually small.

TABLE 1

RATES OF REACTIONS IN THE DECOMPOSITION OF C_3O_2

Reaction	*Rate coefficients*[a]	*Temp.(°K)*	*Ref.*
$C_3O_2 \xrightarrow{1} C_2O(^3\Sigma)+CO$	$k_1 = 1.94\times10^9 \exp-(52{,}000\pm2{,}000/RT)$[b]	900–1100	3
$CO+C_2O(^3\Sigma) \xrightarrow{-1} C_3O_2$	$k_{-1} = 1.5\times10^6 \exp-(0\pm1{,}000/RT)$	1060	3
$2\,C(^3P) \xrightarrow{k_c} C_2$	$k_c = 6.5\times10^9$ [c]	300	13

[a] Units: sec^{-1} (for k_1); $l.mole^{-1}.sec^{-1}$ (for k_{-1} and k_c).
[b] Assuming that the unimolecular reaction is in its first-order region at 1 atm He.
[c] At p = 60 torr He; pressure dependence not investigated.

2.1.2 *Photolysis of* C_3O_2

Although the photolysis of C_3O_2 has been used extensively as a source of C_2O radicals and C atoms, little is known about the photochemistry of this molecule. Absorption and photo-decomposition commence[2,4] at 3300 A and result in the production of CO and a polymer[5]. CO_2 is a minor product at short wavelengths[6]. The careful work of Bayes *et al.*[4] has produced compelling, although indirect evidence for the importance of

$$C_3O_2 + h\nu \rightarrow CO + C_2O \tag{2}$$

as a primary step in the 2300–3300 A region. Their results are most readily interpreted by assuming that C_2O is solely in its ground ($^3\Sigma$) state when generated at $\lambda > 2900$ A, but principally in an excited state, probably $^1\Delta$, for $\lambda < 2400$ A. Flash photolysis experiments[7] have confirmed that triplet C_2O is a primary product of the long-wavelength photolysis; an unidentified transient spectrum has been tentatively ascribed[8] to $C_2O(^1\Delta)$.

Secondary reactions in the long-wavelength photolysis are speculative. According to Cundall *et al.*[9] Φ_{CO} at 2537 A is close to 2, confirming Bayes' deduction[4] that (2) is followed by the overall reaction

$$C_2O + C_3O_2 \rightarrow \text{polymer} + n\,CO$$

with $n = 1$. Forchioni and Willis' results[10] at 2537 A, however, suggest $n = 2.5 \pm 0.5$. Morrow and McGrath[8] have demonstrated the appearance of substantial yields of C_3 at short delays after flash photolysis and conclude that it is not produced from C atoms or excited molecules, but rather from the reactions

$$C_2O(^1\Delta) + C_3O_2 \rightarrow C_5O_3 \rightarrow C_4O_2 + CO$$

$$C_4O_2 \rightarrow C_3O + CO$$

$$C_3O + \begin{cases} C_2O \\ C_3O_2 \end{cases} \rightarrow \begin{cases} C_3 + 2\,CO \\ C_3 + 3\,CO \end{cases}$$

Intermediates such as these react with C_3O_2 to produce large molecules which provide nuclei for thermal polymerization[5].

The assumption of a simple primary dissociation (2) has been cast in doubt by the results of an application of the Cundall technique to the photolysis[9], which suggest that excited-molecule reactions are very important at 2537 A. There is clearly a need here for some careful quantum yield measurements to establish the nature of the primary step.

The primary production of carbon atoms

$$C_3O_2 + h\nu \rightarrow 2\,CO + C \qquad (2a)$$

is thermodynamically feasible at wavelengths shorter than about 2000 A, yet the chemical evidence for the participation of (2a) in the primary process is not very convincing[6, 10, 11]. Braun *et al.*[12] have, however, recently detected carbon atoms spectroscopically in the 1400–1700 A flash photolysis of C_3O_2 and have shown that (2a) is the dominant primary process in that region. Both $C(^3P)$ and $C(^1D)$ atoms were observed as primary products and from the kinetics of their decay estimates were made for the rate coefficients of a number of elementary reactions of carbon atoms. Carbon atoms are also produced in an electrodeless discharge in C_3O_2, and observations[13] on their subsequent decay in a flow system have permitted the calculation of the rate coefficient for their recombination (Table 1).

2.2 CARBON MONOXIDE

2.2.1 *Thermal decomposition of* CO

The decomposition of CO has been studied in shock tubes[14–16] at temperatures in excess of 6000 °K, and in glass vessels[17] at temperatures in the region of 1000 °K. The low-temperature pyrolysis is entirely heterogeneous. Fairbairn's studies[14] have shown that the assumption adopted by previous workers that the decomposition is controlled by

$$CO + M \rightleftharpoons C + O + M \qquad (3, -3)$$

does not account for the time variations of CO, C_2 and C recorded by him. His observation of autoacceleration of the rate of CO removal and of the rates of formation of the intermediates clearly indicates a complex reaction mechanism. Reactions (4)–(7) and their reverse reactions (−4 to −7)

$$C + CO \rightleftharpoons C_2 + O \qquad (4, -4)$$

$$O + CO \rightleftharpoons O_2 + C \qquad (5, -5)$$

$$C_2 + M \rightleftharpoons 2\,C + M \qquad (6, -6)$$

$$O_2 + M \rightleftharpoons 2\,O + M \qquad (7, -7)$$

adequately described the observed concentration profiles, self-acceleration of the rate being the result of decomposition of a steadily increasing concentration of the intermediate C_2. Initial rates of decomposition fitted

$$-\frac{d[CO]}{dt} = k_3[CO][Ar]$$

TABLE 2

RATE COEFFICIENTS OF SOME ELEMENTARY REACTIONS INVOLVED IN CO DECOMPOSITION

Reaction	*Rate coefficient*[a]	*Temp.(°K)*	*Ref.*
$C+O+Ar \xrightarrow{-3} CO+Ar$	$k_{-3} = 7\times10^{7}$	8000	14
$C_2+O \xrightarrow{-4} CO+C$	$k_{-4} = 4\times10^{11}$	8000	14
$C+O_2 \xrightarrow{-5} CO+O$	$k_{-5} = 3\times10^{11}$	8000	14
$C+C+M \xrightarrow{-6} C_2+M$	$k_{-6} = 1\times10^{9}$	8000	14

[a] Units: $l^2.mole^{-2}.sec^{-1}$ (for k_{-3} and k_{-6}); $l.mole^{-1}.sec^{-1}$ (for k_{-4} and k_{-5}). All reactants and products assumed to be in their ground states; uncertainty in the rate coefficients approx. ±50 %.

with $k_3 \cong 6\times10^{10}\, T^{0.5} \exp(-D/RT)$ $l.mole^{-1}.sec^{-1}$ for $T = 6000°$–9000 °K, but could be measured with no greater precision than to within a factor of three. Rate coefficients of earlier workers were appreciably larger than these initial values, but were in excellent agreement with those measured by Fairbairn after the induction period where the reaction rate is no longer solely determined by (3) but by the complete scheme (3)–(7). Estimates for rate coefficients of these elementary steps are given in Table 2.

2.2.2 *Photolysis and radiolysis of* CO

Photolysis of CO has been conducted[18-20] only at wavelengths greater than the threshold for dissociation (~ 1100 A) where reactions of excited molecules must account for the decomposition, *viz.*

$$CO^*+CO \rightarrow CO_2+C \quad (\lambda < 2200 \text{ A}) \tag{8}$$

$$CO^*+CO \rightarrow C_2O+O \quad (\lambda < 1580 \text{ A}) \tag{9}$$

Subsequent reactions of the C and O atoms give rise to the end products CO_2 and C_3O_2, *viz.*

$$C+CO+M \rightarrow C_2O+M \tag{10}$$

$$C_2O+CO(+M) \rightarrow C_3O_2(+M) \tag{11}$$

$$O+CO+M \rightarrow CO_2+M \tag{12}$$

Approximate quantum yields have been measured and discussed in terms of possible contributions to reaction from $CO(A^1\Pi)$ and from various states in the triplet manifold. However, it is difficult to obtain precise quantum yields for a system with such sharp absorption bands, as evidenced by the lack of agreement between the reported values[18]. Even weak impurity lines in the light source, especially the CO emission lines invariably present in discharge lamps[21], may make a dominant but unsuspected contribution to the photolysis in such a case. Evidently extraneous light was responsible for the decomposition by light from an iodine lamp[19] (2062 A) and by Hg (1849 A) sensitization[20], since the results of flash photolysis experiments[22] clearly show that $CO(a^3\Pi)$, the only state accessible at 1849 A and longer wavelengths, does not react at all with the ground-state molecule.

Radiolysis of CO, both in the liquid and gaseous states, also leads to CO_2 and C_3O_2 formation, the latter product appearing mainly as a polymeric solid[23–27]. A scheme of reactions consisting of the initial production of a carbon and an oxygen atom, followed by (10)–(12) and

$$O+C_2O \rightarrow 2\,CO \tag{13}$$

satisfactorily accounts for the observations made in the gas phase. The contribution to the decomposition from reactions of ions is probably small[27]. Reaction (13) is required to explain the decrease in G values with increasing dose rate; a value of $k_{13}k_{12}/k_{11} = 4\times10^{10}$ l.mole^{-1}.sec^{-1} at 300 °K adequately describes the experimental curves of $G(CO_2)$ *versus* dose rate[23]. A chemical determination of the oxygen atom yield has shown that $G[O(^3P)]$ is close to $G(CO_2)$, suggesting that (12) is indeed a major source of CO_2[28].

$C_2O(^3\Sigma)$ and $C(^1S)$ have been observed spectroscopically during pulse radiolysis of CO[29,30]. The failure to observe the lower states of carbon atoms may simply be a reflection of their greater reactivity towards CO[12,30].

2.3 CARBON DIOXIDE

2.3.1 Thermal decomposition of CO_2

Carbon dioxide decomposes behind a shock front in accordance with the kinetics expected of a unimolecular reaction in its low-pressure region[31–37]. The second-order rate coefficients obtained by a number of experimentalists in the temperature range 2500–11000 °K are in reasonable agreement, but there is a considerable spread in the values derived for the Arrhenius activation energy (Table 3). Furthermore, even the highest of these values[31] is much smaller than the endothermicity (D = 125.8 kcal.mole^{-1}) of

$$CO_2+M \rightarrow CO+O(^3P)+M \tag{14}$$

In an attempt to explain the discrepancy in terms of a participation of internal degrees of freedom of the CO_2–M complex in the process of collisional activation, some authors have invoked the classical RRK expression for the rate of energization

$$k_{14} = \frac{pZ}{(s-1)!}\left(\frac{D}{RT}\right)^{s-1} e^{-D/RT}$$

With a suitable choice of s this expression certainly describes the observed temperature dependence very well, but abnormally small steric factors p are implied. A quantum statistical evaluation of the rate of energization is, in any case, more appropriate for CO_2, since its characteristic vibrational temperatures are high. Using quantum statistics and a theory of strong collisions one calculates[38, 39] that at 4000 °K for CO_2 the observed activation energy should be some 11 kcal. mole^{-1} lower than the energy barrier to decomposition. The assumption of a theory of weak collisions, with a consequent depletion of the equilibrium population of CO_2 molecules in the vibrational levels immediately beneath the energy barrier, leads to a further lowering of $\sim$ ***RT*** *i.e.* $\sim$ 8 kcal.mole^{-1} at 4000 °K. Only in the latter case and for the highest reported value of E_A (99 kcal.mole^{-1}) does one approach an energy barrier (118 kcal.mole^{-1}) within ***RT*** of the endothermicity of (14). Troe and Wagner[39] suggest that this result implies a rate-determining step consisting of activation to a bound triplet state whose subsequent dissociation is facile. In such a case the activation energy at high pressure should also be less than ΔE^0 for (14), its exact value depending in a complex fashion on the shapes of the potential surfaces of upper and lower states and on the transition probabilities for crossover in the region of their intersection. A value for E = 110 kcal.mole^{-1} has in fact been reported[38] for the high-pressure ([M = Ar] > 0.5 M), first-order region of the decomposition. Acceptance of this mechanism for dissociation is subject to confirmation of the "high" activation energy obtained by Wagner *et al.*[31, 38]. The much lower values obtained by all the other workers may have arisen through underestimation of a number of effects such as the temperature drop due to dissociation, and the growth of the boundary layer. An estimate of the rate of

$$O(^3P)+CO_2 \rightarrow CO+O_2 \tag{15}$$

obtained from measurements of the reverse reaction[40] shows that secondary reactions probably did not contribute significantly to the measured rate of disappearance of CO_2 during the observation periods employed in the shock-tube experiments. A recent investigation[41], however, suggests much larger rate coefficients for both k_{15} and k_{-15} (Table 3) than the values accepted hitherto, but

TABLE 3

RATES OF REACTIONS IN THE DECOMPOSITION OF CO_2

Reaction	*Rate coefficients*[a]	*Temp.*(°*K*)	*Ref.*
$CO_2+M \xrightarrow{14} CO+O+M$	$k^0_{14,Ar} = 10^{11.7} \exp-(99{,}000 \pm 300)/RT)$	2800–4400	31
	$k^0_{14,Ar} = k^0_{14,N_2}$		
	$k^0_{14,Ar} = 3\times10^8\ T^{\frac{1}{2}} \exp-(86{,}000/RT)$	2550–3000	32
	$k^0_{14,Ar} = 2.88\times10^8\ T^{\frac{1}{2}} \exp-(74{,}700/RT)$	3500–6000	33
	$k^0_{14,N_2} = 2.45\times10^8\ T^{\frac{1}{2}} \exp-(74{,}500/RT)$	3500–6000	33
	$k^0_{14,Ar} = 1.15\times10^8\ T^{\frac{1}{2}} \exp-(68{,}300/RT)$	6000–11000	34
	$k^0_{14,Ar} = 7.11\times10^8\ T^{\frac{1}{2}} \exp-(84{,}500/RT)$	3000–5000	35
	$k^0_{14,N_2} = 5.33\times10^8\ T^{\frac{1}{2}} \exp-(79{,}600/RT)$	3000–5000	35
	$k^0_{14,CO_2} = 1.5\times10^{11} \exp-(87{,}600 \pm 900)/RT)$	3000–5500	36
	$k^0_{14,Ar} = 2.26\times10^8\ T^{\frac{1}{2}} \exp-(71{,}900/RT)$	3300–6000	37
	$k^0_{14,N_2} = 1.18\times10^8\ T^{\frac{1}{2}} \exp-(73{,}200/RT)$	3100–7700	37
	$k^{\infty}_{14} = 10^{11.4} \exp-(110{,}000/RT)$	2800–3700	38
$O(^3P)+CO_2 \xrightarrow{15} CO+O_2$	$k_{15} = 10^{10.2\pm.8} \exp-(32{,}500 \pm 10{,}900)/RT$	2800–3200	41
	$k_{15} = 10^{11} \exp-(61{,}200/RT)$	3500–5500	42
$CO+O_2 \xrightarrow{-15} CO_2+O$	$k_{-15} = 2.5\times10^9 \exp-(48{,}000/RT)$	1500–3000	40
	$k_{-15} = 5.8\times10^9 \exp-(35{,}300 \pm 1800)/RT$	1800–3000	43

[a] Units: l.mole^{-1}.sec^{-1} (for k^0_{14}, k_{15} and k_{-15}); sec^{-1} (for k^{∞}).

shows that while (15) may contribute to the observed rate, it has little or no effect on the measured activation energy.

2.3.2 *Photolysis of* CO_2

In spite of numerous investigations[44–53], many of which were conducted in recent years, the mechanism of photolytic decomposition of CO_2 into CO and O_2 remains uncertain. Many of the proposed reaction schemes contain unsatisfactory features and none of them accounts for all of the experimental facts. There is even considerable disagreement over the correctness of certain observations, *viz.* the absence of $O(^3P)$ during photolysis as indicated by chemiluminescence studies[44–46], the yield of substantial amounts of a gaseous, long-lived CO_3 molecule[47, 48], or the formation of O_3 in quantities sufficient to account for the "lost" oxygen[44, 47, 49]. Accordingly, a lengthy treatise on the kinetics of the photodecomposition is unwarranted at this time. Only the better-established facts will therefore be briefly outlined with a discussion of some reactions of atomic oxygen with carbon oxides which we believe may be important in this system.

Carbon dioxide commences to absorb light in the region of 1750 A. Photochemical studies have been confined to the wavelength region 1200–1700 A where absorption consists of diffuse bands superimposed on a continuum[2]. The upper state, $^1B_2$ or $^1\Delta_u$, is believed to predissociate to yield $O(^1D_2)$ atoms

$$CO_2+h\nu(\lambda < 1670 \text{ A}) \rightarrow CO_2^* \rightarrow CO+O(^1D_2) \quad (16)$$

although at wavelengths shorter than 1286 A the production of $O(^1S_0)$ is energetically feasible. There is certainly both spectroscopic[54] and chemical evidence[55] for the participation of $O(^1D_2)$ in the photolysis, yet we hesitate to conclude that dissociation is the sole fate of CO_2^*. If reactions of electronically excited CO_2 molecules make an important contribution to the photodecomposition, a possibility admitted elsewhere[46, 47], the failure of schemes comprised only of oxygen atom reactions to account for the experimental observations would be understandable.

The fate of an $O(^1D_2)$ atom in a mixture of CO_2, CO and O_2 has been the subject of considerable controversy[46–48, 50, 52, 54, 56–60]. There is, however, agreement on one point[61, 62], that the reaction

$$O(^1D)+CO_2 \rightarrow CO+O_2 \quad (17)$$

makes a negligible contribution to the total rate of destruction of the excited atom by CO_2. If one ignores excited-molecule reactions, CO is then a product of the primary process only and, in the absence of back reactions, its quantum yield is a measure of the efficiency of that process. Most workers[44, 47, 51, 53] agree that Φ_{CO} is close to unity at 1236 A and 1470 A although Ung and Schiff[46] have recently determined $\Phi_{CO} = 0.25$ at 1470 A.

$O(^1D)$ undergoes rapid reactions with CO_2, CO and O_2[54, 56, 61, 64(a)], reactions which we firmly believe result in deactivation of the excited atom to the ground state (at least in the gas phase at conventional pressures), *e.g.*

$$O(^1D)+CO_2 \rightarrow CO_2+O(^3P) \quad (18)$$

Evidence which has recently been obtained[56] in our laboratory strongly suggests that each $O(^1D)$ atom which disappears by reaction with CO_2 or CO is replaced by a ground-state O atom. The same conclusion was reached, although with less certainty, from the results of earlier experiments on the exchange of $O(^1D_2)$ with ^{18}O-labelled CO_2[60, 61] and on the photolysis of O_3/CO mixtures[63]. This finding conflicts with the opinion of some experimentalists[46–48, 57–59] who consider that the reactions of $O(^1D)$ with CO_2 and CO in the gas phase are association reactions resulting in the formation of stable CO_3 and CO_2 molecules respectively. However, Slanger and Black[64(b)] have recently repeated experiments with CO and find that it deactivates $O(^1D_2)$ and does not associate to form CO_2.

CO_3 formation and accumulation in the gas phase[47], or destruction at the wall[48], certainly provide a simple explanation for the most puzzling feature of the photolysis, the loss of oxygen. Yet there is no convincing evidence for its existence in the gas phase beyond a short-lived intermediate in reaction (18). Doubt has even been cast[50] on claims that CO_3 is stable and may be observed spectroscopically in low-temperature matrices[65, 66].

Consideration of the rates of the possible reactions which a ground-state O atom may undergo in this system, such as

$$2\,O+M \rightarrow O_2+M \tag{19}$$

$$O+O_2+M \rightarrow O_3+M \tag{20}$$

$$O+O_3 \rightarrow 2\,O_2 \tag{21}$$

$$O+CO(+M) \rightarrow CO_2(+M) \tag{22}$$

$$O \rightarrow \text{wall} \tag{23}$$

shows that the mixture of CO_2 and its decomposition products is quite unreactive towards $O(^3P)$ at room temperature. For the absorbed light intensities and reaction vessel dimensions usually employed in conventional, slow photolysis, diffusion of atoms to the wall (23) will be competitive with (19) even at pressures of hundreds of torr. In addition to atom recombination at the wall there may well be reaction with the wall itself or with some contaminant, leading to a loss of product oxygen. There is considerable experimental evidence to support this hypothesis of diffusive loss of a reaction intermediate. At finite conversions, (20), (21) and (22) consume O atoms, although at 1 % decomposition calculations suggest that (23) will be competitive with the other reactions even at pressures of tens of torr. Ozone formation (20) may be partly responsible for the departure of the product ratio $[CO]/[O_2]$ from 2, although the observed oxygen deficiency under certain conditions is much too large to be entirely accounted for in this way[46].

2.3.3 *Photosensitized decomposition of* CO_2

Both the resonance lines at 1849 A and at 2537 A from a low-pressure mercury lamp effect the decomposition of CO_2 containing mercury vapour[67–69]. At the longer wavelength decomposition is a very inefficient process resulting in the formation of small yields of CO ($\Phi_{CO} < 0.01$) and HgO (O_2 is not formed)[69]. The low isotopic enrichment of the product HgO shows that the direct abstraction of an oxygen atom from CO_2 by the excited Hg $6(^3P_{1,0})$ atoms does not occur to any appreciable extent[69]. Since the Hg $6(^3P_1)$ atom possesses insufficient energy to break the C–O bond in CO_2 directly, the decomposition must be due to more energetic atoms, such as Hg $7(^3S_1)$, formed from the ^3P state by the secondary absorption of radiation, or alternatively must be due to the stepwise excitation of CO_2.

The photosensitized decomposition of CO_2 at 1849 A produces CO and a solid containing mercuric oxide and oxalates of mercury[68]. Oxalate formation is believed to result from dimerization on the vessel walls of a complex formed between CO_2 and Hg $6(^1P_1)$ atoms.

Excited Xe atoms formed from the ground state by absorption of 1470 A radiation decompose CO_2[70, 71]. A noticeable difference[70] between the results obtained in the Xe-photosensitized experiments and in the direct photolysis studied at the same wavelength is the complete recovery of product O_2 achieved in the former case. Other major differences which have been reported and given mechanistic significance[70] have since been ascribed[72] to effects associated with the intense absorption of the 1470 A resonance radiation by Xe and with the divergence of the radiation.

2.3.4 Radiolysis of CO_2

In contrast to the effects of vacuum ultraviolet radiation, ionizing radiation is remarkably inefficient in decomposing gaseous CO_2[73]. At high dose rates, very small steady-state concentrations of the products CO and O_2 are generated, amounting to no more than a few hundred p.p.m. of the CO_2 concentration. Observations[74, 75] on the exchange reactions of both labelled C and O atoms in irradiated mixtures of CO, CO_2 and O_2 demonstrate that the stability towards ionizing radiation is only apparent and that the small extent of decomposition is due to the establishment of a dynamic equilibrium between decomposition and reoxidation. The steady-state concentrations of products are proportional to (dose rate)$^{\frac{1}{2}}$, suggesting that equilibrium is established between CO formation, which depends linearly on the dose rate, and its destruction by reaction with an entity (or entities) whose rate of formation also depends linearly on the dose rate. The much greater efficiency of the back reaction in radiolysis compared to that in photolysis indicates that intermediates other than those involved in the photodecomposition are responsible for the rapid reoxidation of CO. There is considerable evidence that ionic species effect the rapid reoxidation, although they have not been identified unambiguously[73, 76–78].

Liquid carbon dioxide is decomposed efficiently by ionizing radiation[79]. The decreased radiation stability of the liquid phase compared to the gas phase has been attributed to the much smaller contribution of ion–molecule reactions to radiolysis in the condensed phase, where an efficient geminate charge neutralization process is likely to minimize the occurrence of such processes. Ion–molecule reactions are probably responsible for the rapid reoxidation observed in the gas phase. The yields of CO, O_2 and O_3 from the γ-radiolysis of liquid CO_2 can be

References pp. 132–141

accounted for in terms of the production of ground-state O atoms and their consumption in reactions (19)–(22).

2.4 CARBON DISULPHIDE

2.4.1 Thermal decomposition of CS_2

Shock-tube studies covering the temperature range 1800–3700 °K have shown that the decomposition of CS_2 is a unimolecular reaction[38, 80, 81]. With highly dilute mixtures of CS_2 in Ar it has been possible to measure the rate of the initial dissociation alone[38, 81]

$$M+CS_2(^1\Sigma) \rightarrow CS(^1\Sigma)+S(^3P)+M(\Delta H_0^0 = 92.58 \text{ kcal.mole}^{-1}) \quad (24)$$

For concentrations of $CS_2 > 0.5\,\%$ the reaction

$$S+CS_2 \rightarrow S_2+CS \quad (25)$$

was sufficiently rapid to maintain a stationary S atom concentration during the measurements, and the observed[80] rate coefficient was 2 k_{24}. Arrhenius expressions (Table 4) have been obtained for k_{24} (M = Ar) both for the low-pressure,

TABLE 4

RATE CONSTANTS FOR THE UNIMOLECULAR DECOMPOSITION OF CS_2

Reaction	*Rate coefficients*[a]	*Temp.* (°*K*)	*Ref.*
$CS_2+Ar \xrightarrow{24} CS+S+Ar$	$k_{24} = 10^{12.64} \exp-(81{,}800 \pm 5{,}000/RT)$	2250–3350	80
	$k_{24,CS_2} \sim 20\, k_{24,Ar}$		80
	$k_{24} = 10^{12.56} \exp-(80{,}300/RT)$	1800–3700	81
	$k_{24}^{\infty} = 10^{12.6} \exp-(87{,}000 \pm 2{,}000/RT)$	1950–2800	38

Units: l.mole^{-1}.sec^{-1} (for k_{24}); sec^{-1} (for k_{24}^{∞}).

second-order region ([Ar] $< 10^{-2}$ *M*) and for the limiting high-pressure region ([Ar] > 1 *M*). The values of the activation energy for the decomposition in both limiting regions prove that reaction takes place *via* a singlet–triplet transition (24), and not *via* the spin-allowed process

$$CS_2(^1\Sigma) \rightarrow CS(^1\Sigma)+S(^1D_2) \ (\Delta H_0^0 = 119.0 \text{ kcal.mole}^{-1})$$

Studies[82] of the visible emission from shock-heated CS_2 confirm this: the temperature dependence of the emission shows that the upper state involved lies approximately 74 kcal.mole^{-1} above ground-state CS_2 and is therefore $^3A_2$ (or possibly $^3B_2$). The relative ease of crossing into this triplet state is evidenced by the almost normal pre-exponential factor for k_{24}^{∞}; this is not entirely unexpected in view of the high spin–orbit coupling parameter of the S atom. A high-pressure activation energy appreciably smaller than the energy requirement for dissociation is to be expected for a non-adiabatic reaction[83]. The low value of the activation energy at low pressures may be reconciled with a bond energy of 93 kcal.mole^{-1} by a quantum statistical theory of "weak" collisions[38].

2.4.2 *Photolysis of* CS_2

The photochemistry of CS_2 has been studied in the wavelength region 1900–3600 A, where absorption consists of at least two systems of discrete bands[2, 85]. The intense absorption from 1900–2200 A corresponds to a $\tilde{A}^1B_2 \leftarrow \tilde{X}^1\Sigma_g^+$ transition; photodissociation occurs for $\lambda < 2068$ A[84], the primary products being vibrationally excited ground-state CS and a ground state S atom[85], *viz.*

$$CS_2 + h\nu(\lambda < 3090 \text{ A}) \rightarrow CS(^1\Sigma^+) + S(^3P)$$

Chemical studies[86] have confirmed the spectroscopic observation[85] that the sulphur atoms generated by photolysis in the 1950–2250 A region are entirely $S(^3P)$. It appears that several transitions may be responsible for the moderately intense absorption system from 2800–3600 A, of which only one, the $\tilde{a}^3A_2 \leftarrow \tilde{X}$, has been positively identified[2]. Since S atoms are not formed in this region[86], photodecomposition must be the result of bimolecular reactions of excited molecules. Decomposition at all wavelengths results in the formation of solid sulphur and a solid polymer $(CS)_x$, but quantum yield data are not available. There are indications that the quantum yield of decomposition may be quite small for photolysis in the 2800–3600 A region. Studies[82, 87] of the fluorescence from the upper state (or states) suggest that electronic deactivation occurs at practically every collision, that reactive collisions between CS_2^* and CS_2 are infrequent, and that the contribution to decomposition from reaction between two CS_2^* molecules is only important at the high light intensities employed in flash photolysis.

The CS molecule produced in the primary step is quite long-lived and disappears in a very slow heterogeneous polymerization reaction[88]. Kinetic spectroscopy[85, 89] of S_2 subsequent to the flash photolysis of CS_2 (or COS) shows that $S(^3P)$ atoms do not react with the parent molecule at temperatures in the range 290–370 °K, but recombine *via* the intermediate formation of the stable complex CS_3, *viz.*

$$S+CS_2+M \underset{-26}{\overset{26}{\rightleftharpoons}} CS_3+M \qquad K_{26} \tag{26}$$

$$S+CS_3 \underset{27}{\rightarrow} CS_2+S_2 \qquad k_{27} \tag{27}$$

At low [M] the disappearance of S is first order in S, *viz.*

$$-\frac{d[S]}{dt} = 2k_{26}[CS_2][M][S]$$

whereas at high [M] it is second order in S, *viz.*

$$-\frac{d[S]}{dt} = 2k_{27}K_{26}[CS_2][S]^2$$

Estimates of k_{26} and $k_{27}K_{26}$ have been made[89] from measurements of the effective rate coefficients for recombination at low and high inert gas pressures. A more exact expression for the observed second-order rate coefficient is

$$k_{obs} = \frac{2k_{27}K_{26}[CS_2]}{(1+K_{26}[CS_2])^2}$$

since estimates of [S] were based on initial and final values of $[S_2]$ and the assumption that $[CS_3] \ll [S]$. While the latter assumption held well for small $[CS_2]$, with increasing $[CS_2]$ k_{obs} approached a maximum value. A rough estimate of K_{26} was made from the pressure of CS_2 at which this value was attained. A similar recombination mechanism is envisaged for COS and NO, and estimated rate coefficients for the various chaperon molecules according to the above scheme are summarized in Table 5. In the case of COS it is not clear to what extent $S(^1D)$ reactions contribute to the measured rate coefficients[89].

TABLE 5

RATES OF RECOMBINATION OF SULPHUR ATOMS IN THE PRESENCE OF VARIOUS CHAPERON MOLECULES (Y)

$$Y+S+M \underset{-26}{\overset{26}{\rightleftharpoons}} YS+M \qquad YS+S \overset{27}{\rightarrow} Y+S_2$$

Y	M	k_{26} ($l^2.mole^{-2}.sec^{-1}$)	k_{27}/k_{-26}	$K_{26} = k_{26}/k_{-26}$ ($l.mole^{-1}$)	*Temp.* (°*K*)	*Ref.*
CS_2	Ar	$(3\pm1)\times10^{11}$	$(1.2\pm0.5)\times10^4$	$\sim2\times10^5$	300	89
COS	Ar	$(3\pm1)\times10^9$	$(4.1\pm0.2)\times10^4$	$\sim10^4$	300	89
NO			$k_{27}K_{26} > 2\times10^{14}$ $l^2.mole^{-2}.sec^{-1}$		300	89

2.5 CARBONYL SULPHIDE

2.5.1 Thermal decomposition of COS

Decomposition of gaseous COS proceeds measurably at temperatures in excess of 350 °C. Early investigators[90] showed that in static systems between 350 and 600 °C a heterogeneous reaction

$$2\,COS = CS_2 + CO_2$$

accompanies the homogeneous process generating CO and sulfur. By assuming that the rate of the homogeneous process was controlled by a unimolecular reaction, Partington and Neville[90] were able to deduce an activation energy of 73 kcal.mole^{-1} for reaction in the temperature range 550–600 °C. Recent studies of the decomposition behind a shock front in dilute COS/Ar mixtures have indeed shown that the rate-controlling reaction is unimolecular, *viz.*

$$COS(^1\Sigma) \rightarrow CO(^1\Sigma) + S(^3P)\ (\Delta H_0^0 = 72.1 \text{ kcal.mole}^{-1}) \qquad (28)$$

At low concentrations ($< 10^{-1}$ mole.l^{-1}, 1600–3200 °K), the rate of (28) is determined by the rate of collisional activation to highly vibrationally excited levels of ground-state COS which possess a triplet component by virtue of spin-orbit coupling. Dissociation takes place from the triplet state. At sufficiently

TABLE 6

RATE COEFFICIENTS OF REACTIONS INVOLVED IN THE DECOMPOSITION OF COS

Reaction	*Rate coefficient*[a]	*Temp.(°K)*	*Ref.*
$COS + M \xrightarrow{28} CO + S(^3P) + M$	$k_{28}^{Ar} = 1.5 \times 10^{11} \exp-(60{,}700/RT)$	1600–3100	91
	$k_{28}^{Ar} = 1.11 \times 10^8 T^{0.5} (71{,}500/RT)^{1.87} \exp-(71{,}500/RT)$[b]	2000–3200	92
	$k_{28}^{\infty} = 3.7 \times 10^{11} \exp-(68{,}300/RT)$	1550–2700	91
$S(^3P) + COS \xrightarrow{29} S_2 + CO$	$k_{29} = 6.0 \times 10^8$	2570	92
$S(^3P) + COS \xrightarrow{30} CS + SO$	$E_{30} - E_{29} = 19$ kcal.mole^{-1}	2500–3000	92
$S(^1D_2) + COS \rightarrow CO + S_2$ ($^1\Delta$ and/or $^1\Sigma$) or $\rightarrow S(^3P) + COS$	$k_{31} > 4 \times 10^{10}$	300	96
$S(^1D) + Ar \xrightarrow{k_q} S(^3P) + Ar$	$k_q > 6 \times 10^7$	300	96
$S(^1S_0) + COS \rightarrow$ removal of $S(^1S_0)$	$k_{31s} = (6 \pm 1) \times 10^9$	300	96

[a] Units: l.mole^{-1}.sec^{-1} (for k_{28}^{Ar}, k_{29}, k_{31}, k_q, k_{31s}); sec^{-1} (for k_{28}^{∞}).
[b] The experimental activation energy = 64 kcal.mole^{-1}.

high densities (> 1 M), the rate of dissociation is first order and is controlled by the transition probabilities for intersystem crossing and the shape of the potential surfaces of triplet and ground states in the region of their intersection. The Arrhenius expressions (Table 6) obtained for the second-order rate coefficient k_{28}^{Ar} are consistent with a theory of weak collisions and the currently accepted value for the bond dissociation energy. Hay and Belford's RRKM calculations[92] indicate a low collisional deactivation efficiency of only 1 %.

At the high temperatures attained in shock-tube experiments (28) is followed by

$$S(^3P)+COS \rightarrow S_2+CO \tag{29}$$

which may contribute to the measured rate of COS disappearance at high COS concentrations and low temperatures. Above 2400 °K CS and SO appear in the products in comparable quantities, suggesting a contribution from

$$S(^3P)+COS \rightarrow CS+SO \tag{30}$$

2.5.2 Photolysis of COS

The photolysis of COS has proven to be a most valuable source of S atoms for kinetic studies[93] and considerable effort has been spent in investigating the photochemistry of this molecule. Absorption commences at 2550 A with a continuum extending to 1600 A[2]; photolytic studies in this region[93-95] have been confined to λ > 2200 A where the primary processes

$$COS+h\nu \rightarrow CO+S(^3P) \quad (\Delta H_0^0 = 72.1 \text{ kcal.mole}^{-1})$$

$$COS+h\nu \rightarrow CO+S(^1D_2) \quad (\Delta H_0^0 = 98.5 \text{ kcal.mole}^{-1})$$

are possible, but

$$COS+h\nu \rightarrow CO+S(^1S_0) \quad (\Delta H_0^0 = 135.5 \text{ kcal.mole}^{-1})$$

is not. Photolysis in the region 1400–1600 A, where absorption consists of strong, diffuse bands[2], gives rise to both $^1S_0$ and $^1D_2$ sulphur atoms[96]. The available evidence indicates[96] that for λ > 1400 A the primary production of O atoms is unimportant.

CO and sulphur are the end products of photolysis. The quantum yield of CO is 1.8, independent of COS pressure, for the photolysis at low light intensities of pure, gaseous COS at 2527 A and at 2288 A[94]. This value is exactly halved by the addition of a large excess of olefin, a scavenger for both ground-state and excited sulphur atoms[93], suggesting that the primary quantum yield of S and CO is 0.9

and that in the absence of scavenger the S atoms react according to

$$S+COS \rightarrow S_2+CO \tag{31}$$

The apparent inefficiency of the primary dissociation may well be due to experimental error. An upper limit of 26 % may be set on the percentage of $S(^3P)$ atoms produced in the primary photolytic act from the result that only 74 % of the S atoms can be scavenged by paraffinic hydrocarbons[93, 94]. $S(^1D)$ atoms disappear in the extremely rapid reaction (31) or are quenched, either by COS or by added gases, to the ground state. Reaction (31) is at least 30 times slower for $S(^3P)$ than for $S(^1D)$[94] (Table 6); at the higher [S] encountered in flash photolysis[95], $S(^3P)$ atoms are removed by recombination (see above) rather than by (31). The slight reduction in Φ_{CO} with increasing inert gas pressure for the low-intensity photolysis of COS is possibly due to an enhanced recombination of S atoms[93, 94].

Flash photolysis experiments[95, 97] have revealed the presence of $S_2(^3\Sigma_g^-)$ in its ground vibrational state. There is good reason to believe, however, that the product of reaction between $S(^1D_2)$ and COS is $S_2(^1\Delta_g)$ or $S_2(^1\Sigma_g^+)$ which is collisionally deactivated slowly to the ground state. Vibrational relaxation in the $^3\Sigma_g^-$ state is very rapid. $S_2(^3\Sigma_g^-)$ disappears in a second-order reaction, presumably

$$2\,S_2+M \rightarrow S_4+M \tag{32}$$

Measurements of the rate coefficient for S_2 removal represent upper limits for k_{32}, since kinetic mass spectrometry[93, 95] of S_4, S_5, S_6 and S_8 suggests[85, 93, 95] that reactions of the type

$$S_2+S_m \rightarrow S_{m+2}$$

may contribute significantly to S_2 removal. Because of this and the somewhat unusual dependence of k_{32} on the nature of the chaperon M, estimates[95] of k_{32} should be accepted with considerable reserve.

$S(^3P)$ and $S(^1S_0)$ have been observed spectroscopically[96] to be primary products of photolysis of COS by light in the region 1400–1600 A. That the primary production of $S(^1D_2)$ is also important at these wavelengths has been inferred[96, 97] from observations on the kinetics of formation and removal of $S_2(^1\Delta_g)$, $S_2(^3\Sigma_g^-)$, and sulphur atoms. Estimates have been made from the kinetic spectroscopy of these species of the rate coefficients for the destruction of $S(^1D_2)$ and $S(^1S_0)$. As has been found for the case of oxygen, the more energetic singlet state is, somewhat surprisingly, less reactive and less easily deactivated by collisions than the lowest singlet state.

The $Hg(^3P_1)$ sensitized decomposition of COS produces S atoms exclusively in the ground state, [94] *viz.*

$$Hg(^3P_1)+COS(^1\Sigma^+) \rightarrow Hg(^1S_0)+CO(^1\Sigma^+)+S(^3P)$$

The quantum yield, $\Phi_{CO} = 1.8$, reported[94] for the photosensitized reaction indicates that the primary process

$$Hg(^3P_1)+COS \rightarrow HgS+CO$$

is unimportant.

2.6 CARBON DISELENIDE AND CARBONYL SELENIDE

The results of an investigation[98] of the flash photolysis of CSe_2 and COSe suggest that the decompositions of these molecules follow mechanisms analogous to those established for the sulphides. Photolysis of CSe in the 2300 A region populates two electronically excited states, probably $^3B_2$ and $^3A_2$, in addition to effecting dissociation into CSe and $Se(^3P)$. Ground-state Se atoms are removed by a third-order recombination reaction rather than by reaction with CSe_2. Ultraviolet photolysis of COSe generates $Se(^1D_2)$ atoms which then abstract Se from the parent molecule.

3. Nitrogen oxides and oxyacids

Seven oxides of nitrogen, N_2O, NO, N_2O_3, N_2O_4, NO_2, N_2O_5 and NO_3 have been characterized in all. Of these, NO_3 cannot be isolated as such, although there is well-documented evidence, both spectroscopic and kinetic, for its existence as a transient species in certain reactions. N_2O_3 and N_2O_4 possess very weak N–N bonds and exist free of their dissociation products in the condensed phases only. In the gas phase, the equilibria

$$N_2O_3 \rightleftharpoons NO+NO_2$$

$$N_2O_4 \rightleftharpoons 2\,NO_2$$

are rapidly established and N_2O_3 and N_2O_4 are often only minor components of gaseous mixtures of NO and NO_2[99,100]. Their presence should not be totally ignored in gas-phase investigations, however, especially in photochemical studies where they may make significant contributions[99-101] to the total light absorbed by an equilibrium mixture. Unfortunately, the photochemistry of N_2O_3 and N_2O_4 is not established. In the case of N_2O_3 there is, in fact, no information on the kinetics of its decomposition, apart from vague indications[99,101] that its formation from the association of NO and NO_2 is very rapid. The decomposition of this molecule will therefore not be discussed further in this chapter. For a general account of the chemical and physical properties of N_2O_3 the reader is referred to the review article by Beattie[99].

Nitric acid is the only oxyacid of nitrogen which is at all stable in the free condition. The kinetics of its decomposition are discussed below. Nitrous acid exists in the gas phase, although it cannot be isolated. Information on the kinetics of its decomposition is restricted to an estimate[103] of the half-life of the forward reaction in the rapid gas-phase equilibrium

$$NO+NO_2+H_2O \rightleftharpoons 2\,HNO_2$$

although there are indications[104] that a quantitative kinetic study of its formation and decomposition may soon be forthcoming.

Such a large group of volatile, reactive compounds as the nitrogen oxides has naturally received considerable attention from kineticists since the earliest days of rate measurements and there is now a wealth of rate data pertinent to the decompositions of these compounds. Indeed, it may be said that the oxides of nitrogen have provided a testing ground for the theories of chemical kinetics. The mechanisms by which the nitrogen oxides decompose are, with few exceptions, well established, and rate data derived for many of the proposed elementary reactions can be accepted with some confidence. Such data have proved invaluable in furthering the understanding of the rôles played by nitrogen oxides in the photochemistry of the atmosphere and in the sensitization and inhibition of certain chemical processes.

3.1 NITROUS OXIDE

3.1.1 Thermal dissociation of N_2O

The homogeneous thermal decomposition of N_2O is a well-documented[105] example of a unimolecular reaction. The high-pressure limit of the rate coefficient for the decomposition is experimentally accessible, so that N_2O is one of the simplest molecules whose unimolecular decomposition may be studied over the full pressure range from low- to high-pressure limit. Not surprisingly, therefore, the decomposition of N_2O has been the subject of many experimental investigations[106–109] designed to test the various theories of unimolecular reactions[105, 110–116]. This aspect of research on the N_2O dissociation has perhaps been overemphasized at the expense of a complete understanding of the mechanism of the reaction. Attempts to interpret the pressure and temperature dependences of the rate coefficient for the initial step in terms of modern versions of Kassel's

References pp. 132–141

theory have met with moderate success[109, 110, 112, 113], but the relative importance of the several postulated secondary reactions has not yet been established with certainty. In consequence, the relationship for a given set of experimental conditions between the observed first-order rate coefficient k defined by

$$-\frac{d[N_2O]}{dt} = k[N_2O]$$

and the rate coefficient for

$$N_2O(^1\Sigma) \rightarrow N_2(^1\Sigma)+O(^3P) \qquad (1)$$

which is believed to be the initial step in the decomposition, is not always known. Kineticists have usually adopted one or other of the two extreme assumptions, $k = k_1$ or $k = 2k_1$, often with no experimental justification.

Available experimental data on the thermal decomposition of N_2O were either obtained by the classical methods[105] of gas kinetics in the temperature range 800–1100 °K, or from shock-tube studies[106–109, 117–119] in the range 1500–2500 °K. Johnston[105] has summarized the low-temperature data obtained prior to 1950. By making suitable allowances for a heterogeneous contribution to the dissociation rate he has shown that the results of different workers for the value of k are in fair accord and show behaviour typical of a straightforward unimolecular reaction. Thus, at 888 °K k increases at first linearly with pressure of N_2O up to $p \cong 50$ torr, increases more slowly as the pressure is further raised and finally levels off to a high-pressure limit k^∞ at $\sim$ 70 atm. Arrhenius expressions for k^∞ and k^0 (the limiting second-order rate coefficient at zero pressure) derived from Johnston's review[105] are given in Table 7.

The validity of Johnston's interpretation of the experimental facts in terms of the simple unimolecular dissociation (1) has been questioned by Lindars and Hinshelwood[120] and by Reuben and Linnett[121]. These workers maintain that isothermal plots of k *versus* p are not smooth curves, but consist of a number of straight lines linked by markedly curved portions. To explain such behaviour they incorporate into their mechanism a collision-induced crossover of vibrationally excited N_2O ($^1\Sigma$) to repulsive $^3\Pi$ and $^3\Sigma$ states. While we incline towards the simpler view held by Johnston[105] and others[106–116], we feel that this feature of the decomposition kinetics merits further investigation.

The results of shock-tube investigations are consistent with the low-temperature data and certainly support[106, 109, 110] the view that the decomposition is a simple unimolecular process. Although free from uncertainties regarding heterogeneous contributions to the reaction rate, the shock-tube technique suffers from errors arising from self-heating effects[118]. Olschewski *et al.*[109] point out that for the case of N_2O decomposition such errors are negligible only for concentrations of

TABLE 7

RATE COEFFICIENT DATA FOR N_2O THERMAL DECOMPOSITION

Rate coefficient	*Rate coefficient expression*[a]	*Temp.(°K)*	*Total conc. (mole.l^{-1})*	*Method*	*Ref.*
$k^0(=2k_1^0?)$	$10^{12.71}\exp(-59{,}200/RT)$	900–1050	$[N_2O] \rightarrow 0$	Static	105[b]
$k^0(=2k_1^0)$	$2.5\times10^4\,T^{\frac{1}{2}}(60{,}000/RT)^{5.09}\exp(-60{,}000/RT)$[c]	1800–2500	1–4×10^{-3} (96 % Ar)	Shock	107
$k^0(=2k_1^0)$	$1\times10^{11}\exp(-49{,}500/RT)$	1500–2200	1×10^{-2} (98 % Ar)	Shock	106
$k^0(=2k_1^0)$	$1\times10^{12}\exp(-58{,}000/RT)$	1500–2500	$<6\times10^{-3}$ (>99 % Ar)	Shock	109
$k^0(=2k_1^0)$	$10^{11.8}\exp(-56{,}800/RT)$	1670–1920	3×10^{-2} (94 % Ar)	Shock	118
$k^\infty(=2k_1^\infty)$	$4.4\times10^{11}\exp(-60{,}000/RT)$	900–1050	$[N_2O] \rightarrow \infty$	Static	105
$k^\infty(=2k_1^\infty)$	$10^{11.4}\exp(-59{,}500/RT)$	1400–2100	>1 (>99.8 % Ar)	Shock	109
$k^\infty(=2k_1^\infty)$	$2.3\times10^{10}\exp(-55{,}700/RT)$	1600–2200	0.15 (98 % Ar)	Shock	106

[a] Units: l.mole^{-1}.sec^{-1} (k^0); sec^{-1} (k^∞).
[b] Supported by refs. 117 and 119.
[c] $E = 60$ kcal./mole^{-1} is an assumed value.

less than 1 % of N_2O in Ar carrier. On this basis, results obtained with higher N_2O concentrations, especially those of Bradley and Kistiakowsky[117] who employed 24 % N_2O in Ar, are suspect. Nevertheless, there is fair agreement between the k values reported by several shock-tube operators; the results of the German workers[109] are perhaps the most reliable since they were obtained for very low N_2O concentrations. Table 7 gives Arrhenius expressions for the limiting low- and high-pressure rate coefficients; rate coefficients for intermediate concentrations may be obtained by reference to the original publications[105, 106, 109].

The low pre-exponential factor associated with k^∞ is an indication of a non-adiabatic reaction. Since there is good evidence[122] against a significant extent of thermal dissociation of N_2O to $N+NO$, the alternative dissociation, reaction (1), has been universally accepted as the initial step in the decomposition. The subsequent fate of the ground-state O atoms is unquestionably reaction with N_2O, *viz.*

$$O+N_2O \rightarrow N_2+O_2 \tag{2a}$$

$$O+N_2O \rightarrow 2\,NO \tag{2b}$$

at least at small conversions. Loss of O atoms in a third-order recombination is of no consequence even at the highest pressures used. Atom recombination at the walls

$$O \rightarrow \text{wall} \tag{3}$$

is important in low-temperature experiments[123]. For finite conversions the following reactions must also be considered:

$$O+NO \rightarrow NO_2+h\nu \qquad (4a)$$

$$O+NO+M \rightarrow NO_2+M \qquad (4b)$$

$$NO_2+N_2O \rightarrow N_2+O_2+NO \qquad (5)$$

$$O+NO_2 \rightarrow NO+O_2 \qquad (6)$$

$$NO+N_2O \rightarrow NO_2+N_2 \qquad (7)$$

Studies[106, 124] of the NO-catalyzed decomposition of N_2O show that (7) may be ignored except at large extents of decomposition and at low temperatures. The observation[123] of a green chemiluminescence during the decomposition of N_2O was a clear demonstration of the importance of contributions from reactions (4a) and (4b) even at quite small conversions. Through these reactions the product NO rapidly inhibits its own further formation in the decomposition. Since the marked fall-off in NO formation rate is not accompanied by any irregularities in the loss-of-N_2O *versus* time curves, one is obliged[123] to assume that the NO_2 generated in (4) destroys a molecule of N_2O by (5) rather than reacting by (6). The mechanism of Kaufman *et al.*[123], consisting of steps (1)–(5), satisfactorily accounts for the observed dependences of the NO yield and the chemiluminescence of reacting N_2O on time, temperature, pressure of N_2O and surface/volume ratio. Reuben and Linnett[121] have offered an alternative interpretation of Kaufman's results in terms of a mechanism which distinguishes between translationally "hot" O atoms produced in (1) and O atoms which have been thermalized through collisions. By assuming that the "hot" O atoms only are capable of generating NO by reaction (2b) with N_2O, it is possible to account for the peculiar (and unexplained) order of effectiveness[123] of additives in quenching the initial yield of NO per N_2O molecule decomposed. However, as Fenimore and Jones[125] point out, such an interpretation is not consistent with the more recent determinations of k_{2a} and k_{2b}.

A dispute[123, 125, 126] over the magnitude of k_{2b} (Table 8) seems to have been settled in favour of Fenimore's value by recent observations[107, 109, 127]. The finding of Gutman *et al.*[107] of a stationary state in O atoms for the decomposition of 4 % N_2O in Ar at temperatures between 1800° and 2500 °K certainly supports Fenimore's rate coefficient. Olschewski *et al.*[109] have clearly demonstrated in their shock-tube study that departure from a steady state in O atoms only occurs at very low N_2O concentrations, *viz* for $T > 2000$ °K at 0.02 % in Ar. This observation suggests, in fact, that Fenimore's value of k_{2b} represents a lower limit for the rate of O atom destruction at such temperatures. Evidently all shock-tube measurements to date refer to steady-state conditions, where $k = 2k_1$. For temperatures below 1300 °K, rate measurements obtained by an adiabatic compression technique[128] show that the unimolecular decomposition of N_2O proceeds with an effective rate coefficient $k = k_1$. It has been suggested[128] that this is the result of (4b) and (6) becoming the dominant modes of O atom destruction at finite conversions. This suggestion is not necessarily in conflict with the above conclusion of Kaufman *et al.*[123] that NO_2 formed in the low-temperature decomposition disappears by (5) rather than (6): the steady-state concentration ratio, [O]/[N_2O], in a typical adiabatic compression or shock-tube experiment is several orders of magnitude higher than in a static run of the type carried out by Kaufman *et al.*, thereby enhancing the importance of (6) relative to (5).

Table 8 gives expressions for the rate coefficients of secondary reactions in the thermal decompositions of N_2O, where these have been measured. To our knowledge no determination has been made of k_5.

Theoretical treatments of the initial step (1) are numerous[109–116]. Attention has been largely confined to the limiting rate coefficients (at low and high pressures) k^0 and k^∞. Anharmonicity effects in N_2O are sufficiently large[112] to ensure good intramolecular coupling between all four modes of vibration. Accordingly, the RRK approach is favoured[130] over a Slater treatment in calculating k^0, although the classical version of Kassel's theory is inappropriate in view of the high vibrational temperatures of the oscillators[109, 110, 112]. The classical expression is, nevertheless, a convenient means of summarizing experimental data for k^0. Agreement between RRK theory and experiment is poor[113] unless one assumes a low efficiency of collisional activation. Adoption of a model of "weak collisions" is certainly consistent with the belief that decomposition occurs (through conversion to a $^3\Pi$ state) in a region of widely spaced vibrational energy levels. Using such a model, Olschewski *et al.*[109] calculate from their experimental data for k^0

TABLE 8

RATE CONSTANTS OF SECONDARY REACTIONS IN THE DECOMPOSITION OF N_2O

Reaction	*Rate coefficient*	*Rate coefficient expression (l.mole^{-1}.sec^{-1})*	*Temp(°K)*	*Ref.*
$O+N_2O \xrightarrow{2b} 2\,NO$	k_{2b}	$1\times10^8 \exp(-15{,}500\pm2{,}000/RT)$	876–1031	123[a]
	k_{2b}	$(1\pm1)\times10^{11}$ $\exp(-28{,}000\pm3{,}000/RT)$	1400–2000	125
$O+N_2O \xrightarrow{2a} N_2+O_2$	k_{2a}/k_{2b}	$0.3 \exp(1{,}000\pm2{,}000/RT)$	876–1031	123
	k_{2a}/k_{2b}	0.83 ± 0.14	1800–2500	107
	$k_{2a}+k_{2b}$	$(2.3\pm0.4)\times10^{10}$ $\exp(-25{,}000\pm800/RT)$	1700–2300	127
$NO+N_2O \xrightarrow{7} NO_2+N_2$	k_7	$2.5\times10^{11} \exp(-50{,}000/RT)$	924–1028	124
	k_7	$2.0\times10^{11} \exp(-50{,}000/RT)$	1600–2200	106

[a] Kaufman[129] now prefers an E of $\sim$ 21 kcal.mole^{-1}.

that the threshold energy for decomposition is 63 kcal.mole^{-1} and that the average vibrational energy removed per N_2O^*/Ar collision is ~ 0.1 kcal.mole^{-1}.

Both the Slater and the RRKM treatments are inappropriate for calculations of k^∞, since the dissociation is not characterized by a critical extension of one bond, but rather by the transition from one potential surface to another. In such a case the observed activation energy at high pressures will be lower than the energy threshold for reaction[110]. From their high-pressure data Olschewski *et al.*[109] calculate that $E_0 = 63$ kcal.mole^{-1} and that the transition matrix–element is ~ 100 cal.mole^{-1}, which is in good agreement with the spin-orbit interaction term for O atoms.

3.1.2 Photolysis and radiolysis of N_2O

Ionizing radiation decomposes gaseous nitrous oxide[131–136] into N_2, O_2, and nitrogen oxides by a series of secondary reactions following primary formation of N_2O^* and N_2O^+. The secondary processes include reactions of neutral fragments and of ionic species and are of considerable complexity. The former are similar to those which occur in the photolytic decomposition. Reactions of ionic species[131–136] will not be discussed in this article.

Light absorption by N_2O is detectable[137, 138] at all wavelengths shorter than 3070 A. The absorption spectrum consists of diffuse bands superimposed on an underlying continuum which exhibits a number of broad maxima. These features are characteristic of highly dissociative upper states, but the spectroscopic information by itself is insufficient to permit a distinction between the many possible modes of dissociation (Table 9) at a given wavelength. Duncan[138b] has ascribed a single electronic transition to each region of intense absorption, although such a simplification is not borne out by the results of photochemical experiments We will, however, retain the original terminology of Duncan[138b] in designating the regions of absorption.

The products of photolysis at all wavelengths are N_2, O_2 and NO; NO_2 may be formed through oxidation of NO by O_2 or O.

Region A; $\lambda > 2100$ A. This is a region of very weak, continuous absorption. There have been no attempts to determine quantum yields in this region, but studies[139] of the photolysis of mixtures of N_2O and hydrocarbons at 2139 A and 2288 A suggest that (b) (Table 9) is an important primary process. Sponer and Bonner[137] have attributed a weak maximum in the absorption (at 2900 A) to a transition to a $^3\Pi$ state which dissociates *via* (a), but it could equally well correspond to an upper $^1\Delta$ state which dissociates *via* (b).

Region B; 1600–2100 A. Quantum yield data for the photolysis of pure N_2O in this wavelength region are summarized in Table 10. The widely accepted Φ_{N_2} value of 1.44 is certainly not sufficiently well established to warrant quotation

TABLE 9

POSSIBLE MODES OF DISSOCIATION OF N_2O OF PHOTOCHEMICAL INTEREST

Photolyzing wavelength(A)		*States of primary products*				*Threshold*	
		N_2	O	NO	N	*(eV)*	*(A)*
1849	a	$^1\Sigma^+$	^3P			1.67	7425
	b	$^1\Sigma^+$	^1D			3.64	3405
	c			$X^2\Pi$	^4S	4.93	2514
	d	$^1\Sigma^+$	^1S			5.86	2115
1470	e			$X^2\Pi$	^2D	7.31	1696
	f	$A^3\Sigma_u^+$	^3P			7.89	1571
	g			$X^2\Pi$	^2P	8.50	1458
1236	h	$B^3\Pi_g$	^3P			9.06	1368
		$A^3\Sigma_u^+$	^1D			9.86	1257
		$B'^3\Sigma_u^-$	^3P			9.89	1253
		$a^1\Pi_g$	^3P			10.26	1208
				$A^2\Sigma^+$	^4S	10.38	1194
				$B^2\Pi_r$	^4S	10.62	1167
		$B^3\Pi_g$	^1D			11.03	1124
		$B'^3\Sigma_u^-$	^1D			11.86	1045
		$A^3\Sigma_u^+$	^1S			12.08	1026
		$a^1\Pi_g$	^1D			12.23	1013
				$A^2\Sigma^+$	^2D	12.76	971

to the second decimal place. A quantum yield for N_2 greater than unity implies dissociation to oxygen atoms which are capable of further reaction with N_2O to generate N_2, *viz.*

$$N_2O \rightarrow N_2 + O^*$$

$$O^* + N_2O \rightarrow N_2 + O_2 \tag{8a}$$

Since the rate of reaction of ground-state O atoms with N_2O is very slow at room temperature, excited O atoms are necessarily involved. Studies[139, 140, 144] of the photolysis of mixtures of N_2O, hydrocarbons and inert gases have clearly demonstrated the participation of electronically excited O atoms in the photolysis at 1849 A. Furthermore, a comparison[145] of relative rate data for reactions of the excited atoms with data for $O(^1D_2)$ strongly suggests that the latter species is dominant in this wavelength region. The production of small amounts of NH_3 from the photolysis of N_2O/H_2 mixtures[146] and the inhibition of NH_3 formation by NO additions suggests some contribution from the spin-forbidden dissociation process (c) (Table 9). Unquestionably N atoms are removed, in the absence of additives, by

$$N + NO \rightarrow N_2 + O(^3P) \tag{9}$$

References pp. 132–141

since the reaction of N with N_2O is extremely slow in comparison (Table 11). From a study of the isotopic distribution in product N_2 from the photolysis[147] of $N_2O/^{15}NO$ and $^{15}NNO/^{15}NO$ mixtures it would appear that (c) contributes between 15 and 25 % to the primary dissociation yield, with approximately one third of the N atoms arising through expulsion of the central N from N_2O †. Since any NO arising from (c) is lost by (9), the observed yield of NO must be the result of a reaction of $O(^1D)$ with N_2O

$$O^* + N_2O \rightarrow 2\,NO \tag{8b}$$

Quenching[146] of the NO yield by added H_2 is then understandable.

In principle, one should be able to calculate k_{8a}/k_{8b} from the observed quantum yields. In view of uncertainties in the NO quantum yield and the primary yield of O*, however, it is not possible at the present time to obtain a reliable estimate for this ratio.

An investigation[148] of 1849 A photolysis of aqueous N_2O has shown that not more than 10 % of quanta absorbed by N_2O in solution gives other than singlet oxygen atoms. This result conflicts with the gas-phase data[147].

Region C; 1380–1600 A. The absorption spectrum is highly structured in this region[138], so that significant participation of electronically excited N_2O is a distinct possibility. Quantum yield data (Table 10), however, suggest that the

TABLE 10

PRODUCT QUANTUM YIELDS (Φ) FROM N_2O PHOTOLYSIS

Wavelength (A)	*Temp(°C)*	Φ				*Ref.*
		$-N_2O$	N_2	O_2	NO	
1850–2000	26[a]		$1.44^b \pm 0.05$		0.82	140
1849	50	1.79 ± 0.10	1.44^c	0.51 ± 0.03	p.d.*	141
1850–2000	25	1.7	1.4^d	0.6	0.5	142
1849	25	2.0	1.44^d	0.5	1.0	143
1849	25		1.32^e			144
1470	25[f]		1.40	0.58	0.78	149
1470(+1295)	25[f]	1.8	1.40		0.79^g	150
1470[h]	50	1.70	1.44	0.50	p.d.*	141
1236	25[f]		1.34			147
1236	25[f]	1.46	1.18		0.55^g	150
1236	50[i]	1.0^j		0.25^j	p.d.*	153

[a] One run at 105 °C gave the same result. [b] Assuming $\Phi_{N_2} = 1$ for N_2O/C_2H_6 photolysis. [c] Assumed equal to Φ_{N_2} at 1470 A[141]. [d] Assumed. [e] Assuming $\Phi_{N_2} = 1$ for large CO_2, Xe additions. [f] Temperature not stated, probably close to 25 °C. [g] $NO + NO_2$. [h] See, however, ref. 150. [i] Temperature not stated, probably 50 °C. [j] Relative to $\Phi_{N_2} = 1.0$. * p.d. = pressure dependent.

† More recent experiments of the same type (K. F. Preston and R. F. Barr, *J. Chem. Phys.*, in press) show, however, that (c) does not occur, at least not in the region 1849–2288A. The contrary carlier results[147] were apparently due to complications arising from photolysis of NO.

1849 A mechanism is also operative at 1470 A. The measurements of Doering and Mahan[147], which show a marked dependence of Φ_{N_2} on N_2O and CO pressures and are therefore suggestive of the involvement of N_2O^*, are in conflict with the bulk of experimental data and have not been confirmed in later work[149]. It is tempting to conclude from the good agreement between quantum yields measured at 1849 A and 1470 A that the mechanisms of photolysis at the two wavelengths are identical. Certainly recent investigations[151] show that the possible additional primary processes (e) and (f) (Table 9) for 1470 A make only minor (although finite) contributions to the photolysis, and that photodissociation leads chiefly to electronically excited O atoms. Using the chemiluminescent emission from NO_2^* as a measure of $O(^3P)$ concentration in the photolysis of N_2O/NO mixtures containing various additives, Young *et al.*[151] obtained the following estimates of primary dissociation yields at 1470 A

$O(^1S)$	0.5	
$O(^1D)$	0.55	(including all or part of $O(^1S)$)
$O(^3P)$	0.08	(assumed equal to $N_2(A^3\Sigma)$ yield)
$N(^2D)$	$>3\times10^{-4}$	
$N_2(A^3\Sigma)$	0.08	

The results of experiments[147] with N^{15}-labelled N_2O and NO suggest that between 10 and 16 % of absorbed quanta at 1470 A led to N+NO, but do not permit a distinction between $N(^2D)$ and $N(^4S)$. Ejection of the internal N atom of N_2O is less important at 1470 than at 1849 A.

At wavelengths shorter than 1520 A photolysis of N_2O is accompanied by fluorescence[152] from the $B^2\Pi_r$ state of NO (β emission). Added or product NO quenches this β emission and enhances fluorescence of the γ bands from $NO(A^2\Sigma^+)$. The fluorescent states of NO must[152] arise in secondary reactions and these are believed[151, 152] to be, for β emission,

$$N(^2D)+N_2O \rightarrow N_2+NO(B^2\Pi)$$

with some contribution from

$$N_2(A^3\Sigma)+NO \rightarrow N_2+NO(B^2\Pi),$$

and, for γ emission,

$$N_2(A^3\Sigma)+NO \rightarrow N_2+NO(A^2\Sigma)$$

Available kinetic data for reactions of the intermediates with N_2O and the photolysis products are summarized in Table 11. Quenching cross sections for the electronically excited species are given elsewhere[139, 144, 145, 154, 155]. Contrary to an earlier expressed opinion[141, 153], $N_2(A^3\Sigma)$ does not induce decomposition of N_2O[151, 154, 159].

Region D: 1215–1380 A. Product quantum yields at 1236 A (Table 10) are noticeably smaller than at longer wavelengths, suggesting a reduced yield of excited oxygen atoms in this region. The primary dissociation to N atoms is not appreciably higher at this wavelength, amounting[147] to ~ 12 % of the primary processes producing N_2. The smaller quantum yields may be the result of an increased importance of (f) and (h). Young *et al.*[154] have detected both the $A^3\Sigma_u^+$ and the $B_3\Pi_g$ states of N_2 in the photolysis at 1236 A and have estimated that the primary quantum yield of the *B* state is only ~ 1 % of that of the *A* state. Unfortunately, the absolute quantum yield of $N_2(A^3\Sigma_u^+)$ is not known. Since (h)

TABLE 11

RATES OF REACTIONS OF INTERMEDIATES IN N_2O PHOTOLYSIS

Reaction[a]	*Rate coefficient* ($l.mole^{-1}.sec^{-1}$)	*Temp.* (°*K*)	*Ref.*
$O(^3P)+N_2O$	See Table 8		
$O(^3P)+O_2+M \rightarrow O_3+M$	See Table 22		
$O(^3P)+NO+M \rightarrow NO_2+M$	$(1.44\pm0.20)\times10^9 \exp+(1{,}930\pm100)/\boldsymbol{RT}$[b]	300–500	251, 158
$O(^3P)+NO \rightarrow N+O_2$	See Table 12		
$O(^3P)+NO_2 \rightarrow NO+O_2$	See Table 18		
$N(^4S)+N_2O \rightarrow NO+N_2$	$<10^5$	553	156
$N(^4S)+NO \rightarrow N_2+O$	$3\times10^{10} \exp(-200/\boldsymbol{RT})$	466–755	157
$N(^4S)+O_2 \rightarrow O+NO$	$1.48\times10^5\, T^{\frac{3}{2}} \exp(-5{,}683/\boldsymbol{RT})$	400–5000	158
$O(^1D)+N_2O$	1.1×10^{11}	~300	164
$+O_2$	4×10^{10}	~300	164
$+NO$	9×10^{10}	~300	64
$+N_2$	3×10^{10}	~ 300	64
	5×10^{10}	~ 300	164
$O(^1S)+N_2O$	6.6×10^9	298	165
$+NO$	4.8×10^{10}	298	165
$+O_2$	2.2×10^8	298	165
$+N_2$	$< 1\times10^5$	298	165
$N_2(A^3\Sigma)+N_2O$	3.9×10^9	~ 300	154
$+O_2$	2.3×10^9	~ 300	154
$+NO$	4×10^{10}	~ 300	154
$+N_2$	$< 4\times10^6$	~ 300	154
$N_2(B^3\Pi)+N_2O$	9.6×10^{10}	~ 300	154
$+N_2O$	3.5×10^{10}	~ 196	159
$+O_2$	6.6×10^{10}	~ 300	154
$+NO$	$\sim 1.4\times10^{11}$	~ 300	154
$+N_2$	1.6×10^{10}	~ 300	154

[a] Where the products are not indicated the rate coefficient refers to total removal of the electronically excited species.
[b] Units: $l^2.mole^{-2}.sec^{-1}$, M = N_2; for other M see quoted literature.

is apparently much less important than was originally suggested[153], the quenching effects observed by Doering and Mahan[147] at 1236 A must relate to excited oxygen atoms, rather than to $N_2(A^3\Sigma_u^+)$.

3.1.3 Mercury-photosensitized decomposition of N_2O

The mercury-photosensitized decomposition of N_2O was studied first by Manning and Noyes[160] and later in greater detail and more quantitatively by Cvetanović[161, 162]. The primary reaction step is

$$\text{Hg } 6(^3P_1)+N_2O \rightarrow N_2(^1\Sigma)+O(^3P)+\text{Hg } 6(^1S_0)$$

and not

$$\text{Hg } 6(^3P_1)+N_2O \rightarrow NO(^2\Pi)+N(^4S)+\text{Hg } 6(^1S_0)$$

Monoisotopic mercury photosensitization shows[163] that HgO is not formed in the primary quenching reaction.

3.2 NITRIC OXIDE

3.2.1 Thermal decomposition of NO

Because of its relevance to the chemistry of air at elevated temperatures the homogeneous decomposition of nitric oxide has received considerable attention from gas kineticists. References to early studies are given in the more recent work discussed below. The mechanisms for the decomposition and for the reverse reaction, the formation of NO from air, are well established and good quantitative data (Table 12) are available for the rate coefficients of the elementary steps.

In pure nitric oxide the bimolecular dissociation

$$2\,NO \underset{-10}{\overset{10}{\rightleftharpoons}} N_2+O_2 \qquad (10, -10)$$

is the dominant reaction[166, 167] at low temperatures, the contribution from the "unimolecular" step

$$NO+M \underset{-11}{\overset{11}{\rightleftharpoons}} N+O+M \qquad (11, -11)$$

becoming significant at low NO concentrations and very high temperatures

References pp. 132–141

TABLE 12

RATE COEFFICIENTS FOR ELEMENTARY STEPS IN THE THERMAL DECOMPOSITION OF NO

Reaction	*Rate coefficient expressions (l.mole⁻¹.sec⁻¹)*	*Temp.(°K)*	*Ref.*
$2\,NO \xrightarrow{10} N_2+O_2$	$k_{10} = 1.3\times10^{9}\exp(-63{,}800\pm600/RT)$	1370–1535	166
	$k_{10} = 2.4\times10^{20}\,T^{-\frac{5}{2}}\exp(-85{,}800/RT)$	1370–4300	166, 168
	$k_{10} = 1.5\times10^{10}\exp(-63{,}100\pm3{,}200/RT)$	1700–2100	176
$NO+M \xrightarrow{11} N+O+M$	$k_{11,Ar} = 7\times10^{12}\exp(-150{,}000/RT)$	3000–4300	168
	$k_{11,Ar} = 7\times10^{7}\,T^{\frac{1}{2}}(D/RT)^2\exp(-D/RT)^a$	3000–8000	169
	$k_{11,Ar} = 20\,k_{11,Ar}$	3000–8000	169
$O+NO \xrightarrow{12} N+O_2$	$k_{12} = 3.6\times10^{9}\exp(-39{,}500\pm1{,}000/RT)$	1500–1700	167
	$k_{12} = 3.2\times10^{6}\,T\exp(-39{,}100/RT)$	450–5000	167, 169 158
$N+O_2 \xrightarrow{-12} NO+O$	$k_{-12} = 1.48\times10^{5}\,T^{\frac{3}{2}}\exp(-5{,}683/RT)$	400–5000	158
	$k_{-12} = (1.41\pm0.07)\times10^{10}$ $\exp(-7{,}900\pm200/RT)$	300–910	172
	$k_{-12} = 3.8\times10^{9}\exp(-7{,}000/RT)$	450–600	173
$N+NO \xrightarrow{13} N_2+O$	$k_{13} = 3\times10^{10}\exp(-200/RT)$	476–755	157
$O+N_2 \xrightarrow{-13} N+NO$	$k_{-13} = 5\times10^{10}\exp(-75{,}500/RT)$	2000–3000	174, 175
	$k_{-13} = 3.7\times10^{10}\exp(-70{,}500/RT)$	2000–2900	171

D is the bond dissociation energy of NO.

(> 3000 °K) only[168, 169]. Thus, in a static system between 1370° and 1530 °K, the reaction was found[166] to be cleanly second order in NO and the rate was uninfluenced by changes in the vessel surface/volume ratio or by the addition of inert gases; below 1370 °K a heterogeneous contribution to the decomposition in quartz vessels cannot be ignored[166, 176, 177]. Values of k_{10} deduced from shock-tube studies[168, 169] are in fair agreement with values obtained by extrapolation of the low-temperature results from the static system. The low-temperature results by themselves lead to an activation energy of 63.8 kcal.mole^{-1}, whereas the combined results, covering a temperature range of 3000 °K, lead to a value of 85.5 kcal.mole^{-1}. The latter value definitely dictates against decomposition *via* the intermediate formation of N_2O,

$$2\,NO \rightarrow N_2O+O \quad \Delta H_0^0 = +36.5 \text{ kcal.mole}^{-1} \qquad (-2b)$$

since the largest value reported for the activation energy of the reverse reaction (2b) is only 28 kcal.mole^{-1} (see Table 8). This possibility cannot be dismissed, however, if 63.8 kcal.mole^{-1} is accepted as the activation energy for the decomposition.

At finite conversions, particularly above 1600 °K, an atomic mechanism[167–171] contributes to the decomposition

$$O+NO \underset{-12}{\overset{12}{\rightleftharpoons}} N+O_2 \qquad (12, -12)$$

$$N+NO \underset{-13}{\overset{13}{\rightleftharpoons}} N_2+O \qquad (13, -13)$$

Atomic oxygen is believed to be in thermal equilibrium with O_2 under these conditions

$$O_2+M \underset{-14}{\overset{14}{\rightleftharpoons}} 2\,O+M, \qquad (14, -14)$$

equilibrium being reached through

$$O_2+NO \rightleftharpoons NO_2+O$$

$$2\,NO_2 \rightleftharpoons 2\,NO+O_2$$

rather than through (14). Reaction (11) is a further source of atoms at very high temperatures. The contribution of this Zel'dovich mechanism[171] to the rate is

$$\frac{-d[NO]}{dt} = 2k_{12}K_{14}[NO][O_2]^{\frac{1}{2}} \bigg/ \left(1+\frac{k_{-12}[O_2]}{k_{13}[NO]}\right)$$

where K_{14}, the equilibrium constant of reaction (14, −14) is defined[167] as $K_{14} = [O]/[O_2]^{\frac{1}{2}}$. Thus, small additions of O_2 should have a catalytic effect, large additions an inhibitive effect. Kaufman and Decker[167] have clearly demonstrated this phenomenon and have achieved a quantitative interpretation of their rate measurements on O_2-catalyzed NO decomposition in terms of the Zel'dovich mechanism[171]. Their estimate for k_{12} is in accord with the high-temperature value derived from a shock-tube study[169] and with the low-temperature values for k_{-12}[158, 172, 173]. The observation of Fenimore and Jones[170] that O_2 impedes NO decomposition in the burnt gas from N_2O/H_2 flames over a wide range of $[O_2]/[NO]$ ratios is at variance with the simple Zel'dovich mechanism, and presumably is a consequence of unsuspected reactions of NO with flame components.

Nitric oxide is thermochemically the least stable of the nitrogen oxides, and while it is quite stable at room temperature at low pressures, it decomposes at a measurable rate at pressures of ~ 100 atm. Melia[178] has shown that, in the temperature range 30–50 °C and for pressures up to 400 atm, the NO decomposition is represented approximately by

$$3\,NO = N_2O+NO_2$$

(apart from traces of N_2O_5 product) and follows a third-order rate law

$$\frac{-\mathrm{d}[\mathrm{NO}]}{\mathrm{d}t} = k[\mathrm{NO}]^3$$

with $k = (7.2 \pm 0.9) \times 10^{-9}$ $1^2.\mathrm{mole}^{-2}.\mathrm{sec}^{-1}$, at 30 °C and 200 atm. A pre-equilibrium step followed by a slow reaction of dimer with NO, *viz.*

$$2\,\mathrm{NO} \rightleftharpoons (\mathrm{NO})_2$$

$$\mathrm{NO} + (\mathrm{NO})_2 \rightarrow \mathrm{N_2O} + \mathrm{NO_2}$$

can account for the observed rate law and the weak temperature dependence of k, but there is no guarantee that the process is not totally heterogeneous.

3.2.2 *Photolysis and radiolysis of* NO

Nitric oxide commences to absorb light at wavelengths just shorter than 2300 A; dissocation is energetically possible with radiation below 1906 A and photo-ionization begins at 1340 A[179]. A multitude of bound doublet states of NO is accessible by direct light absorption[179]. In addition, the lowest excited state, NO ($a^4\Pi_i$), may be produced by mercury ($^3P_1$) photosensitization[180].

Considerable effort has been devoted to the determination of cross-sections for quenching of fluorescence from some of the lower states, but apart from the first two excited states $a^4\Pi_i$ and $A^2\Sigma^+$, little attention has been paid to their photochemistry.

The NO ($a^4\Pi_i$) state produced by mercury photosensitization

$$\mathrm{Hg}(^3P_1) + \mathrm{NO} \rightarrow \mathrm{NO}(a^4\Pi_i) + \mathrm{Hg}(^1S_0) \tag{15}$$

reacts with ground-state NO to produce a short-lived dimer[181]

$$\mathrm{NO}^* + \mathrm{NO} \rightarrow (\mathrm{NO})_2^* \tag{16}$$

but is also deactivated very efficiently to the ground state

$$\mathrm{NO}^* + \mathrm{NO} \rightarrow 2\,\mathrm{NO} \tag{17}$$

There is some evidence[182-184] to suggest that the formation of HgO in the primary quenching step is insignificant and that quenching to Hg (6 $^3P_0$) is perhaps unimportant. Reaction (15) would appear to be the process responsible for the large quenching cross-section[185] of NO for Hg (6 $^3P_1$), and the surprisingly small quantum yield[181] ($\Phi_{N_2}+\Phi_{N_2O} \sim 10^{-3}$) of product formation is a consequence of the inefficiency of (16) relative to (17). $(NO)_2^*$ decomposes

$$(NO)_2^* \rightarrow N_2+O_2 \quad \text{or} \quad N+NO_2 \tag{18}$$

or reacts with a further molecule of NO

$$(NO)_2^*+NO \rightarrow N_2O+NO_2 \tag{19}$$

to yield the products N_2, N_2O and NO_2. In accordance with this scheme the ratio of quantum yields Φ_{N_2O}/Φ_{N_2} increases linearly with [NO], with $k_{19}/k_{18} = 77.5$ l.mole^{-1} at 30 °C[181]. In the region of complete quenching ($P_{NO} > 4$ torr) $\Phi_{N_2}+\Phi_{N_2O}$ is pressure independent and equal to 1.9×10^{-3} at 30 °C; this is the value of $k_{16}/(k_{16}+k_{17})$ if the outlined mechanism is correct.

In more recent work, Polanyi *et al.*[186] have observed infrared emission of vibrationally excited NO molecules (NO^{vib}) formed in the quenching of both Hg ($^3P_1$) and Hg ($^3P_0$) atoms, suggesting, by analogy with CO, that the primary quenching reactions are

$$Hg(^3P_1)+NO \rightarrow HgNO^* \rightarrow Hg+NO^{vib}$$

and

$$Hg(^3P_0)+NO \rightarrow HgNO^0 \rightarrow Hg+NO^{vib}$$

rather than reaction (15). However, it was not possible from their results to determine the potential importance of reaction (15) followed by reaction (17) which may be written as

$$NO(^4\Pi_i)+NO \rightarrow NO^{vib}+NO^{(vib)}$$

NO is decomposed by the 2265 A and 2144 A lines from a Cd arc[187] which selectively excite the ground and first vibrational levels, respectively, of the $A^2\Sigma^+$ state. There is some evidence[188] that the ground vibrational state of NO ($A^2\Sigma^+$) may be unreactive, but this fact has not been definitely established. The following scheme accounts for the experimental observations to date

$$NO+h\nu \rightarrow NO^*(A^2\Sigma^+) \tag{20}$$

$$NO^* \rightarrow NO+h\nu' \tag{21}$$

$$NO^*+M \rightarrow NO+M \tag{22a}$$

$$NO^*+NO \rightarrow N_2+O_2 \text{ or } N+NO_2 \tag{22b}$$

$$NO^*+NO \rightarrow N_2O+O \tag{22c}$$

The quantum yields of N_2 ($\Phi_{N_2} = 0.19$) and of N_2O ($\Phi_{N_2O} = 0.096$) are independent of temperature in the range 20–200 °C and of [NO] above 20 torr[189, 190]. The lack of dependence of Φ_{N_2} on pressure above a few torr is consistent with measurements of quenching of the γ-band fluorescence of NO itself[179, 191]. Relative rate coefficients for $k_{22a,NO}$, k_{22b} and k_{22c} calculated from Heicklen's quantum yield data[190], are summarized in Table 13 along with a few values for k_{22} ($= k_{22a}+k_{22b}+k_{22c}$) obtained from studies of fluorescence quenching. (A comprehensive list of k_{22} values is available elsewhere[179].) The proposed reaction for the formation of N_2O from NO ($A^2\Sigma^+$) is to be contrasted with the method of formation from the less energetic $a^4\Pi_i$ state which involves the intermediate $(NO)_2^*$.

Because of unfavourable Franck–Condon factors absorption from ground-state NO $X^2\Pi_r$ into the $B^2\Pi_r$ state (β bands) is very weak. In addition, many of the β bands are partially or completely overlapped by strong δ and γ bands[192]. It is not surprising, therefore, that no systematic study has been made of the photodecomposition of NO at a wavelength where the B state is exclusively excited. Some early work of Macdonald[193] on the photolysis of NO by light from an aluminium spark possibly[179] involved the exclusive production of the 5th vibrational level of NO($B^2\Pi_r$). At this energy level dissociation to atoms is not possible and the observed products, N_2, O_2, NO_2 and N_2O, presumably arose by way of a scheme similar to that proposed for decomposition *via* the $A^2\Sigma^+$ state. The invariance of the product ratio $[N_2]/[N_2O]$ found by Macdonald is evidence for such a mechanism, and suggests a value of ~ 9 for k_{22b}/k_{22c} when NO* is the B state.

The $C^2\Pi$ state of NO is predissociated for all vibrational levels above the ground-vibrational state and is even weakly predissociated in the latter. Predissociation to ground-state atoms proceeds *via* the $a^4\Pi_i$ state, but is not sufficiently rapid to cause detectable broadening of the rotational fine structure of the δ absorption bands. Flory and Johnston[195] clearly demonstrated that photolysis of NO at 1832 A, the wavelength corresponding to the $\delta(1, 0)$ band, proceeds through dissociation. Callear and Smith[196] have determined rate coefficients for the loss of $C^2\Pi$ ($v' = 0$)

$$NO \rightarrow NO\,(C^2\Pi) \rightarrow N+O \qquad (21a)$$

by fluorescence, quenching and predissociation (Table 13). Reaction (21a) is presumably followed by

$$N+NO \rightarrow N_2+O \qquad (9)$$

$$O+NO+M \rightarrow NO_2+M \qquad (4b)$$

TABLE 13

NO PHOTOLYSIS

Reaction	*Rate coefficient*[a]	*Temp.* (°*K*)	*Ref.*
$NO(a^4\Pi_i)+NO \xrightarrow{16} (NO)_2^*$	k_{16}		
	$k_{16}/k_{17} = 1.9\times10^{-3}$	303	181
$NO(a^4\Pi_i)+NO \xrightarrow{17} 2\,NO$	k_{17}		
$(NO)_2 \xrightarrow{18} N_2+O_2$ (or $N+NO_2$)	k_{18}		
	$k_{19}/k_{18} = 77.5$	303	181
$(NO)_2^*+NO \xrightarrow{19} N_2O+NO_2$	k_{19}		
$NO(A^2\Sigma^+) \xrightarrow{21} NO+h\nu'(\gamma)$	$k_{21}{}^b = 4.5\times10^6$	300	191
$NO(A^2\Sigma^+)+NO \xrightarrow{22a} 2\,NO$	$k_{22a}/k_{22} = 0.71$	300–470	190
$NO(A^2\Sigma^+)+NO \xrightarrow{22b} N_2+O_2$ (or $N+NO_2$)	$k_{22b}/k_{22} = 0.19$	300–470	190
$NO(A^2\Sigma^+)+NO \xrightarrow{22c} N_2O+O$	$k_{22c}/k_{22} = 0.10$	300–470	190
$NO(A^2\Sigma^+)+M \xrightarrow{22}$ loss of $NO(A^2\Sigma)$[c]	$k_{22,NO} = 9.2\times10^{10}$	300	191
	$k_{22,CO_2} = 2.7\times10^{11}$	300	191
	$k_{22,Ar} = \sim 6\times10^7$	300	191
$NO(B^2\Pi_r) \xrightarrow{21'} NO+h\nu'(\beta)$	$k'_{21}{}^b = \sim 4\times10^5$	r.t.[d]	194
$NO(B^2\Pi_r)+NO \xrightarrow{22b'} N_2+O_2$(or $N+NO_2$)			
	$k'_{22b}/k'_{22c} = 9$	r.t.[d]	193
$NO(B^2\Pi_r)+NO \xrightarrow{22c'} N_2O+O$			
$NO(C^2\Pi_{v=0}) \xrightarrow{21''} NO+h\nu(\delta)$	$k_{21} = 2.2\times10^7$	293	196
$NO(C^2\Pi_{v=0}) \xrightarrow{21a} N+O$	$k_{21a} = 6.6\times10^8$	293	196
$NO(C^2\Pi_{v=0}) \xrightarrow{21b} NO(A^2\Sigma^+)+h\nu$	$k_{21b} \sim 10^7$		197
$NO(C^2\Pi_{v=0})+M \rightarrow$ loss of NO^*[c]	$k_{NO} < 8\times10^{11}$	293	196
	$k_{Ar} < 3\times10^{10}$	293	196
	$k_{N_2} = 3.0\times10^{11}$	293	196
	$k_{CO} = 4.5\times10^{11}$	293	196
	$k_{CO_2} = 6.3\times10^{11}$	293	196
$NO(D^2\Sigma^+) \xrightarrow{21'''} NO+h\nu(\varepsilon)$	$k_{21}' = 3.8\times10^7$	293	196
$NO(D^2\Sigma^+)+M \rightarrow$ loss of NO^*[c]	$k_{Ar} = 5.8\times10^{10}$	293	196
	$k_{CO} = 8.7\times10^{10}$	293	196
	$k_{N_2} = 1.1\times10^{11}$	293	196
	$k_{CO_2} = 5.4\times10^{11}$	293	196
	$k_{He} = 5.8\times10^{10}$	293	196

[a] Units: l.mole^{-1} (for k_{19}/k_{18}); sec^{-1} (for k_{21}, k'_{21}, k''_{21}, k_{21a}, k_{21b}, k_{21}); l.mole^{-1}. sec^{-1} (for $k_{22}, k_{22a}, k_{22b}, k_{22c}, k_M$); dimensionless (for $k_{16}/k_{17}, k_{22a}/k_{22}, k_{22b}/k_{22}, k_{22c}/k_{22}, k'_{22b}/k'_{22c}$).

[b] $v' = 0, 1, 2$.

[c] For comprehensive list of k_{22} values, see ref. 179.

[d] r.t. = room temperature.

Photodecomposition in the region 1650–1470 A is also believed to proceed through atom formation[198]. The pressure dependence of the quantum yields of products (N_2, NO_2, N_2O) suggests that in this wavelength region a bound state ($D^2\Sigma^+$ or $E^2\Sigma^+$) of NO is produced which is subject to collisionally induced predissociation. Further evidence for this supposition is provided by the observation[196] that the quenching of ε-band fluorescence from $D^2\Sigma^+$ is not accompanied by population of lower excited states. Collisional quenching of $C^2\Pi$, by contrast, inhibits predissociation. N_2O is a product[198] of NO photolysis at 1470 A, but it is probably formed from the reaction of N atoms with product NO_2, rather than from a reaction of type (22c).

Somewhat surprisingly, the pressure dependence of quantum yields from the photolysis of NO at 1236 A, where the primary process is almost entirely photoionization[199], is very similar to that observed for 1470–1650 A[198]. Clearly, recombination of the ionic fragments must lead to an excited state of NO which predissociates upon collision. However, the neutralization reaction

$$NO^+ + e^- \rightarrow \text{neutrals} \tag{23}$$

is definitely dissociative and cannot be the source of NO*, as proposed by Leiga and Taylor[198(a)]. At the pressures used in the photolysis study electron loss occurs mainly through three-body attachments[200, 201]

$$\left.\begin{array}{ll} e^- + NO + NO & \rightarrow NO^- + NO \\ \text{or} & \rightarrow (NO)_2{}^- \\ \text{or} & \rightarrow NO_2{}^- + N \end{array}\right\} \tag{24}$$

and (23) is relatively unimportant.

NO* is presumably a product of

$$NO^+ + NO^- \rightarrow \text{neutrals} \tag{25}$$

or of

$$NO^+ + NO_2{}^- \rightarrow \text{neutrals} \tag{26}$$

probably the latter since there is good reason[203] to believe that $NO_2{}^-$ is the principal negative ion present in photoionized NO. Mahan and Person[203] have measured the rate coefficient for (26) and have shown that neutralization also occurs by a termolecular process involving an inert gas molecule. Rate coefficients of reactions (23), (24) and (26) are summarized in Table 14.

Radiolysis[206] of NO has been studied by Mund and Gillerot[207], by Harteck and Dondes[208, 209] and, recently by Hochanadel *et al.*[101] who used the pulse

TABLE 14

RATE COEFFICIENTS OF REACTIONS INVOLVED IN THE PHOTOIONIZATION OF NO

Reaction	*Rate coefficient*[a]	*Temp(°K)*	*Ref.*
$NO^+ + e^- \xrightarrow{23} N + O$	$k_{23} = 6.0 \times 10^{14}$	196 ($T^{-1.2}$ dependence)	202
	$k_{23} = 2.8 \times 10^{14}$	298 ($T^{-1.2}$ dependence)	202
	$k_{23} = 2.1 \times 10^{14}$	358 ($T^{-1.2}$ dependence)	202
	$k_{23} = 2.4 \times 10^{13}$	2900	204
	$k_{23} = 1.8 \times 10^{18}\ T^{-1.5}$	4300–5000	205
	$k_{23} = 5 \times 10^{18}\ T^{-1.5}$	5000–7800	179
$e^- + 2\,NO \xrightarrow{24} \begin{Bmatrix} NO^- + NO \\ (NO)_2^- \\ NO_2^- + N \end{Bmatrix}$	$k_{24} = 2.5 \times 10^{11}$	196 (T^{-3} dependence)	201
	$k_{24} = 8.0 \times 10^{10}$	298 (T^{-3} dependence)	201
	$k_{24} = 4.0 \times 10^{10}$	358 (T^{-3} dependence)	201
$NO^+ + NO_2^- \xrightarrow{26}$ neutrals[b]	$k_{26} = 1.3 \times 10^{14}$	300	203

[a] Units: l.mole^{-1}.sec^{-1} (for k_{23}, k_{26}); l^2.mole^{-2}.sec^{-1} (for k_{24}).
[b] Also occurs by termolecular process.

radiolysis technique. The products formed are N_2, O_2 and nitrogen oxides. It is believed that their formation occurs through reactions of oxygen and nitrogen atoms, although ionic species are also produced initially.

3.3 NITROGEN DIOXIDE ($2\,NO_2 \rightleftharpoons N_2O_4$)

3.3.1 Thermal decomposition of NO_2 *and* N_2O_4

(a) Thermal decomposition of NO_2

At temperatures in excess of 200 °C nitrogen dioxide decomposes at a measurable rate into nitric oxide and oxygen in accordance with the stoichiometry

$$2\,NO_2 = 2\,NO + O_2$$

The classical study of Bodenstein and Ramstetter[210] established the second-order rate law

$$-\frac{d[NO_2]}{dt} = 2\,k_d[NO_2]^2$$

Subsequent investigations[211–214] have confirmed this finding and have shown that at least up to 750 °C the reaction is completely homogeneous and the rate is independent of the pressure of added inert gases. The interpretation of the kinetic data in terms of the elementary molecular process

$$NO_2+NO_2 \rightarrow NO+NO+O_2 \quad (27)$$

was generally accepted until recently when Ashmore *et al.*[212, 213] showed the presence of a free radical contribution

$$NO_2+NO_2 \rightarrow NO_3+NO \quad (28)$$

$$NO_3+NO \rightarrow 2\,NO_2 \quad (29)$$

$$NO_3+NO_2 \rightarrow NO+O_2+NO_2 \quad (30)$$

Steps (29) and (30) are those involved also in the decomposition of N_2O_5. Initial rates of decomposition are apparently higher than at later stages in the reaction because, initially, the free-radical reaction is not inhibited by the fast step (29). When sufficient nitric oxide is present, either initially added or formed by NO_2 decomposition, the free-radical reaction path is suppressed. Ashmore *et al.*[212, 213] found indeed that the value of the second-order rate coefficient of decomposition k_d, depends on the $[NO]/[NO_2]$ ratio in agreement with the relation

$$k_d = k_{27}+\frac{k_{28}k_{30}[NO_2]}{k_{29}[NO]+k_{30}[NO_2]}$$

derived from a steady-state treatment of steps (27–30). These findings have been confirmed quantitatively by Graham[214], who used a similar experimental technique. Table 15 gives the Arrhenius parameters obtained in the two studies for k_{27}, k_{28} and $k_{28}k_{30}/k_{29}$ and those for k_{27} obtained by other workers. The latter values have been adjusted when necessary[213] to allow for the free-radical contribution to the rate. Included in the table are the Arrhenius parameters for k_{29}, k_{30} and k_{-30}, derived from the experimental rate coefficients and the thermodynamic data in the literature[1] for the equilibria

$$2\,NO_2 \rightleftharpoons NO_3+NO \quad K_{28,29} = k_{28}/k_{29}$$

$$2\,NO_2 \rightleftharpoons 2\,NO+O_2 \quad K = k_{28}k_{30}/k_{29}k_{-30}$$

The derived values of k_{29}, k_{30} and k_{-30} are in fair agreement with rate coefficients obtained by independent methods. Thus, for example, the derived value of k_{-30} at 298 °K of 142 $l^2.mole^{-1}.sec^{-1}$ should be compared with 66 $l^2.mole^{-2}.sec^{-1}$ obtained by Ray and Ogg[215] from a study of NO oxidation at high NO_2/NO ratios. Similarly, the derived value of k_{30} at 820 °K of 1.3×10^7 $l.mole^{-1}.sec^{-1}$ compares well with the k_{30} value of 1.5×10^7 $l.mole^{-1}.sec^{-1}$ from the data of Schott and Davidson[226] obtained in a shock-tube study of N_2O_5 pyrolysis. The derived value of k_{29} is an order of magnitude smaller than the other estimates listed in Table 15.

TABLE 15

RATE COEFFICIENTS FOR THE THERMAL DECOMPOSITION OF NO_2

Reaction	Rate coefficient	Arrhenius parameters: A[a]	Arrhenius parameters: E(kcal.mole^{-1})	Temp.(°K)	Ref.
$2\,NO_2 \xrightarrow{27} 2\,NO+O_2$	k_{27}	$(1.99\pm0.19)\times10^9$	26.9 ± 0.1	470–820	213
	k_{27}	1.99×10^9	26.9	630–1020	211
	k_{27}	2.45×10^9	27.1	590–650	210[b]
	k_{27}	2.23×10^9	27.0 ± 0.2	470–670	214
$2\,NO_2 \xrightarrow{28} NO+NO_3$	k_{28}	$(3.9\pm2.3)\times10^8$	23.9 ± 0.6	470–700	213
	k_{28}	2.81×10^8	23.4 ± 0.5	470–670	214
$NO_3+NO \xrightarrow{29} 2\,NO_2$	k_{29}	$(1.82\pm1.1)\times10^9$	0.7 ± 1.0	470–700	213
	k_{29}	6.0×10^{10}	1.4 ± 2.5	566–576	226, 227
	k_{29}[c]	1.32×10^9	0.2 ± 1.0	470–670	214
	k_{29}[d]	3.4×10^9		573	216
$NO_3+NO_2 \xrightarrow{30} NO+O_2+NO_2$	k_{30}	$(5.13\pm2.3)\times10^7$	2.25 ± 1.0	470–700	213
	k_{30}[c]	1.14×10^8	3.0 ± 1.0	470–670	214
	k_{-30}	15.8 ± 7	$-(1.3\pm1.0)$	470–700	213
	k_{-30}[c]	35.2	$-(0.5\pm1.0)$	470–670	214
$Ar+NO_2 \xrightarrow{31} Ar+NO+O$	k_{31}	$6.9\times10^{18}\,T^{-\frac{3}{2}}$	71.9	1400–2300	222
	k_{31}	$2.3\times10^{18}\,T^{-\frac{3}{2}}$	71.9	1830–2200	223[e]
	k_{31}	$6.2\times10^{18}\,T^{-\frac{3}{2}}$	71.9	1400–2000	224
$2\,NO_2 = 2\,NO+O_2$	k_d^f	1.25×10^{10}	25 ± 5	1400–2300	222
	k_d^f	4.5×10^9	25.7	1400–2000	224
	$k_{28}k_{30}/k_{29}$	$(1.10\pm0.30)\times10^7$	25.4 ± 0.3	470–700	213
	$k_{28}k_{30}/k_{29}$	2.45×10^7	26.2 ± 0.9	470–670	214

[a] Units: l.mole^{-1}.sec^{-1} (for k_{27}, k_{28}, k_{29}, k_{30}, k_{31}, k_d, and $k_{28}k_{30}/k_{29}$); l^2.mole^{-2}.sec^{-1} (for k_{-30}).
[b] Recalculated in ref. 213.
[c] Derived values, as explained in the text.
[d] k_{29} value at 300 °C.
[e] Fishburne *et al.*[224] have recalculated Hiraoka and Hardwick's data[223] erroneously.
[f] Rate coefficients for the overall decomposition, including the free radical path.

The agreement between rate coefficients for reaction of NO_3 derived from the decomposition of N_2O_5 and NO_2 has led to the suggestion that the same NO_3 intermediate is involved in both reactions. This is believed to be the symmetrical (D_{3h}) nitrate radical because of the agreement between calculated and experimentally determined values of (*i*) the entropy of a D_{3h} nitrate radical[226] and (*ii*) the equilibrium constant k_{30}/k_{-30}[215, 226].

A transition state formulation[217]

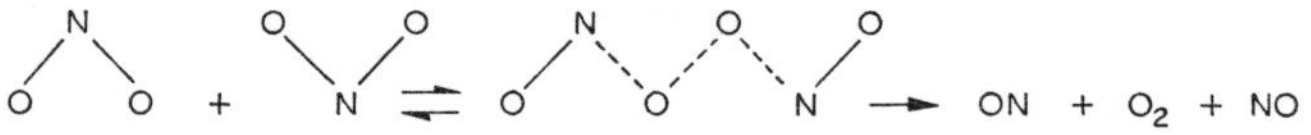

of reaction (27) has led to a calculated A factor for this reaction at 570 °K of 4.5×10^9 l.mole^{-1}.sec^{-1} in good agreement with the experimental value. Reaction (28) presumably proceeds *via* a complex of the following structure[213]

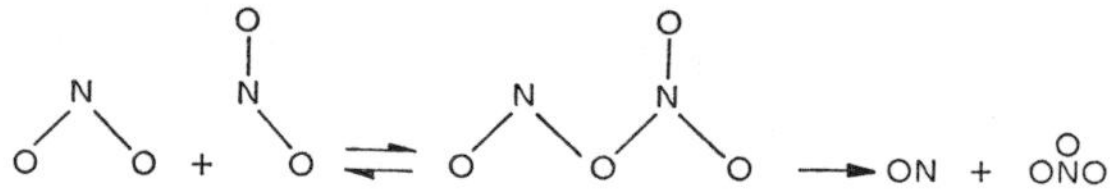

There have been suggestions[218, 219] that reaction (27) and its reverse involve the intermediate formation of a peroxy NO_3 radical and there is some confirmation of this view in the results of Ogg[220] for the NO-catalyzed isotopic exchange between NO_2 and $^{18}O_2$ and in the results of infrared absorption studies[221] of reacting mixtures of NO and O_2.

The thermal decomposition of NO_2 has been studied[222–224] in the temperature range 1400–2300 °K by the shock-tube technique. Changes in the concentration of NO_2 in shock-heated argon-diluted NO_2 were monitored by visible absorption[222] or visible emission[224] spectrophotometry. The data fitted a complex rate law of the form

$$-\frac{d[NO_2]}{dt} = k_a[Ar][NO_2]+2\,k_b[NO_2]^2$$

clearly suggesting that the bimolecular mechanism obtaining at low temperatures is accompanied at higher temperatures by unimolecular dissociation of NO_2 at its low-pressure limit, *viz.*

$$Ar+NO_2 \rightarrow Ar+NO+O \qquad (31)$$

$$O+NO_2 \rightarrow NO+O_2 \qquad (32)$$

with $k_a = 2k_{31}$. It was not possible to separate k_b into its molecular and free-radical components, so that the values given in Table 15 represent the overall rate coefficient k_d. Huffman and Davidson's rate coefficients[222] should be accepted with caution since they were calculated on the basis of an assumed activation energy for the unimolecular dissociation. Their value for k_d is appreciably higher than that obtained by extrapolation of the low-temperature rate coefficients whereas later data[224] show better agreement. The Arrhenius parameters for the unimolecular rate coefficients given in Table 15 are based on the classical version of the RRK formulation[225] for the low-pressure, second-order dissociation rate coefficient. The difference of 6 kcal.mole^{-1} between the measured[223] activation energy for the low-pressure dissociation and the bond energy[1] is consistent with the assumption that three effective oscillators contribute to the reaction.

(*b*) *Thermal decomposition of* N_2O_4

The rate of dissociation of gaseous N_2O_4 has been determined near room temperature from studies of the relaxation of non-equilibrium mixtures of NO_2 and N_2O_4 which were produced in several ways, such as the passage of a sound wave[228–230] or a weak shock wave[231] through an equilibrium mixture, or the high-velocity flow through a perforated diaphragm[232] or past an obstruction[233]. Shock-tube studies[231] have provided the most reliable information (Table 16) on the kinetics of the decomposition showing that the reaction

$$N_2O_4+M \rightarrow 2\,NO_2+M \tag{33}$$

is "unimolecular" and in its second-order region at pressures less than 1 atm.

TABLE 16

RATE CONSTANT OF THERMAL DECOMPOSITION OF N_2O_4

Reaction	*Rate coefficient expressions*[a]	*Temp.(°K)*	*Pressure*	*Ref.*
$N_2O_4+M \xrightarrow{33} 2\,NO_2+M$	$k^0_{33,N_2} = 2.0\times10^{14}\exp-(11{,}000/\boldsymbol{R}T)$	250–300	<1 atm N_2	231
	$k^0_{33,CO_2} \sim k^0_{33,N_2}$			
	$k^\infty_{33} = 1\times10^{16}\exp-(13{,}100/\boldsymbol{R}T)$	250–300	∞	231[b]
	$k^0_{33,NO_2} = 10^{14.14}\exp-(10{,}240/\boldsymbol{R}T)$	270–320	<1 atm.	228[c]
	$k^0_{33,N_2O_4} = 4.5\times10^6$	298.2	<1 atm.	230
	$k^\infty_{33} = 1.7\times10^5$	298.2	∞	230[d]
	Efficiencies for activating N_2O_4 relative to N_2O_4 = 1.0: NO_2 1.0; CO_2 1.0; N_2 0.5; Ar 0.3			230

[a] Units: l.mole^{-1}.sec^{-1} (for k^0_{33}); sec^{-1} (for k^∞_{33}).
[b] Limiting first-order rate coefficient at infinite pressure estimated[114] from effect of pressure on k^0_{33} at pressures well below the high pressure limit and from the assumption $E^\infty_a = \Delta E^0_0 =$ 13,100 cal.mole^{-1}.
[c] We assume that M is the equilibrium mixture of NO_2 and N_2O_4, although this is not clear in the Chemical Abstract.
[d] Obtained by a long extrapolation; an order of magnitude estimate only.

Rate coefficients obtained by the other relaxation methods[228, 229, 232, 233] are less reliable (perhaps with some exceptions[228, 230]) because of the lengthy computations involved and because of uncertainties in the underlying assumptions. Nevertheless, estimates of k_{33} obtained by these techniques are in excellent accord with the shock-tube values of Carrington and Davidson[231] (Table 16). Curvature in the plot of the first-order rate coefficient for (33) against concentration is noticeable above 2 atm, although the high-pressure limit is not attained even at 7 atm. Estimates[230, 231] have been made of k_{33}^{∞} from the pressure dependence of k_{33} and a relationship derived by Johnston[114] (Table 16). The high value of the pre-exponential factor associated with k_{33}^{∞} is an indication that the activated complex for the dissociation is "loose", *i.e.* resembles two freely rotating NO_2 molecules in close proximity more closely than a rigid N_2O_4 molecule.

3.3.2 *Photolysis of* NO_2

Nitrogen dioxide possesses an extensive and extremely complex electronic absorption spectrum which, with the exception of a few bands in the region 3700–4600 A[234], at 2500 A[235] and at 1500 A[236], has not been successfully analyzed. The spectrum shows vibrational and rotational structure, although at wavelengths shorter than 4000 A the bands are superposed on underlying continua and many of them have only diffuse rotational structure. Photodecomposition occurs at wavelengths shorter than 4050 A[237] and photoionization[238] below 1270 A.

Much of the research on NO_2 photolysis has been confined to the 2700–4500 A spectral region and has been concerned with establishing the importance of the reaction

$$NO_2(^2A_1)+h\nu \rightarrow NO(X^2\Pi_{\frac{1}{2}})+O(^3P) \tag{34}$$

as a primary process and determining the subsequent fate of the ground-state oxygen atoms, $O(^3P)$. Relatively few studies have been made at wavelengths shorter than 2500 A, although recently attention has been focussed[145] on the importance of

$$NO_2+h\nu \rightarrow NO+O(^1D_2) \tag{34a}$$

as a primary process at wavelengths shorter than 2450 A.

In gaseous NO_2 the equilibrium fractions of the dimer, dinitrogen tetroxide, are usually relatively small but N_2O_4 absorbs strongly[102, 238, 239] below 4000 A and substantial corrections to light absorption measurements are necessary at shorter wavelengths. Furthermore, relatively little information is available on the photochemical behavior of N_2O_4. In an early study, Holmes and Daniels[239] established that N_2O_4 decomposed at 2650 A with a quantum efficiency of 0.4 but was not measurably decomposed at 3130 and 3660 A although it absorbs light at these wavelengths. Further and more detailed studies of the photochemical behavior of N_2O_4 are highly desirable.

(a) *Photolysis of* NO_2 *in the* 2500–4500 *A region*

Significant quantum yields for the photodecomposition of NO_2 into O_2 and NO have been found at 4047 A and at shorter wavelengths. In the absence of foreign gases the quantum yields are independent of NO_2 pressure at pressures greater than 2 torr. The existing data for such limiting quantum yields are summarized in Table 17.

Very small amounts of decomposition are observed at 4358 A and it has been suggested[244] that they are due to the presence of some light of shorter wavelengths

TABLE 17

LIMITING QUANTUM YIELDS OF O_2 FORMATION IN THE PHOTOLYSIS OF NO_2 AT HIGH NO_2 CONCENTRATION AND WITHOUT OTHER GASES ADDED

Wavelength(A)	*Temp. (°K)*	Φ	*Ref.*
4358	296	0.005	244
4358	344	0.013	244
4358	406	0.018	244
4358	496	0.032	244
4358	298	0	240
4358	295	0.005	256
4358	273	0	239
4047	296	0.36	244
4047	344	0.41	244
4047	406	0.51	244
4047	496	0.71	244
4047	566	0.91	244
4047	298	0.39	246
4047	298	0.36	247
4047	298	0.37	240
4047	295	0.36	256
4047	273	0.25	239
3800	297	0.82	244
3660	296	0.93	244
3660	344	0.98	244
3660	406	0.98	244
3660	496	1.00	244
3660	298	0.96	246
3660	298	1.05	240
3660	295	0.77	256
3660	273	0.92	239
3130	273	0.97	239
3130	297	0.97	244
3130	298	0.96	247
3160–2700	298	1.03	240

in the filtered light sources used. However, measurements[245] of ^{18}O scrambling in $NO_2/^{18}O_2$ mixtures show that a very slight decomposition of NO_2 does occur at wavelengths > 4200 A but this takes place by a "molecular" mechanism

$$NO_2^* + NO_2 \rightarrow 2\,NO + O_2$$

while at 4047 A and shorter wavelengths the decomposition involves free oxygen atoms generated in reaction (34).

The onset of rotational diffuseness in the absorption spectrum of NO_2 at 3979 A coincides almost exactly[234] with the energy requirement for reaction (34) strongly suggesting predissociation to ground-state oxygen atoms. The weakening of fluorescence[240-242, 248] and the sudden appearance of products characteristic of $O(^3P)$ atom reactions[243] from the photolysis of NO_2–olefin mixtures, on passing from 4358 A to 4047 A, are also indications of a dissociative primary process at and below 4047 A. Quantitative data[244, 245] on the extent of isotopic scrambling in irradiated NO_2–$^{18}O_2$ mixtures show that photo-decomposition at wavelengths from 4047 A to 2537 A indeed proceeds entirely by way of an oxygen-atom mechanism. An earlier report[247] of no scrambling at 4047 A is contrary to the more recent findings[244-246].

The wavelength dependence of Φ_{O_2} in Table 17 reflects a variation in the efficiency of the primary photodissociation process. The high efficiency of the room-temperature photodissociation at 4047 A, where the photon possesses some 1.3 kcal.mole^{-1} energy less than the accepted (JANAF[1]) bond dissociation energy of NO_2 of 71.9 kcal.mole^{-1}, is surprising. Pitts *et al.*[244] have explained Φ_{O_2} at 4047 A and its temperature dependence by assuming that most of the rotational energy of excitation was available for bond rupture in a bound upper electronic state of NO_2. A lower efficiency of internal conversion of rotational energy perhaps combined with a somewhat reduced bond dissociation energy could equally well account for the observed results. Another possibility is that the energy (of about 1 kcal) which is deficient is supplied by collisional activation of the upper bound state ($^2B_2$). This was considered unlikely[244] because even at 500 °K the extent of dissociation at 4358 A was only about 3–4 % of that observed at 4047 A. However, the difference in photon energies at the two wavelengths is 5 kcal.mole^{-1}. At 4047 A the energy deficit is very small and it is possible that collisional activation may be efficient. At 4358 A, on the other hand, collisional deactivation may occur preferentially.

Although considerable time and effort have been devoted to the study of the further reactions of $O(^3P)$ atoms produced in reaction (34) with NO_2, the precise mechanism has not yet been established with certainty. The derived rate coefficients for the proposed individual elementary reactions should therefore be accepted with great caution. The low Φ_{O_2} values[253] at very small NO_2 concentrations in an atmosphere of N_2 and their dependence[244, 246, 247, 249] on inert gas pressure at NO_2 pressures in the 1 torr range have been explained by Ford and Endow[249, 253] by the following competing reactions

$$O+NO_2 \rightarrow NO_3^* \quad (35)$$

$$O+NO_2 \rightarrow NO+O_2 \quad (6)$$

$$NO_3^* \rightarrow NO_2+O \quad (-35)$$

$$NO_3^*+M \rightarrow NO_3+M \quad (36)$$

$$NO_3^*+NO_2 \rightarrow NO_2+NO+O_2 \quad (30^*)$$

$$NO_3+NO_2 \rightarrow NO_2+NO+O_2 \quad (30)$$

$$NO_3+NO \rightarrow 2\,NO_2 \quad (29)$$

The involvement of NO_3 in the photolysis has been clearly demonstrated by its positive spectroscopic identification[250] in flash photolysis of NO_2.

At room temperature reaction (30) is slow and (29) is sufficiently fast[250] to be effective even at very small conversions of NO_2 to NO. Reaction (30*) is rapid and explains the increase in Φ_{O_2} with increasing $[NO_2]$ at constant total pressure. It may be regarded[249] as a decomposition of NO_3^* promoted by a rapid conversion of symmetrical NO_3 to the less stable peroxy form

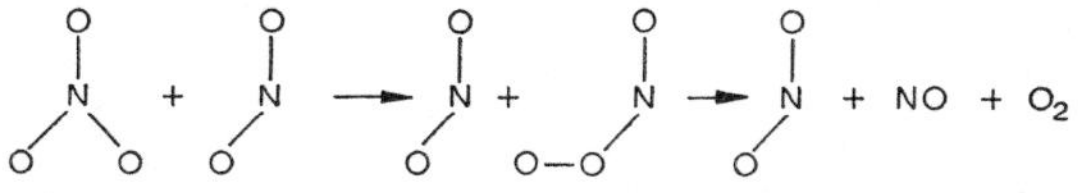

The existence of two modes of reaction of O atoms with NO_2, reactions (35) and (6) may be explained similarly

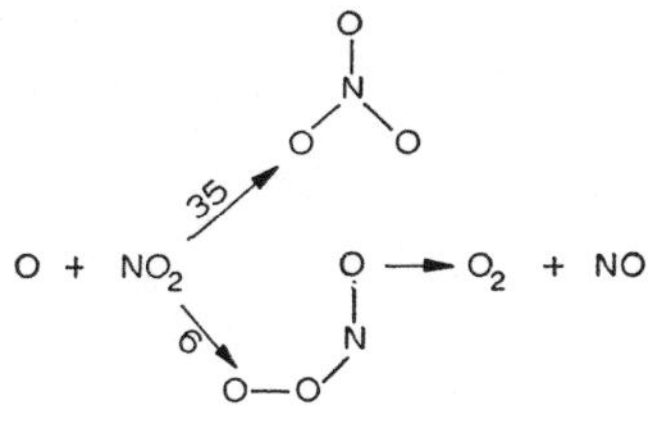

A simpler assumption, favoured by some workers[251, 252] is that only one intermediate NO_3^* is involved and that reaction (6) is the result of (35) followed by

$$NO_3^* \rightarrow NO+O_2 \quad (37)$$

The existing kinetic data are not sufficiently extensive and precise to differentiate between the two mechanisms and, moreover, neither mechanism can explain all

TABLE 18

RATE COEFFICIENTS RELEVANT TO NO_2 PHOTOLYSIS

Reaction	*Rate coefficients*[a]	*Temp(°K)*	*Ref.*
$O+NO_2 \xrightarrow{35} NO_3^*$	$k_{35} = 6.3\times10^9$	310	251
$O+NO_2+M \xrightarrow{35'} NO_3+M$	$k'_{35} = 2.88\times10^{11}$ (for $[M] = 1$ atm. N_2]	300	253
$O+NO_2 \xrightarrow{6} NO+O_2$	$k_6 = 6.2\times10^9$	300	253
	$k_6 = (1.95\pm0.61)\times10^{10}\exp-(1{,}060\pm200/RT)$	278–374	251
	$k_6 = (1.5\pm0.4)\times10^9$	298	257
$NO_3^*+M \xrightarrow{36} NO_3+M$	$k_{36}/k_{30*} = 9.3\times10^{-3}$ ($M = CO_2$)	297	244
$NO_3+NO_2 \xrightarrow{30*} NO+O_2+NO_2$	$k_{36}/k_{30*} = 3.3\times10^{-3}$ ($M = CO_2, N_2$)	298	246
	$k_{36}/k_{30*} = 4.1\times10^{-3}$ ($M = CO_2$)	298	247
	$k_{36}/k_{30*} = 5\times10^{-2}$ ($M = CF_2Cl_2$)	298	247
	$k_{36}/k_{30*} = 1.7\times10^{-2}$ ($M = C_2H_6$)	298	247
	$k_{36}/k_{30*} = 2.3\times10^{-2}$ ($M = C_3H_8$)	298	247
	$k_{36}/k_{30*} = 8.3\times10^{-2}$ ($M =$ iso-C_4H_{10})	298	247

[a] Units: $l.mole^{-1}.sec^{-1}$ (for k_{35} and k_6); $l^2.mole^{-2}.sec^{-1}$ (for k'_{35}); k_{36}/k_{30*} ratios are dimension less.

the experimental observations. In particular, the fall-off[246] of Φ_{O_2} with decreasing $[NO_2]$ below 2 torr (in the absence of additives) and the lack of dependence[253, 254] of Φ_{O_2} on $[NO_2]$ at NO_2 pressure below 0.1 torr in the presence of 1 atm of inert gas are not accounted for.

The formation and subsequent redissociation of an NO_3^* intermediate of finite lifetime is consistent with the observed rapid incorporation[251] of ^{18}O into NO_2 by exchange with ^{16}O atoms. The value of the elementary rate coefficient k_{35} (Table 18) was determined by Herron and Klein[251] from the rate of incorporation of ^{18}O assuming that only (35) and (−35) were responsible for the exchange.

Steady-state treatment of reactions (27), (35), (−35), (6) and (32) gives

$$\Phi_{O_2}^{-1} = \phi_\lambda^{-1}\left\{1+\frac{k_{35}k_{36}[M]}{k_6k_{-35}+k_6k_{36}[M]+k_{30*}(k_{35}+k_6)[NO_2]}\right\} \tag{a}$$

where ϕ_λ is the primary quantum yield of reaction (34) at a given wavelength (λ). At NO_2 pressures larger than 2 torr it may be assumed that $k_{36}[M] \ll k_{30*}[NO_2]$ and the effect[244, 246, 247] of $[NO_2]$ on Φ_{O_2} at fixed [M] suggests further that k_6k_{-35} can also be neglected in comparison with $k_{30*}(k_{35}+k_6)[NO_2]$, so that under these conditions

$$\Phi_{O_2}^{-1} = \phi_\lambda^{-1}\left\{1+\frac{k_{35}k_{36}[M]}{k_{30*}(k_{35}+k_6)[NO_2]}\right\} \tag{a'}$$

Using trace concentrations of NO_2 ($\sim 10^{-7}$ M) in an atmosphere of N_2, Ford and Endow[253] determined Φ_{-NO_2} at 3660 A. Under these conditions it appears justified to assume that k_{30*} $[NO_2] \ll k_{36}$ [M] and with the additional elementary steps

$$O+O_2+M \rightarrow O_3+M \tag{38}$$

$$O+NO+M \rightarrow NO_2+M \tag{39}$$

$$O_3+NO \rightarrow NO_2+O_2 \tag{40}$$

the steady-state treatment gives

$$\Phi^{-1}_{-NO_2} = 0.5\phi_\lambda^{-1}\left\{1+\frac{k'_{35}[M]}{k_6}+\frac{k_{38}[O_2][M]}{k_6[NO_2]}+\frac{k_{39}[NO][M]}{k_6[NO_2]}\right\}$$

where $k'_{35} = k_{35}k_{36}(k_{-35}+k_{36}[M])^{-1}$ is the effective rate coefficient for the association of O and NO_2. From the slopes and intercepts of the linear plots at fixed [M] (1 atm N_2) but at variable $[O_2]/[NO_2]$ or $[NO]/[NO_2]$ the values of k'_{35}, k_6, and k_{39} were obtained using known values of k_{38} and ϕ_{3660} (assumed to be unity). These coefficients have been recalculated with a more recent[255] value of k_{38} and are listed in Table 18. With k_{35} and k_6 known, it was possible to use eqn. (a′) to calculate the values of k_{36}/k_{30*} in Table 18. The latter should be regarded only as order of magnitude values in view of the uncertainties in the mechanism and the approximations in eqn. (a′).

(*b*) *Photolysis of* NO_2 *below* 2450 *A*

The photochemistry of NO_2 at wavelengths shorter than about 2450 A, *i.e.* in the region where there is enough energy for reaction (34a), has been investigated very little. However, definitive chemical evidence has been obtained for the photochemical generation of $O(^1D_2)$ atoms. These atoms react very rapidly with a number of substances, including NO_2, and are readily deactivated by collision with such gases as O_2, N_2, CO_2, Xe, but not by SF_6, He, etc. Relative rates of these processes have been measured[145] at 2288 A and have been found to agree with the values obtained using other sources of $O(^1D_2)$. Formation of $O(^1D_2)$ in this spectral region is in agreement with the predissociation observed[2] at about 2490 A, although no prior evidence that reaction (34a) actually occurs was available.

3.3.3 *Radiolysis of* NO_2

Decomposition of NO_2 irradiated by neutrons[258] and by fission fragments[208] of ^{235}U has been studied by Harteck and Dondes. Reaction intermediates appear to be NO_2^+ ions, excited NO_2 molecules, oxygen atoms and nitrogen atoms. Attack on NO_2 by O atoms produces $NO+O_2$. Attack by N atoms on NO gives N_2+O, and attack on NO_2 gives, respectively 2 NO, N_2O+O, N_2+O_2 and $N_2+2\,O$. Decomposition of NO_2 by γ-radiation[259, 260] is basically similar. The products are NO, O_2, N_2O and N_2 and the principal reactions responsible for their formation are[260]

$$NO_2+O \rightarrow NO+O_2 \tag{6}$$

$$NO_2+N \rightarrow 2\,NO \tag{41a}$$

$$NO_2+N \rightarrow N_2O+O \tag{41b}$$

$$NO_2+N \rightarrow N_2+2\,O \tag{41c}$$

$$NO_2+N \rightarrow N_2+O_2 \tag{41d}$$

and also reactions involving intermediate formation of NO_3, discussed above in connection with photolysis of NO_2. From their initial yields of the products and using rate measurements of Clyne and Thrush[261], Dmitriev *et al.*[260] estimate the rate ratios for reactions (41a), (41b), (41c) and (41d) as 2.5 : 12 : 1 : 5.

3.4 NITROGEN PENTOXIDE

3.4.1 Thermal decomposition of N_2O_5

(*a*) *Gas phase decomposition of* N_2O_5

The thermal decomposition of nitrogen pentoxide proceeds according to the stoichiometric equation

$$\begin{array}{l} N_2O_5 = 2NO_2+\tfrac{1}{2}O_2 \\ \qquad\qquad \rightleftharpoons \\ \qquad\qquad N_2O_4 \end{array}$$

and is measurable at room temperature. Following the discovery[262] that the reaction is homogeneous and obeys a first-order kinetic law, a large number[263–271] of experimentalists investigated the kinetics of the decomposition for the temperature range 0–120°C and for pressures ranging from 5×10^{-4} torr to 700 torr. Decomposition rates were measured in static systems by following the pressure change[262, 265, 268–271] or by monitoring NO_2 formation colorimetrically[263, 269, 272] and in flow systems[264, 266] by chemical analysis. Above 10^{-1} torr of N_2O_5 the reaction in the neighbourhood of room temperature is first-order, the first-order rate coefficient being independent of pressure of added inert gases[265, 266],

NO_2[263–265], O_2[265, 266], surface area[262], P_2O_5[266], dust[266], Pt[266]. The results of different workers[263–272] agree remarkably well and we find that they all fit the following expression for the experimental first-order rate coefficient (k^{∞}_{exp})

$$k^{\infty}_{exp} = \frac{-1}{[N_2O_5]}\frac{d[N_2O_5]}{dt} = (3.21 \pm 0.47) \times 10^{13} \exp -(24{,}499 \pm 97)/RT$$

sec^{-1}

with a standard deviation of only 1 % at 332 °K.

A fall-off[270, 271, 273, 274] in the first-order rate coefficient is observed at pressures below 5×10^{-2} torr, although because of the experimental difficulties of working in this low pressure region and the contribution of a heterogeneous reaction[271, 273, 274] there is considerable scatter among the results of different workers. The results of Linhorst and Hodges[273] are perhaps the most reliable and agree fairly well with those of Schumacher and Sprenger[270] and of Ramsperger and Tolman[274]. Using a 22 l vessel to limit the importance of heterogeneous reactions, Linhorst and Hodges[273] followed the decline in first-order rate coefficients between 10^{-1} and 10^{-4} torr for temperatures between 35 and 65 °C. They found that plots of $1/k_{exp}$ *versus* $1/p$ were linear and that the reaction order with respect to N_2O_5 approached 2 at the lowest pressure used. From their data and the expression above for k^{∞}_{exp}, we deduce for the experimental bimolecular (zero pressure limit) rate coefficient for N_2O_5 decomposition (k^{0}_{exp, N_2O_5})

$$k^{0}_{exp, N_2O_5} = 10^{15.97 \pm 1.02} \exp -(18{,}393 \pm 1464)/RT \text{ l.mole}^{-1}.\text{sec}^{-1}$$

For a molecule of the complexity of N_2O_5, quasiunimolecular rate theories predict a fall-off in the first-order rate constant of decomposition in a region of pressure appreciably higher than that where fall-off is actually observed. This anomaly remained unexplained until Ogg proposed[275] that decomposition of N_2O_5 was not a true unimolecular reaction but involved, instead, the transient intermediate NO_3 in a more complex process, which, however, was still first-order in N_2O_5. The following currently accepted mechanism is basically that suggested by Ogg[275].

$$N_2O_5 \underset{-42}{\overset{42}{\rightleftarrows}} NO_2 + NO_3 \quad (K_{42} = k_{42}/k_{-42}) \tag{42, −42}$$

$$NO_2 + NO_3 \rightarrow NO_2 + NO + O_2 \tag{30}$$

$$NO + NO_3 \rightarrow 2\,NO_2 \tag{29}$$

$$2\,NO_3 \rightarrow 2\,NO_2 + O_2 \tag{43}$$

$$NO_2 + O_3 \rightarrow NO_3 + O_2 \tag{44}$$

Johnston[276] has shown that with the inclusion of steps (43) and (44) the scheme can quantitatively account for the kinetics of the decomposition of N_2O_5 alone and in the presence of NO, the N_2O_5-sensitized decomposition of O_3, and the formation of N_2O_5 from NO_2 and O_3. The important feature of the mechanism is the rapid pre-equilibration in which recombination of NO_2 and NO_3, reaction (−42), competes effectively with the alternate step (30). Ogg *et al.*[277] have clearly demonstrated the validity of this assumption by establishing that the rate of isotopic scrambling between $^{15}N_2O_5$ and NO_2 is many orders of magnitude greater than the rate of N_2O_5 decomposition. The scrambling was first order in N_2O_5 and zero order in NO_2, and had a rate coefficient at a given total pressure identical with that for the NO-sensitized decomposition of N_2O_5. Schott and Davidson[278] observed, by spectrophotometric measurements of both NO_2 and NO_3, a state of near equilibrium in the shock-tube decomposition of N_2O_5, and Katan[279] unambiguously identified the NO_3 intermediate and directly observed the pre-equilibrium in the pyrolysis of N_2O_5 in a flow system.

For a steady state in NO and NO_3, reactions (42), −(42), (30) and (29) give for the experimental first-order rate coefficient ($k_{N_2O_5}$) for decomposition of pure N_2O_5, the expression

$$k_{\text{exp}, N_2O_5} = k_{42}(1+k_{-42}/2\,k_{30})^{-1}$$

k_{42} and k_{-42} are, of course, pressure-dependent, but k_{exp, N_2O_5} is pressure independent in the region where $k_{-42} \gg 2\,k_{30}$, *i.e.*

$$k_{\text{exp}}^{\infty} = 2\,k_{30}(k_{42}/k_{-42}) = 2\,k_{30}\,K_{42}$$

being simply the product of an equilibrium constant and a bimolecular rate coefficient. Evidently, the inequality $k_{-42} \gg 2\,k_{30}$ holds for pressures of $N_2O_5 > 0.1$ torr. The experimental second-order rate coefficient at zero pressure will be

$$k_{\text{exp}, N_2O_5}^{0} = \left\{\frac{k_{\text{exp}, N_2O_5}}{[\text{M}]}\right\}_{[\text{M}]\to 0} = \frac{k_{42}}{[\text{M}]}$$

The addition of nitric oxide markedly increases[280] the rate of N_2O_5 decomposition. In terms of the accepted mechanism, NO removes NO_3 in the very rapid reaction (29), thereby preventing reassociation. The stoichiometric equation is now

$$N_2O_5 + NO = 3\,NO_2$$

and the assumption of a steady state in $[NO_3]$ for steps (42), (−42), (30) and (29) leads to the experimental first-order rate coefficient of N_2O_5 decomposition in the presence of NO

$$k_{\text{exp},(N_2O_5+NO)} = \frac{-1}{[N_2O_5]}\frac{d[N_2O_5]}{dt} = k_{42}(1+k_{-42}[NO_2]/k_{29}[NO])^{-1}$$

when $k_{29}[NO] \gg k_{30}[NO_2]$.

Hisatsune *et al.*[281], using infrared spectrophotometry to monitor N_2O_5 and NO_2, have confirmed the form of this rate law and have deduced Arrhenius parameters for k_{42} and the ratio $k_{42}k_{29}/k_{-42}$ (Table 19). The initial rate of the reaction is simply $k_{42}[N_2O_5]$, so that studies of the decomposition of N_2O_5 in the presence of NO, rather than of N_2O_5 alone, yield genuine unimolecular decomposition rate coefficients (k_{42}).

In a series of careful investigations, Johnston *et al.*[282–285] have measured the initial rates of the N_2O_5+NO reaction as a function of temperature and of pressure of various additives. They have successfully interpreted the pressure dependence of the first-order rate coefficient, k_{42}, in terms of the expression

$$k_{42} = \sum_i \frac{c_i \sum_M a_{Mi}[M_M]}{\sum_M b_{Mi}[M_M]+c_i}$$

where *a* and *b* refer to the energy transfer processes

$$N_2O_5+M \xrightarrow{a_{Mi}} (N_2O_5)_i^*$$

$$(N_2O_5)_i^*+M \xrightarrow{b_{Mi}} N_2O_5$$

and *c* to the decomposition of the energized molecules

$$(N_2O_5)_i^* \xrightarrow{c_i} NO_2+NO_3$$

A decline in k_{42} with decreasing pressure was evident even at 10^4 torr, and at 10^{-1} torr the reaction was clearly in its second-order region. The bimolecular rate coefficients $\Sigma_i\, a_{Mi}$ were computed[284, 285] for a number of gases at 323.7 °K and are summarized in Table 19. (The efficiency factors for collisional activation of N_2O_5 by different gases have also been estimated[284, 285] but are not reproduced in this article.)

In an elegant study of shock waves in dilute N_2O_5/Ar mixtures, Schott and Davidson[278] have measured the low-pressure limit of the rate coefficient for pure N_2O_5 decomposition, k_{42}^0, by monitoring NO_3 and NO_2 formation behind the shock front. By subsequently following the decay of NO_3 they were able also to compute values of k_{30} and k_{43}. These values, however, should be accepted with caution since there was some difficulty in separating the individual contributions of steps (30) and (43) to the total rate of NO_3 destruction. However, their rate

TABLE 19

RATE COEFFICIENTS FOR ELEMENTARY REACTIONS IN N_2O_5 DECOMPOSITION

Reactions	*Rate coefficients*[a]	*Temp(°K)*	*Ref.*
$N_2O_5 \rightarrow 2\,NO_2+\frac{1}{2}\,O_2$	$k_{exp}^{\infty} = (3.21\pm0.47)\times10^{13}$ $\exp-(24{,}499\pm97)/RT)$[b]	273–393	263–272[b]
	$k_{exp,N_2O_5}^{0} = 10^{15.97\pm1.02}$ $\exp-(18{,}393\pm1{,}464)/RT$[b]	308–338	273[b]
$N_2O_5 \underset{-42}{\overset{42}{\rightleftharpoons}} NO_2+NO_3$	$K_{42} = k_{42}/k_{-42} = 9.26\times10^4$ $\exp-(20{,}100\pm1{,}100)/RT$	450–550	278
	$K_{42} = 1.37\times10^{-6}$	356.2	279
	$k_{42}/[M] = 4.8\times10^{13}$ $\exp-(16{,}500\pm700)/RT$[c]	450–550	278
$NO_2+NO_3 \xrightarrow{30} NO+O_2+NO_2$	$k_{30} = 2.26\times10^8 \exp-(4{,}420\pm700)/RT$	550–1100	278
$2\,NO_3 \xrightarrow{43} 2\,NO_2+O_2$	$k_{43} = 2.63\times10^9 \exp-(7{,}700\pm1{,}000)/RT$	550–1100	278
	$k_{43} = 1.9\times10^8 \exp-(5{,}710/RT)$[d]	293–313	291
	$k_{43} = 9.5\times10^5 \exp-(2{,}750/RT)$[e]	293–309	290
$N_2O_5+NO \rightarrow 3\,NO_2$	$k_{exp,M}^{0} = k_{42}/[M] = 1.28\times10^{16}$ $\exp-(19{,}270\pm630)/RT$[f]	300–444	283
	$k_{exp,N_2O_5}^{0} = k_{42}/[N_2O_5] = 22.5\times10^2$	323.7	284
	$k_{42}^{\infty} = 6.62\times10^{14}$	273–300	282
	$\exp-(21{,}100\pm2{,}000)/RT$	273–300	282
	$k_{42}^{400} = 1.3\times10^{14}$ $\exp-(21{,}000/RT)$[g]	293–303	281
	$k_{42}^{57} = 1.3\times10^{13}$ $\exp-(20{,}000/RT)$[h]	293–303	281
	$K_{42}k_{29} = 2.0\times10^{16}$ $\exp-(22{,}400/RT)$	293–303	281
$NO_2+O_3 \xrightarrow{44} NO_3+O_2$	$k_{44} = 5.9\times10^9$ $\exp-(7{,}000\pm600)/RT$	290–302	294

Additive (M)	N_2O_5	He	Ne	Ar	Kr	Xe	N_2	NO	CO_2	CCl_4	SF_6
$k_{exp,M}^{0}/k_{exp,N_2O_5}^{0}$[i]	1.00	0.124	0.090	0.135	0.159	0.147	0.234	0.300	0.40	0.551	0.32

[a] Units: sec^{-1} (for k_{exp}^{∞}, k_{42}^{∞}, k_{42}^{400}, k_{42}^{57}, $K_{42}k_{29}$); $l.mole^{-1}.sec^{-1}$ (for k_{exp}^{0}, $k_{42}/[M]$, k_{30}, k_{43}, k_{44}); $mole.l^{-1}$ (for K_{42}).

[b] Deduced from quoted literature as explained in the text.

[c] $[M] = [Ar] = 7.6\times10^{-3}$ $mole.l^{-1}$.

[d] Calculated from K_{42} at 300 °K[278], k_{44}[294] and [291] $(2\,k_{43})^{\frac{1}{2}}(K_{42}k_{44})^{\frac{1}{2}} = 1.13\times10^{13} \exp(-20{,}570/RT)$ $l^{\frac{1}{2}}.mole^{-\frac{1}{2}}.sec^{-1}$.

[e] Calculated from K_{42} at 300 °K[278], k_{44}[294] and [290] $(2\,k_{43})^{\frac{1}{2}}(K_{42}k_{44})^{\frac{1}{2}} = 1.93\times10^{12}$ $\exp(-19{,}583/RT)$ $l^{\frac{1}{2}}.mole^{-\frac{1}{2}}.sec^{-1}$.

[f] $[M] = [NO]+[N_2O_5]$.

[g] The value of the unimolecular rate coefficient k_{42} at 400 torr N_2.

[h] The value of the unimolecular rate coefficient k_{42} at 57 torr N_2.

[i] Relative values for different additives of the experimental rate coefficients in the limit of zero pressure at 323.7 °K[284].

data for the temperature range 450–1100 °K are consistent with the room-temperature data for N_2O_5 decomposition and related reactions.

(b) *Liquid phase decomposition of* N_2O_5

Lueck[286] studied the kinetics of N_2O_5 decomposition in CCl_4 and $CHCl_3$ solvents in the temperature range 25–55 °C, showing that the reaction was first order in N_2O_5 and that the rate coefficients differed only slightly from the gas-phase value at a given temperature. Subsequently, Eyring and Daniels[287] extended Lueck's investigation to include a large number of solvents. Their results are summarized along with Lueck's in Table 20. For all but two of the solvents,

TABLE 20

RATE PARAMETERS FOR N_2O_5 DECOMPOSITION IN VARIOUS SOLVENTS

	Arrhenius parameters			
Solvent	*A* (*sec*$^{-1}$)	*E* (*kcal.mole*$^{-1}$)	*Temp.* (°*C*)	*Ref.*
N_2O_4	1.44×10^{14}	25.0	15–45	287
CH_3CHCl_2	1.17×10^{14}	24.9	15–45	287
$CHCl_3$	0.593×10^{14}	24.6	15–45	287
$CHCl_3$	0.473×10^{14}	24.5	35–50	286
CH_2ClCH_2Cl	0.365×10^{14}	24.4	15–45	287
CCl_4	0.256×10^{14}	24.2	15–45	287
CCl_4	2.0×10^{14}	25.5	25–55	286
C_2HCl_5	0.945×10^{14}	25.0	15–45	287
Br_2	0.166×10^{14}	24.0	15–45	287
CH_3NO_2	0.282×10^{14}	24.5	15–45	287
$C_2F_3Cl_3$	0.40×10^{14}	24.5	13.6–20	288
HNO_3	5.37×10^{14}	28.3	15–45	287
Propylene chloride	3.21×10^{14}	27.0	20–25	287

nitric acid and propylene chloride, the first-order rate coefficients for the decomposition differ little from the gas-phase values. The rate coefficients show a slight increase[287] with increasing concentration of N_2O_5, but there is no deviation from the first-order kinetic law. The addition of polar molecules, H_2O and dinitrobenzene, had no influence on the rate coefficients. In view of Ogg's findings[289] we can justifiably assume that the decomposition of N_2O_5 in solution proceeds according to the complex mechanism outlined above for the gas phase in which the rate coefficient is given by

$$k_{\text{exp}}^{\infty} = 2\, K_{42}\, k_{30}$$

Accordingly, the lack of solvent effect relates either to the absence of an effect on both K_{42} and k_{30} or to compensatory solvent effects on K_{42} and k_{30}.

(c) N_2O_5*-sensitized decomposition of* O_3

It is convenient to deal with this reaction here since the elementary reactions proposed above for N_2O_5 decomposition are involved.

Ozone decomposition in the gas phase may be effected at room temperature by the addition of small amounts of N_2O_5[290, 291]. Although the N_2O_5 concentration remains virtually unchanged during the reaction[291], the detection of strong absorptions in the visible[291, 292] and in the infrared[293] attributed to the nitrate radical NO_3 strongly suggests that the equilibrium of reactions (42) and (−42) plays an important role in the decomposition. A combination of steps (42), (−42), (43) of the N_2O_5 decomposition mechanism with step (44) and the assumption of stationary-state concentrations of NO_3 and NO_2 leads to the expression for the rate of ozone depletion

$$-\frac{\mathrm{d}(O_3)}{\mathrm{d}t} = (K_{42}\,k_{44})^{\frac{2}{3}}(2\,k_{43})^{\frac{1}{3}}[N_2O_5]^{\frac{2}{3}}[O_3]^{\frac{2}{3}}$$

and for the NO_3 stationary concentration

$$[NO_3] = \left(\frac{K_{42}\,k_{44}}{2\,k_{43}}\right)^{\frac{1}{3}}[N_2O_5]^{\frac{1}{3}}[O_3]^{\frac{1}{3}}$$

which are satisfied by the experimental data[290, 291, 293]. The high concentration of NO_3 encountered in this reaction is a result of the slow removal of the nitrate radical in step (43). The alternative, rapid destruction of NO_3 *via* (29) is negligible since NO is absent: rate (44) $\gg$ rate (30). Using available data for K_{42}[278] and k_{44}[294] we have derived from the observed rate coefficient[290, 291] for the sensitized ozone decomposition two expressions for k_{43} (Table 19, footnotes d and e).

3.4.2 Photolysis and radiolysis of N_2O_5

In contrast to the numerous investigations of the thermal decomposition of N_2O_5, very few studies of its photochemical and radiolytic decomposition have been made. Nitrogen pentoxide absorbs below 3000 A[295, 296] and decomposes (into NO_2, N_2O_4, and O_2) with a quantum yield[296] of about 0.6 at 2800 A and 2650 A. The primary photolytic step is not known, although the following reaction

$$N_2O_5 + h\nu \rightarrow N_2O_5^* \rightarrow (N_2O_4 \rightleftharpoons 2\,NO_2) + O$$

has been considered[295, 296], together with a possible decomposition into N_2O_3

and O_2. To this one should perhaps also add a possible primary decomposition into NO_2 and NO_3, although all three alternatives are entirely speculative.

Direct radiolysis of N_2O_5 does not seem to have been investigated. Castorina and Allen[297] have observed that N_2O_5, a product in the γ-irradiation of liquid N_2O_4, is stabilized against thermal decomposition by the γ-rays, which cause the products O_2 and N_2O_4 to reform N_2O_5 in N_2O_4 solution.

3.5 NITRIC ACID

3.5.1 Thermal decomposition of HNO_3

Nitric acid vapour pyrolyzes at a measurable rate above 100 °C, decomposition proceeding in accordance with the stoichiometric equation

$$2\,HNO_3 = 2\,NO_2 + H_2O + \tfrac{1}{2}\,O_2$$

At high temperatures, product NO_2 decomposes further to NO and O_2. Below 300 °C the reaction is predominantly heterogeneous; above 400 °C it is essentially homogeneous[298] in a vessel of a surface/volume ratio $< 6\ \text{cm}^{-1}$. The homogeneous decomposition has been investigated for a static system in the temperature region 300–480 °C by optical analysis for both HNO_3 and NO_2[298–301]. The kinetics are complex and incompletely understood, particularly at low total pressures[300]. Johnston *et al.*[298–300] have shown that at 400 °C and 1 atm the *initial* rate of decomposition is first order in HNO_3. NO_2 inhibits the fast initial rate and slows the reaction as its concentration builds up in the course of the decomposition. These observations and the estimate[298] of $\sim$ 40 kcal.mole^{-1} for the activation energy of the process suggest that the unimolecular decomposition

$$HNO_3 + M \underset{-45}{\overset{45}{\rightleftharpoons}} OH + NO_2 + M \quad \Delta H_0^0 = +47.6\ \text{kcal.mole}^{-1} \qquad (45, -45)$$

(in its first-order region) and its reverse control the rate of decomposition. Nitric oxide sustains the initial rate of decomposition. With excess NO the reaction is first order in HNO_3 and zero order in NO. The overall reaction is then

$$2\,HNO_3 + NO = 3\,NO_2 + H_2O$$

The following elementary steps in combination with (45, −45) represent a mechanism which satisfactorily describes observations at 1 atm:

$$OH + HNO_3 \rightarrow H_2O + NO_3 \qquad (46)$$

$$NO_3+NO_2 \rightarrow NO_2+NO+O_2 \tag{30}$$

$$NO_3+NO \rightarrow 2\,NO_2 \tag{29}$$

$$OH+NO+HNO_3 \rightarrow H_2O+2\,NO_2 \tag{47}$$

Assuming a steady-state concentration of NO_3, the reaction scheme leads to the rate expression

$$-\frac{d[HNO_3]}{dt} = \frac{2\,k_{45}[HNO_3]^2\{k_{46}+k_{47}[NO]\}}{k_{-45}[NO_2]+k_{46}[HNO_3]+k_{47}[NO][HNO_3]}$$

There is good evidence[250, 302] that OH reacts rapidly with HNO_3 in accordance with (46). Reactions (30) and (29) are well established, and NO_3 has been observed both in thermally shocked[302] and flash-photolyzed HNO_3[250]. The exact nature of the OH-removing reaction (47) is uncertain and its ability to compete efficiently with the rapid[250] reaction (46) may be open to question. However, it offers one possible explanation of the observed effect of nitric oxide additions. (It should be noted here that at low temperatures there is an additional, very rapid, heterogeneous reaction between NO and HNO_3[299, 303].) The rapidity of (29) ensures that [NO] is stationary and that $k_{46} \gg k_{47}[NO]$ in the absence of added NO; NO only appears as a detectable product towards the end of the reaction when the slow decomposition of NO_2 can compete with (45), (46), (30) and (29). Reducing agents such as H_2, CO and hydrocarbons sustain the initial rate, in a similar manner to NO, by intercepting OH radicals from (45).

The expected fall-off in first-order rate coefficient, k_{45}, with decreasing pressure has been observed by Johnston *et al.*[300]. They have shown that the initial rate of decomposition at 400 °C for nitric acid pressures $<$ 10 torr is second-order in HNO_3. However, they point out that the magnitude of the observed effects of inert gases on the empirical first-order rate coefficient was such that their measurements did not refer to the truly second-order region of a unimolecular reaction. Side reactions involving H_2O, NO, NO_2 and HNO_2 contribute substantially to the decomposition rate at low pressures, so that a simple interpretation of the second-order dependence of the rate on HNO_3 is not permissible. Fréjacques[301] deduced from the time dependence of $[HNO_3]$ that in the region 280–480 °C the rate showed a second-order dependence on HNO_3 for pressures $<$ 100 torr. His deduction is, however, inadmissible since the inhibiting effect of product NO_2 on the rate gives rise to reactant–time curves suggestive of second-order decay. A shock-tube investigation[302] has shown that at 1000 °K and 10^{-2} *M* (principally Ar) reaction (45) is close to its second-order limit and has an activation energy of $\sim$ 31 kcal.mole^{-1}. The difference between this value and the HO–NO_2 bond energy is reasonable for a pentatomic molecule.

Rate coefficients of elementary steps in the thermal decomposition of HNO_3 are summarized in Table 21.

TABLE 21

RATE COEFFICIENTS OF ELEMENTARY STEPS IN THE DECOMPOSITION OF HNO_3

Reaction	Rate coefficients[a]	Temp.(°K)	Ref.
$M+HNO_3 \xrightarrow{45} M+OH+NO_2$	$k_{45}^{Ar} = 10^{12.2\pm1.0}$ $\exp-(30{,}600\pm1{,}800)/RT$	800–1200	302
	$k_{45}^{HNO_3} = 5.6\times10^{14} \exp-(38{,}300/RT)$[b]	650–700	300
	$E_A^{\infty} > 40$ kcal.mole^{-1}		298
	$k_{45}^{N_2=1\text{ atm}} = 0.16$	670	299
$OH+HNO_3 \xrightarrow{46} H_2O+NO_3$	$k_{46} = 1\times10^8$	300	250
	$k_{46} = (1.4\pm0.6)\times10^7$	~300	307
$OH+NO_2+M \xrightarrow{-45} HNO_3+M$	$k^{N_2}_{-} = 1.3\times10^8 \exp(7{,}000/RT)$	673	304
	$k_{-45}^{N_2}/k_{46} = 209$	670	299
	$k_{-45}^{Kr}/k_{46} = 13$	~ 300	306

[a] Units: l.mole^{-1}.sec^{-1} (for k_{45}^{Ar}, $k_{45}^{HNO_3}$, k_{46}); sec^{-1} (for $k_{45}^{N_2=1\text{ atm}}$); l.mole^{-1} (for k_{-45}/k_{46}); l^2.mole^{-2}.sec^{-1} (for $k_{-45}^{N_2}$). Superscripts to rate coefficients (Ar, HNO_3, N_2, Kr) indicate the predominant third body (M) in third-order reactions or activating gas in the second-order region of unimolecular reactions.

[b] Probably not measured in truly second-order region (see text).

3.5.2 Photolysis and radiolysis of HNO_3

Because anhydrous nitric acid is of low thermal stability even at room temperature, very few photochemical and radiolytic studies have been carried out with it. A flash photolytic study of nitric acid vapour by Husain and Norrish[250] has provided evidence for a primary photochemical split into OH and NO_2, *viz.*

$$HNO_3+h\nu \rightarrow OH+NO_2$$

followed by

$$OH+HNO_3 \rightarrow H_2O+NO_3 \tag{46}$$

A kinetic study of the photolysis at low conversions by Bérces and Förgeteg[305] leads to the conclusion that OH is the sole reactive species formed in the primary process. At 2650 A and 2537 A, respectively, the quantum yields are 0.1 and 0.3. Secondary reactions are expected to be similar to those involved in the thermal decomposition of HNO_3, as discussed in the preceding section.

Simultaneous photolysis[306] of HNO_3 and NO_2 leads to a room temperature value of $k_{-45}/k_{46} = 13$ l.mole^{-1} with M = Kr and to the estimates $k_{30} = 10^7$ l.mole^{-1}.sec^{-1} and $k_{48} = 10^{11}$ l.mole^{-1}.sec^{-1}, both at room temperature, where k_{48} is for the reaction

$$O+HNO_3 \rightarrow OH+NO_3 \qquad (48)$$

The estimated value of k_{30} is about two orders of magnitude larger than the value obtained by extrapolation to room temperature of Schott and Davidson's rate expression[278] (Table 19). The above k_{-45}/k_{46} value for M = Kr appears to be grossly inconsistent[306] with the value of Johnston *et al.*[299] for M = N_2 (Table 21) approximately extrapolated to room temperature or with the ratio calculated from Burnett's expression[304] for k_{-45} and Husain and Norrish's estimate[250] for k_{46} (Table 21). Bérces *et al.*[307] measured indirectly k_{46} and obtained a value 7 times smaller than the value of Husain and Norrish in Table 21. All these inconsistencies point out the need for further work with these kinetically very complex systems.

Breakdown to NO_2, O_2 and H_2O of gaseous anhydrous nitric acid under γ-irradiation has been studied at 25 °C by Dmitriev *et al.*[308] who also irradiated samples of liquid and solid nitric acid over the temperature interval from −196 to +32 °C. The ESR spectrum of solid HNO_3 showed the presence of NO_3 and NO_2 radicals. The radiational breakdown of HNO_3 appears to be principally through rupture of the weakest HO–NO_2 bond.

4. Ozone

A knowledge of the kinetics of the decomposition of ozone is essential for the understanding of the chemistry of some important processes which occur in earth's atmosphere. Yet, in spite of numerous studies and the structural simplicity of ozone, the mechanism of its ultraviolet photolysis is still uncertain. Electronically and vibrationally excited species are involved in ozone decomposition and the current knowledge of the chemical behavior of such intermediates is still in its infancy.

4.1 THERMAL DECOMPOSITION OF O_3

The mechanism for the thermal decomposition of O_3 is well established[309–312]. Both low-temperature studies (in clean glass vessels) and high-temperature shock-tube studies suggest that a simple mechanism, involving ground-state O atoms,

$$O_3+M \underset{-1}{\overset{1}{\rightleftharpoons}} O_2+O+M \qquad (1, -1)$$

$$O+O_3 \rightarrow O_2+O_2 \qquad (2)$$

is operative over the temperature range 300–900 °K. There is a heterogeneous contribution[309, 313] to the decomposition rate at low temperatures, but it is unimportant for unpacked, clean glass vessels. The occurrence of a direct bimolecular reaction in the gas phase

$$2\,O_3 \rightarrow 3\,O_2 \qquad (3)$$

is doubtful. Benson and Axworthy[309] have suggested that the abnormally high rates of decomposition measured by Glissman and Schumacher[314] are to be attributed to self-heating rather than to a contribution from (3). Pshezhetskii *et al.*[313], on the other hand, have reported a progressive lowering of the activation energy for the decomposition with increasing O_3/O_2 ratio and maintain that (3) is the sole reaction pathway in the decomposition of 100 % O_3. Their belief that O_3 has zero efficiency as a third body in (1) contradicts the experimental evidence[309, 310]. More importantly, the results of Zaslowsky *et al.*[311] for nearly pure ozone do not confirm Pshezhetskii's findings and show no departure from the atomic mechanism.

Rate data for the decomposition yield values for k_1 (Table 22) and, in some instances, k_1k_2/k_{-1}. With the aid of thermodynamic data for $K_1 = k_1/k_{-1}$ one may calculate k_{-1} and k_2 (Table 22). Values of k_2 obtained by direct measurement in a flow system are, generally speaking, too high because of contributions to O_3 removal by H atoms and "hot" molecules[325, 327].

Reaction (1) is a unimolecular reaction which, for the temperature and pressure ranges investigated, is close to its second-order limit. (The first-order rate coefficient for the decomposition of O_3 dissolved in CCl_4 is possibly a measure of k_1^{∞}, or $2\,k_1^{\infty}$, but may alternatively refer to a complex reaction between O_3 and the solvent[328].) The pre-exponential factor of k_1^0 is appreciably smaller than the A factors reported for other spin-allowed dissociations of triatomic molecules. Formulation of the rate coefficients according to the classical RRK theory leads to an expression[310, 315] which suggests that inefficient collisional activation of O_3 may be the cause of a small A_1. Classical theory is, however, inappropriate for dissociation with such a small critical energy. The modification of Wieder and Marcus[113] of RRK theory gives a more satisfactory expression for k_1^0 and indicates that the efficiency of collisional activation approaches unity; the small A factor is a result of the low density of vibrational levels at the dissociation limit of O_3.

Energy-rich oxygen molecules are expected from the very exothermic reaction (2), and vibrationally excited ground-state molecules have indeed been observed[310, 316, 317]. Both the thermal[309] and the long-wavelength photochemical data[329] show, however, that energy chains are unimportant in the decomposition of ozone when $O(^3P)$ is an intermediate. The results of a study of reaction (2) in a flow system support this conclusion[326]. Evidently the transfer of vibrational energy between vibrationally "hot" O_2 and O_3 in amounts sufficient to break the O_2–O bond is an inefficient process.

References pp. 132–141

TABLE 22

RATE COEFFICIENTS FOR ELEMENTARY REACTIONS IN O_3 DECOMPOSITION

Reaction	*Rate expressions*[a]	*Temp.(°K)*	*Ref.*
$M+O_3 \xrightarrow{1} M+O_2+O$	$k_1^{O_3} = (4.61\pm0.25)10^{12} \exp-(24{,}000/RT)$	340–390	309, 314
	$k_1^{O_3} : k_1^{O_2} : k_1^{N_2} : k_1^{CO_2} : k_1^{He} = 1.00 : 0.44 : 0.41 : 1.06 : 0.34$	340–390	309, 314
	$k_1^{N_2} = (5.8\pm0.6)\times10^{11} \exp-(23{,}150\pm300)/RT$	300–900	309, 310, 314
	$k_1^{N_2} = (1.54\pm0.17)k_1^{Ar}$	700–900	310
	$k_1^{O_3} = 8\times10^{12} \exp-(24{,}300/RT)$	388–403	311
$O+O_2+M \xrightarrow{-1} O_3+M$	$k_{-1}^{Ar} = 8.9\times10^6 \exp+(1{,}800\pm400/RT)$	188–373	315
	$k_{-1}^{Ar} = 1.45\times10^8$	300	255
	$k_{-1}^{Ar} : k_{-1}^{He} : k_{-1}^{O_2} : k_{-1}^{N_2} : k_{-1}^{CO_2} : k_{-1}^{N_2O} : k_{-1}^{CF_4} : k_{-1}^{SF_6} : k_{-1}^{H_2O} = 1.0 : 1.0 : 1.6 : 1.4 : 3.8 : 3.8 : 4.0 : 8.5 : 15.0$	300	255
	$k_{-1}^{O_2} = 2.3\times10^8$	297	318
	$k_{-1}^{Ar} = 0.83\times10^8$	296	319, 320
	$k_{-1}^{Ar} : k_{-1}^{CO_2} : k_{-1}^{N_2O} : k_{-1}^{He} = 1.0 : 5.0 : 4.2 : 0.8$	296	319, 320
	$k_{-1}^{O_2} = 1.47\times10^8$	300(?)	321
	$k_{-1}^{O_2} : k_{-1}^{CO_2} : k_{-1}^{N_2O} : k_{-1}^{CO} = 1.0 : 2.5 : 2.2 : 1.1$	300(?)	321
	$k_{-1}^{O_3} : k_{-1}^{He} : k_{-1}^{Ar} : k_{-1}^{N_2} : k_{-1}^{O_2} : k_{-1}^{CO_2} = 1.0 : 0.34 : 0.25 : 0.39 : 0.44 : 0.96$	291	329
	$k_{-1}^{O_2} = 2.98\times10^7 \exp+(892/RT)$	340–390	309
	$k_{-1}^{O_2} = 1.33\times10^8$	300	309
	$k_{-1}^{Ar} = (1.7\pm0.2)\times10^7 \exp+(1{,}680\pm100)/RT$	213–386	323
	$k_{-1}^{CO_2} = (8.4\pm1.1)\times10^7 \exp+(1{,}450\pm140)/RT$	213–386	323
	$k_{-1}^{O_2} : k_{-1}^{He} : k_{-1}^{Ar} : k_{-1}^{CO_2} = 1.0 : 0.74 : 0.90 : 3.1$	213–386	323
$O+O_3 \xrightarrow{2} 2\,O_2$	$k_2 = 3.35\times10^{10} \exp-(5{,}708/RT)$	340–390	1, 309, 314
	$k_2 = 2.36\times10^{10} \exp-(5{,}364/RT)$	420–560	324, 1
	$k_2 = 2.7\times10^9 \exp-(3{,}270/RT)$	400–500	312, 1
	$k_2 = 4.7\times10^6$	300	325, 255
	$k_2 = 1.9\times10^{10} \exp(-4{,}500/RT)$	270–340	350
	$k_2/k_{-1}^{Ar} = 4\times10^2 \exp-(5{,}440\pm1{,}000/RT)$	280–300	329
$O(^1D)+O_3 \xrightarrow{6} 2\,O_2$	$k_6 = (2\pm1)\times10^{11}$	200–300	335
$O_2(^1\Delta_g)+O_3 \xrightarrow{9} 2\,O_2+O$	$k_9 = (1.7\pm0.5)\times10^6$	298	346
	$k_9 = (2.1\pm0.6)\times10^6$	?	346[b]
	$k_9 = (1.5\pm0.3)\times10^6$	298	347
$O_2(^1\Sigma_g^+)+O_3 \xrightarrow{9a} 2\,O_2+O$	$k_{9a} = 3.6\times10^8$	298	348
	$k_{9a} = (4.3\pm0.5)\times10^9$	294	349

[a] Units: l.mole^{-1}.sec^{-1} (for k_1, k_2, k_6, k_9, k_{9a}); l^2.mole^{-2}.sec^{-1} (for k_{-1}); mole.l^{-1} (for k_2/k_{-1}).

[b] The result quoted in ref. 346.

Many measurements of k_{-1} have been made from direct observations of the decay of [O] in mixtures of O and O_2. Only recently[315, 255] however, was it realized how important were the contributions to the measured rate of recombination from catalysis by H atoms and from redissociation of O_3 by $O_2(^1\Sigma^+)$ and by states of O_2 of higher energy. Table 22 summarizes values of k_{-1} obtained by experimenters[315, 255, 318–321, 323] who successfully eliminated or corrected for these extraneous effects. Agreement between the various values is excellent when one considers the variety of experimental techniques employed; furthermore, the value of k_{-1} calculated from the pyrolysis results for k_1 and thermochemical data[1] is close to (although perhaps significantly smaller than[323]) the observed rate coefficient. Hochanadel *et al.*[318] have followed the relatively slow relaxation of the vibrationally excited ozone produced in the initial association

$$O+O_2 \underset{-4}{\overset{4}{\rightleftharpoons}} O_3^* \qquad (4, -4)$$

$$O_3^*+M \rightarrow O_3+M \qquad (5)$$

and conclude that $\sim 10^4$ collisions are required on average for deactivation of O_3* to the ground state. Their observations suggest that rate coefficients k_{-1} obtained[319–321] by monitoring the ultraviolet absorption at 2600 A of product O_3 from (−1) are somewhat too low on account of the significant reduction in extinction coefficient at that wavelength with increasing vibrational energy content of O_3. Such a slow relaxation of O_3* is difficult to reconcile with calculations of the rate of collisional activation of O_3 based on the modern theory of unimolecular reactions.

Studies[251, 322] of the isotopic exchange reaction between ^{18}O and O_2 and of the rate of O_3 formation at very high pressures[319] have shown that $k_4 \sim 10^9$ l.mole^{-1}.sec^{-1} at 300 °K.

4.2 PHOTOLYSIS AND RADIOLYSIS OF O_3

Ozone undergoes decomposition when irradiated with α-particles[352, 353] or with γ-rays[354]. The decomposition may proceed through a number of distinct reaction sequences[354], such as negative ion chain decomposition reactions in all-glass vessels and involvement of impurities in vessels fitted with fluorocarbon-lubricated stopcocks.

Photodecomposition of ozone has been studied both in the region of its weak absorption[329–331] centered at 6000 A and in the strong ultraviolet absorption band[316, 317, 332–338] centered at 2550 A. A quantum yield of $\Phi_{-O_3} = 2$ observed for photolysis with visible light under oxygen-free conditions has been interpreted[329] in terms of the primary process (a) (Table 23), proceeding with unit

TABLE 23

POTENTIAL PRIMARY PROCESSES IN THE PHOTOLYSIS OF OZONE

	ΔH_0^0*(kcal.mole*$^{-1}$*)*	*Threshold wavelength (A)*
(a) $O_3(^1A) \rightarrow O_2(^3\Sigma_g^-)+O(^3P_2)$	24.25	11788
(b)† $\rightarrow O_2(^3\Sigma_g^-)+O(^1D_2)$	69.61	4107
(c)† $\rightarrow O_2(^3\Sigma_g^-)+O(^1S_0)$	120.85	2365
(d)† $\rightarrow O_2(^1\Delta_g)+O(^3P_2)$	46.78	6110
(e)† $\rightarrow O_2(^1\Sigma_g^+)+O(^3P_2)$	61.76	4629
(f) $\rightarrow O_2(^1\Delta_g)+O(^1D_2)$	92.14	3102
(g) $\rightarrow O_2(^1\Sigma_g^+)+O(^1D_2)$	107.11	2669
(h) $\rightarrow O_2(^1\Delta_g)+O(^1S_0)$	143.38	1994
(i) $\rightarrow O_2(^1\Sigma_g^+)+O(^1S_0)$	158.35	1805
(j) $\rightarrow O_2(^3\Sigma_u^+)+O(^3P_2)$	126.34	2263
(k) $\rightarrow O_2(^3\Sigma_u^-)+O(^3P_2)$	165.35	1729

† Spin-forbidden processes.

efficiency, followed by reaction (2). The observed effects of temperature and inert gas additives on the quantum yield in the visible are quantitatively consistent with the rate parameters deduced from thermal decomposition data for the scheme (1), (−1) and (2) (Table 22).

Quantum yield values reported for ultraviolet photolysis in the region 2000–3100 A are not in good agreement, but are all in excess of 2. Norrish and Wayne[332] obtained a value as high as 17 for the photolysis of dry O_3 at 2537 A, which implies a reaction mechanism involving short chains. Other workers[336−338] have reported values of Φ_{-O_3} no larger than 6–7, which can be explained without necessarily invoking a chain mechanism.

In spite of differences of opinion over the nature of the secondary processes in the ultraviolet photolysis, there is now abundant evidence[339−341, 344, 345] that at wavelengths shorter than 3200 A $O(^1D_2)$ atoms are produced in the primary process and disappear in a very rapid[335] reaction

$$O(^1D)+O_3 \rightarrow 2\,O_2^\dagger \tag{6}$$

It has been established[316, 317, 333, 334] that the exothermicity of (6) appears, at least in part, as vibrational excitation of product $O_2(^3\Sigma_g^-)$ molecules (levels up to $v'' = 23$ have been observed), and there have been suggestions[309, 332, 343] that one of the product molecules is electronically excited. The "hot" products of (6) decompose further molecules of O_3 in reactions which are appreciably slower[334] than (6)

$$O_2^\dagger+O_3 \rightarrow 2\,O_2+O(^3P) \tag{7}$$

$$O(^3P)+O_3 \rightarrow 2\,O_2^\dagger \tag{2a}$$

Studies of the flash photolytic decomposition by Bair *et al.*[333−335] clearly show that decomposition occurs in two stages: a very fast initial reaction with a quantum yield of 2, which presumably comprises the primary photolytic act and (6), followed by a much slower secondary decomposition. The quenching effect of inert gas additions on the quantum yield for the initial decomposition was consistent with the previously established[139, 145] relative efficiencies of the same inert gases for the deactivation of $O(^1D_2)$ to the ground state. Their effect on the slow, secondary reaction is related to the relaxation of vibrationally hot $O_2^\dagger$ and to their relative effectiveness in (−1).

In connection with the possible existence of short energy chains in O_3 photolysis, a significant observation[333, 334] has been the appearance of vibrationally hot O_2 early in the flash photolysis and the observation of a rapid transfer of vibrational energy between $O_2^\dagger$ and O_3. If the average vibrational temperature of O_3 is sufficiently enhanced, for example at high densities of absorbed radiation, short energy chains involving (2a) and (7) could perhaps be maintained. (In thermally equilibrated ozone, such a chain reaction evidently does not occur (see above).) The necessary high density of absorbed energy may have been attained even in the conventional low-intensity photolysis experiments at 2537 A where the extinction coefficient for O_3 is large.

Available experimental evidence[333, 334] dictates against a chain mechanism[332] involving $O(^1D)$ atoms

$$O(^1D)+O_3 \rightarrow O_2^*+O_2$$

$$O_2^*+O_3 \rightarrow 2\,O_2+O(^1D) \qquad (8)$$

where O_2^* is either vibrationally or electronically excited. Reaction (8) is exothermic for O_2 in vibrational levels higher than the 17th, yet there is no discontinuity[333] in the population of levels from $v'' = 13$–23 of $O_2(^3\Sigma_g^-)$ formed in the flash photolysis of O_3; furthermore, the calculated fraction of O_2^* molecules formed with $v'' \gg 17$ in (6) is too small to maintain such a chain. Of the electronically excited states of O_2 capable of participating in (8), the $B(^3\Sigma_u^-)$ state may be eliminated as a possibility since its formation in (6) would be 2 kcal.mole^{-1} endothermic, whereas the experimental data indicate that the reaction of $O(^1D)$ with O_3 is temperature independent[334, 342]; production of $O_2(A^3\Sigma_u^+)$ in (6) is only sufficiently exothermic to excite the other product O_2 molecule to $v'' \leqq 8$ of the ground state, whereas states $v'' \gg 17$ are observed *early* in the photolysis[334]. As a result of their recent study of low-intensity conventional photolysis of pure O_3 using light in the region 2980–3220 A from a high-pressure Hg lamp, Castellano and Schumacher[337] conclude that the decomposition does not proceed *via* a chain mechanism. They account for their quantum yield of 6, observed for oxygen-free ozone and found to be independent of temperature, light intensity and $[O_3]$, in terms of primary process (f) (see Table 23) followed by

$$O_2(^1\Delta_g)+O_3 \rightarrow 2\,O_2+O \tag{9}$$

$$O(^1D)+O_3 \rightarrow O_2^{\dagger}+O_2 \tag{6a}$$

$$O_2^{\dagger}+O_3 \rightarrow 2\,O_2+O$$

and reaction (2). $O_2^{\dagger}$ was considered to be a vibrationally excited, ground-state molecule. The quenching effect of inert gas additions, which reduced Φ_{-O_3} to 4, was attributed to

$$O(^1D)+M \rightarrow O(^3P)+M \tag{10}$$

Significant deactivation of $O_2(^1\Delta_g)$ would not have taken place even at the highest $[M]/[O_3]$ ratios used. Reduction of Φ to 2 for large inert gas additions, as observed by Norrish and Wayne[332] at 2537 A, could be interpreted to imply the production of $O_2(^1\Sigma_g^+)$ in the primary process at that wavelength, and its subsequent deactivation to the ground state. However, Gauthier and Snelling[355] have now obtained experimental evidence that $O_2(^1\Delta_g)$ and not $O_2(^1\Sigma_g^+)$ is a primary photolysis product at 2537 A.

The finding of Gauthier and Snelling supports reaction (f) as the primary process in the experiments of Castellano and Schumacher[337]. Unfortunately, the photon energy corresponding to the wavelength region chosen by the latter authors for photolysis is barely sufficient to effect (f) and there may therefore be some uncertainty regarding the primary process. Furthermore, there is good evidence to suggest[342], at least for O_3 dissolved in liquid Ar, that for $\lambda > 3000$ A the primary quantum yield of $O(^1D)$ is less than the value of unity assumed by Castellano and Schumacher. Also, the relative values for k_{10} deduced by them are in poor quantitative agreement with other data in the literature[139, 145] and suggest that reaction (10) is not the sole deactivation step which determines the observed quenching effects. In view of the considerable dissension over the nature of the primary process and the secondary reactions in the UV photolysis, there is clearly a need for renewed experimentation, preferably at wavelengths in the region 2700–2900 A where the primary quantum yield of $O(^1D)$ is unity and ε_{O_3} is not too large.

5. Sulphur oxides

SO_2 and SO_3 are the only compounds considered in this section. S_2O is a well-established[356] oxide of sulphur, although there is apparently no information on the kinetics of its decomposition. Sulphur monoxide, SO, is a reactive intermediate which features in the decompositions of SO_2 and SO_3.

5.1 SULPHUR DIOXIDE

5.1.1 Thermal decomposition of SO_2

SO_2 is a very stable oxide and its thermal decomposition is only measurable at the very high temperatures attained in a shock tube. A study[357] of the time-dependence of light emission from shock-heated SO_2/Ar mixtures in the region of 3000 °K has shown that SO_2 is removed in accordance with a sigmoid-shaped concentration–time curve typical of a chain or autoaccelerated reaction. The induction period observed[357] prior to the onset of detectable decomposition corresponded closely with the time for the formation of a fixed concentration of O (or SO) calculated from the rate expression (Table 24) for the unimolecular decomposition

$$SO_2+M \rightarrow SO+O+M \qquad (1)$$

obtained in an independent investigation[358]. Empirical rate expressions[357–359] obtained for the maximum rate of decomposition show that the rate is first order in SO_2 but complex in total pressure, and suggest that decomposition occurs by at least two pathways. The observed activation energy of $\sim$ 70 $kcal.mole^{-1}$ is certainly reasonable for a chain reaction initiated by (1) and terminated by a reaction second order in chain carrier concentration; the nature of the individual steps remains obscure, however.

Emission from shock-heated SO_2 arises from three electronically excited states $^3B_1$, $^1B_2$, $^1B_1$, populated by collisions from the $^1A_1$ ground state and, for finite extents of dissociation, by the recombination of O and SO[360]. Equilibrium between the emitting states and the ground state

$$SO_2+M \rightleftharpoons SO_2^*+M$$

is achieved immediately behind the shock front[360]; the contribution of chemical quenching to the removal of SO_2^* by, *e.g.*

$$SO_2^*+SO_2 \rightarrow SO_3+SO$$

is negligible compared to physical quenching.

In addition to providing a detailed mechanism for the initial process of dissociation, the shock-tube studies have yielded a value[360] for the absolute intensity of the sulphur dioxide afterglow

$$SO+O \rightarrow SO_2+h\nu \qquad (2)$$

TABLE 24

ELEMENTARY REACTIONS IN THE DECOMPOSITION OF SO_2

Reaction	*Rate coefficients*[a] *and expressions*	*Temp.(°K)*	*Ref.*
$SO_2+M \xrightarrow{1} SO+O+M$	$k_1 = 10^{11.4}$ $\exp-(110{,}000\pm3{,}000)/RT$	4500–7500	358
$SO+O \xrightarrow{2} SO_2+h\nu$	$I_0^{Ar} = 1.5\times10^5[O][SO]$	298	361
	$I_0^{Ar} = (1.4\pm0.4)\times10^4[O][SO]$	3500	360
	$I_0^{N_2} = 1.9\times10^5\ (300/T)^{1.6}[O][SO]$	300–1500	362
	$I_0 = 4\times10^5[O][SO]$	300	363
	$I_0 = 3\times10^6[O][SO]$	300	364
$SO+O+M \xrightarrow{-1} SO_2+M$	$k_{-1}^{Ar} = (3.2\pm0.4)\times10^{11}$	300	361
$SO+O_2 \xrightarrow{-4} SO_2+O$	$k_{-4} = 10^{10.3\pm0.3}$ $\exp-(10{,}000\pm1{,}000)/RT$	440–530	366
	$k_{-4} = 5.2\times10^{11}$ $\exp-(19{,}300/RT)$	750–1100	367
$SO+SO \xrightarrow{5} ?$	$k_5 < 4\times10^6$	300	361
	$k_5 < 10^7$	300	368
$O+SO_2+M \xrightarrow{-3} SO_3+M$	$k_{-3}^{O_2} = (2.7\pm0.5)\times10^9$	299	371
	$k_{-3}^{Ar} = (2.4\pm0.15)\times10^9$	299	371
	$k_{-3}^{SO_2} = (10\pm4)\times10^9$	299	371
	$k_{-3}^{O_2} = k_{-3}^{Ar} = (4.7\pm0.8)\times10^9$	300	361
	$k_{-3}^{O_2} = 4.9\times10^9$	300	372
	$k_{-3}^{N_2} = 4.5\times10^9$	300	372
	$k_{-3}^{O_2} \sim 3\times10^{10}$	300	373
	$k_{-3}^{H_2} \not< 4\times10^9$	784	374
	$k_{-3}^{SO_2+NO_2} = 1.4\times10^{10}$	297	375
	$E_A \sim -2000$ cal/mole		366
$O+SO_2 \xrightarrow{-3a} SO_3^*$	$k_{-3a} = (1.1\pm0.1)\times10^9$	297	375
$SO_3^*+M(=SO_2+NO_2) \xrightarrow{3b} SO_3+M$	$k_{3b}/k_{3a} = 13$	297	375
$S+O_2 \xrightarrow{7} SO+O$	$k_7 = (1.2\pm0.3)\times10^9$	298	389

[a] Units: $l.mole^{-1}.sec^{-1}$ (for k_1, I_0, k_{-4}, k_5, k_{-3a}, k_7); $l^2.mole^{-2}.sec^{-1}$ (for k_{-1}, k_{-3}); $l.mole^{-1}$ (for k_{3b}/k_{3a}).

which compares favorably (Table 24) with room-temperature measurements. Rate coefficients for (2) and a number of other possible secondary reactions in the decomposition of SO_2 have been deduced from flame studies and from room-temperature investigations of the sulphur dioxide afterglow. There have been now several determinations[360–364] of the absolute intensity of (2) and, apart from one rather high value[364], they are in good quantitative agreement. The reported[361, 363] lack of pressure dependence of I_0 does not necessarily imply that emission is the result of a two-body radiative collision[363]; it is more likely[361] that the emitting states are held in pseudo-equilibrium by formation and destruction processes that are both pressure dependent, *viz.*

$$O+SO+M \rightarrow SO_2^*+M \quad (2a)$$

$$SO_2^*+M \rightarrow SO_2+M \quad (2b)$$

$$SO_2^* \rightarrow SO_2+h\nu \quad (2c)$$

When $k_{2c} \ll k_{2b}(M)$, I_0 will be pressure independent, as observed by Halstead and Thrush[361] for pressures > 0.2 torr. However, at the low pressures employed by Rolfes *et al.*[363] one would expect[361] a pressure dependence, *viz.* $I_0 = k_{2a}$ [O][SO][M], which was not observed by those workers. The fall-off to a third-order dependence was perhaps masked by emission from SO_2^* generated at the reaction cell surface[365].

The third-order recombination reactions

$$O+SO+M \rightarrow SO_2+M \quad (-1)$$

and

$$O+SO_2+M \rightarrow SO_3+M \quad (-3)$$

are the most important secondary reactions following the dissociation of SO_2 at room temperature; they are likely to play an important role even at high temperatures since measurements[366, 367] of the reverse of

$$O+SO_2 \rightarrow SO+O_2 \quad (4)$$

combined with thermochemical data[1], show that this alternative path for O atom removal has an activation energy of at least 20 kcal.mole^{-1}. Halstead and Thrush[361] have determined k_{-1} at room temperature from observations of the decay of the sulphur dioxide and air afterglows in discharged SO_2 containing small amounts of NO. Their observations strongly suggest that (−1) is the dominant process responsible for SO removal and that any contributions from

$$SO+SO \rightarrow SO_2+S \quad (5)$$

or

$$\rightarrow S_2O_2$$

or other reactions of SO, are negligible. Hoyermann *et al.*[368] and Donovan *et al.*[369] reached the same conclusion and set similar upper limits to the rate coefficient for (5) as a result of their studies of the reaction of O+COS[368] and flash photolysis of $SOCl_2$[369]. The rapid removal of SO observed by Sullivan and Warneck[370] in an independent study of the latter reaction does not necessarily imply a fast reaction (5), as suggested by those authors, but more likely[361] an alternative, four-centre reaction

$$SO + COS \rightarrow CO_2 + S_2$$

The rate of the three-body recombination (−3) has been determined from observations[361, 371–373] of the decay of O atoms in a discharge-flow tube, by measurement[375] of the quantum yield of the photolysis of NO_2 containing SO_2, and from the inhibiting effect[374] of SO_2 on the ignition limits of H_2–O_2 mixtures. We have chosen to reject estimates of k_{-3} deduced[376] from a very involved analysis of reactions occurring in flames containing sulphur oxides in favour of the more directly determined values. Jaffe and Klein's study[375] of ^{18}O exchange with SO_2 has provided rate coefficients (Table 24) for the individual steps

$$O + SO_2 \underset{3a}{\overset{-3a}{\rightleftharpoons}} SO_3^*$$

$$SO_3^* + M \overset{3b}{\rightarrow} SO_3 + M$$

comprising the overall reaction (−3). Their results suggest that at room temperature the recombination (−3) will remain third order up to ~ 100 torr. The rapidity of the recombination suggests that (−3a) is a spin-allowed process yielding a triplet state of SO_3[371, 374]. If electronically excited SO_3 molecules participate in further reactions with the products of discharged O_2, *e.g.*

$$SO_3^* + O_2(^1\Delta_g \text{ or } ^1\Sigma_g^+) \rightarrow SO_2 + O_2 + O$$

which regenerate O atoms, the rate data obtained by the discharge-flow technique would be vitiated. However, the excellent agreement (Table 24) between values of k_{-3} obtained by the discharge-flow method and from the photolysis of SO_2/NO_2 mixtures argues against such complications. There are indications[373] from gas phase studies that the reaction

$$O + SO_3 \rightarrow SO_2 + O_2 \tag{6}$$

is too slow at room temperature to influence determinations of k_{-3}. The reaction

$$SO + SO_3 \rightarrow 2\,SO_2$$

is also believed to be very slow at room temperature[361]. However recent[377] measurements of oxygen yields from the γ radiolysis of SO_2/SO_3 solutions support a value for k_6 in liquid SO_2 at 266 °K of ~ 5×10^9 l.mole^{-1}.sec^{-1} (based on Jaffe and Klein's k_{-3a}[375]). Such a high value of k_6 would imply that the values of k_{-3} deduced from discharge-flow experiments have been overestimated by a factor of 2.

5.1.2 *Photochemical decomposition of* SO_2

The increasing awareness of the health hazard posed by the pollution of our atmosphere with ever-increasing quantities of SO_2 has stimulated research activity into its photodecomposition and photooxidation; yet the chemistry of these processes is still only very poorly understood.

The processes accompanying the absorption of ultraviolet light by SO_2 have been investigated, but principally from the point of view of characterizing the upper electronic states involved in the transitions[2] and establishing their lifetimes with respect to the various decay processes of fluorescence, phosphorescence, internal conversion and intersystem crossing[378, 379]. Absorption commences at 3880 A and consists of a series of clearly separated bands which increase in intensity towards shorter wavelengths. Table 25 summarizes the spectroscopic data.

TABLE 25

EXCITED STATES AND SPECTRAL CHARACTERISTICS OF SO_2 IN ULTRAVIOLET ABSORPTION[2]

Wavelength region (A)	*Upper state*	ΔE^a *(kcal.mole*$^{-1}$*)*	*Band structure*	*Remarks*
3900–3400	$\tilde{a}(^3B_1)$	74	Weak, discrete	Phosphoresces; decomposes (?)
3400–2400	$\tilde{A}(^1B_1)$	85	Strong, discrete	Fluoresces; $\Phi_{SO_3} = 8\times10^{-2}$ (ref. 384)
2400–1800	$\tilde{C}, \tilde{D}$, possibly others		Strong, discrete, evidence for underlying continuum at $\lambda < 2280$ A; diffuseness at $\lambda < 1950$ A	Fluoresces at $\lambda > 2000$ A; $\Phi_{SO_3} = 0.5$ at 1849 A
1600–1400	$\tilde{E}$		Weak, diffuse bands superimposed on strong continuum commencing at $\lambda < 1680$ A.	

^{a} Approximate energy of the (0, 0) transition.

Stable end products of the photodecomposition of SO_2 are S and SO_3; spectroscopic observations have revealed the transient presence of SO and S_2O[356, 380, 381]. Photodissociation is energetically possible only for $\lambda < 2200$ A, so that excited molecule reactions must be responsible for the decomposition accompanying absorption into the $\tilde{A}(^1B_1)$ state. The persistence of fluorescence and of discrete structure in the absorption spectrum is an indication of the continuing importance of excited molecule reactions at $\lambda < 2200$ A. Rate data for the various processes of removal of $^1B_1$ and $^3B_1$ SO_2 molecules, following their formation by absorption in the $\tilde{A}$–$\tilde{X}$ system, are summarized in Table 26. The more recent and more direct determinations of the lifetime of $^3SO_2$, obtained by flash photolytic and laser excitation techniques[382], give a value (7.9×10^{-4} sec) an order of magnitude lower than earlier determinations[378, 383], but do agree with

TABLE 26

SECONDARY PROCESSES FOLLOWING ABSORPTION INTO $SO_2\ \tilde{A}(^1B_2)$ AT 2875 A AND 25 °C

Reaction	*Rate coefficient*[a]	*Ref.*
$^1SO_2 \xrightarrow{8} SO_2+h\nu_f$[b]	$k_8 = (5.1\pm4.0)\times10^3$	378, 379
$^1SO_2 \xrightarrow{9} SO_2$	$k_9 = (1.7\pm0.4)\times10^4$	378, 379
$^1SO_2 \xrightarrow{10} {}^3SO_2$[c]	$k_{10} = (1.5\pm0.8)\times10^3$	378
$^1SO_2+SO_2 \xrightarrow{11} (2\ SO_2)$	$k_{11} = (2.0\pm1.0)\times10^{10}$[d]	378, 379
$^1SO_2+SO_2 \xrightarrow{12} {}^3SO_2+SO_2$	$k_{12} = (0.18\pm0.08)\times10^{10}$	378, 379
$^3SO_2 \xrightarrow{13} SO_2+h\nu_p$	$k_{13} = (1.0\pm0.6)\times10$[e]	378
$^3SO_2 \xrightarrow{14} SO_2$	$k_{14} = (1.3\pm0.2)\times10^2$ $k_{13}+k_{14} = (1.3\pm0.3)\times10^3$	378, 383 382
$^3SO_2+SO_2 \xrightarrow{15} (2\ SO_2)$	$k_{15} = (1.4\pm0.7)\times10^7$[d]	378, 383
$^3SO_2+SO_2 \xrightarrow{15'} SO_3+SO(^3\Sigma)$	$k_{15'} = 4\times10^8$	382

[a] Units: sec^{-1} (for $k_8, k_9, k_{10}, k_{13}, k_{14}$); $l.mole^{-1}.sec^{-1}$ (for k_{11}, k_{12}, k_{15}).
[b] $^1SO_2 = \tilde{A}(^1B_1)$. For wavelength dependence of τ_f see ref. 387.
[c] $^3SO_2 = \tilde{a}(^3B_1)$.
[d] Total rate, including chemical quenching; for quenching rate coefficients by gases other than SO_2 see refs. 378, 379, 386.
[e] Much larger values are suggested in refs. 360, 365, 379.

some solid-state measurements (5×10^{-4} sec)[379] and an estimate (7.6×10^{-4} sec) based on studies of the chemiluminescence from O+SO recombination[365]. The quantum yield of decomposition, Φ_{-SO_2}, in the 2400–3400 A region is very small, but Calvert *et al.*[384] estimate that practically all $^3SO_2$ formed react with SO_2 to form SO_3 and SO

$$^3SO_2+SO_2 \rightarrow SO_3+SO(^3\Sigma^-) \tag{15'}$$

The inefficiency of photodecomposition of SO_2 in this spectral region is then ascribed to the small quantum yield of $^3SO_2$.

Warneck *et al.*[385] have presented evidence that absorption in the 1800–2400 A range consists of discrete structure superimposed on a continuum which they attribute to the dissociation to ground state SO and O. The overlying bands are predissociated at $\lambda < 1900$ A, but the nature and states of the resulting fragments are not established. On the basis of their measured quantum yield, $\Phi_{SO_3} = 0.5$ at 1849 A, Driscoll and Warneck[388] have discussed the relative importance of the possible primary processes. There is as yet, however, insufficient information for the evaluation of quantitative rate data for the photolysis in this region.

5.2 DECOMPOSITION OF SO_3

In spite of their relevance to air pollution and to the combustion of sulphur-contaminated fuels, kinetic data for the homogeneous decomposition of SO_3 are practically non-existent. The thermal decomposition is dominated by a heterogeneous component, even in clean silica vessels[390]; reactions of SO_3 at a glass surface are evident even at room temperature[380]. The kinetics of approach to the equilibrium

$$2\,SO_2+O_2 \rightleftharpoons 2\,SO_3$$

in the presence of a catalyst have been investigated, but there appears to be no quantitative information whatsoever on the reaction rates in the complete absence of catalyst[391].

SO_3 is decomposed to SO_2 and O_2 upon photolysis. Absorption, which commences at 3000 A, consists of broad, diffuse bands superimposed on a continuous background[392]. The relative contributions of the reactions

$$SO_3+h\nu \rightarrow SO_2+O(^3P) \qquad (\lambda < 3500\ \text{A})$$

$$\rightarrow SO+O_2 \qquad (\lambda < 3050\ \text{A})$$

$$\rightarrow SO+O(^1D_2) \qquad (\lambda < 2250\ \text{A})$$

to the primary process of dissociation are not established, however. Norrish and Oldershaw[380] have discussed the implications upon the photolysis mechanism of their observation of vibrationally excited SO as a transient intermediate. A number of the secondary reactions which they proposed have been discussed above.

6. Halogen oxides and oxyacids

The halogens form a number of stable oxides, but information on their thermal and photochemical decompositions in the gas phase is limited to two oxides of fluorine and several oxides of chlorine. Little research has been done on the decompositions of bromine, iodine or mixed-halogen oxides.

The oxyacids of iodine are solids and will not be discussed here. Perchloric acid is the only oxyacid of bromine or chlorine which is stable in the pure state. There have been a few studies of its pyrolysis in the vapour phase.

References pp. 132–141

6.1 FLUORINE OXIDES

Only two stable fluorine oxides are known. Except for an early investigation of the decomposition of F_2O_2, studies of decomposition kinetics have been confined to F_2O. There still remains considerable doubt concerning the mechanisms by which these two compounds decompose to their elements. Unlike its chlorine analogue, FO has defied spectroscopic detection in the fluid phases, although it is considered an important intermediate in many of the reactions of F_2O.

6.1.1 Decomposition of F_2O

The thermal decomposition of F_2O has been studied in both static[393, 394] and flow[395] systems and also by the shock-tube method[396, 397]. All investigators of this reaction are agreed that the kinetics for the overall reaction

$$F_2O+M \xrightarrow{k_1} F_2+\tfrac{1}{2}\,O_2+M \tag{1}$$

are well represented by

$$-\frac{d[F_2O]}{dt} = k_1[F_2O][M]$$

the rate law characteristic of a unimolecular decomposition in its second-order region; the investigations cover a temperature range of 500–1300 °K and pressures up to 20 atm. In seasoned nickel, aluminum, magnesium, or even quartz vessels there is apparently[393–395] no heterogeneous contribution to the decomposition rate. The net reaction in metal vessels is

$$2\,F_2O \rightarrow 2\,F_2+O_2$$

but in glass vessels the F_2 is consumed by reaction with silica. Agreement between different workers' k_1 values (Table 27) is quite good; the more recent determinations of the Arrhenius parameters are more reliable since in the earliest study[393] only a small temperature range was employed. The simplest interpretation of the experimental results is, of course, that k_1 indeed represents the rate of a unimolecular reaction

$$F_2O+M \rightarrow FO+F+M \tag{2}$$

or possibly twice the rate of (2), depending on the rapidity of secondary reactions. The experimental value of the pre-exponential factor for k_1 is entirely reasonable

TABLE 27

OVERALL RATE COEFFICIENTS OF F_2O DECOMPOSITION ($F_2O+M \xrightarrow{k_1} F_2+\frac{1}{2}O_2+M$)

Rate coefficient[a]	*Temp.(°K)*	*Ref.*
$k_1^{F_2O} = 10^{14.44} \exp-(39{,}000\pm3{,}000)/RT$	523–543	393
$k_1^{F_2O} = (4.7\pm4.3)\times10^{10} \exp-(30{,}100\pm1{,}000)/RT$[b]	563–634	394
$k_1^{Ar} = 10^{12.1} \exp-(34{,}200/RT)$	820–1240	396
$k_1^{Ar} = 10^{11.2} \exp-(31{,}500/RT)$	860–1300	397
$k_1^{Ar} = 10^{12.8} \exp-(37{,}800/RT)$	770–1000	400
$k_1^{He} = 10^{12.2\pm0.2} \exp-(32{,}200\pm600)/RT$	773–973	395
$k_1^{M}/k_1^{F_2O} = 0.96$ (O_2); 0.96 (N_2); 0.87 (SiF_4); 0.68 (He); 0.43 (Ar); 0.99 (F_2).		393
$k_1^{M}/k_1^{F_2O} = 1.0$ (O_2); 1.0 (N_2); 0.80 (CF_4); 0.75 (He); 0.52 (Kr); 0.64 (F_2).		394

[a] Units: $l.mole^{-1}.sec^{-1}$ (for k_1); dimensionless (for $k_1^{M}/k_1^{F_2O}$).
[b] Rate expression calculated from the first-order rate coefficients at 800 torr.

for the dissociation of a triatomic molecule, as is the bond energy derived for F–OF. However, recent evidence has raised the possibility that the decomposition may be autocatalytic, *viz.* the observation[394] of induction periods prior to reaction in a static system, and the autoacceleration evident[397] in the decomposition of shocked F_2O/Ar mixtures of high F_2O content. The shock-tube values for k_1, in our opinion, almost certainly refer to the initial step (2), and their proximity to the other values in Table 27 suggests that the decomposition kinetics are not complex. There is clearly a need for further investigation of this reaction to account for the unexpected phenomena mentioned above. Meanwhile, it may be assumed, though with some reserve, that $k_1 = k_2$ (or possibly $2\,k_2$).

Because of the paucity of rate data for the reactions of F and FO one can only speculate on the course of the reaction subsequent to the initial dissociation. The results of photochemical studies[398, 399] give some guidance. The quantum yield Φ_{-F_2O} of photodecomposition is 1.0 at 3650 A, independent of temperature in the range 15–45 °C, pressure of F_2O and pressure of oxygen[398]; the primary step is almost certainly as in (2)[398, 399]. Thus, at room temperature at least, any contribution from

$$F+F_2O \rightarrow F_2+FO \tag{3}$$

is negligible compared to radical and atom loss by

$$F+F+M \rightarrow F_2+M \tag{4}$$

$$2\,FO \rightarrow F_2+O_2 \tag{5}$$

or at the wall. This is in accord with the belief that (3) is endothermic. However, the conclusion of Gatti *et al.*[398] that $E_3 > 15$ kcal.mole^{-1}, a limit based on the observed lack of pressure change during the photolysis of F_2 containing 15 % F_2O, is difficult to accept. Under such conditions the reverse of (3) was perhaps fast enough to prevent detectable decomposition of F_2O. At the higher temperatures necessary for thermal decomposition, (3) may well make an important contribution to the reaction. Blauer and Soloman[397] have proposed a chain mechanism consisting of (2), (3), (4) and

$$2\,FO \rightarrow 2\,F + O_2 \tag{5a}$$

to account for their observations of self-acceleration of the rate of thermal decomposition and of inhibition of the rate, *via* (−3), by F_2.

In a recent publication Lin and Bauer[400] have pointed out that the reverse of reactions (2) and (4) cannot be neglected even in the early stages of a shock-initiated decomposition. For temperatures below 1000 °K their overall rate constant (k_1^{Ar}) is in fair agreement with previous determinations (Table 27). They attribute the marked curvature in their Arrhenius plot for the overall rate coefficient at higher temperatures to the approach of reaction (2) to equilibrium during the period of observation. For an assumed reaction scheme consisting of steps (2), (−2), (5a), (4) and (−4) they computed values for k_2 ($k_2 = 10^{14.3\pm1.0}\exp-(42{,}500\pm4{,}100)/\boldsymbol{RT}$) l.mole^{-1}.sec^{-1}) and k_{5a} ($k_{5a} = 10^{9.10\pm0.12}$ l.mole^{-1}.sec^{-1}) from their measurements of the overall rate coefficient and the oxygen yields. Their calculations suggest that (3) and (−3) make no significant contributions towards the decomposition in a shock tube except at high concentrations of added F_2 when (−3) (and (−4)) may be sufficiently rapid to retard the reaction.

6.1.2 *Decomposition of* F_2O_2

The thermal decomposition of F_2O_2 was studied by Schumacher and Frisch[401] who found it to be a homogeneous, unimolecular reaction with the first-order rate coefficient given by

$$k^{\infty} = 5.0\times10^{12}\exp-(17{,}300/\boldsymbol{RT})\ \text{sec}^{-1}$$

in the temperature range −60 to −25 °C. The fall-off of the rate coefficient with pressure in the range 400–2 torr is, however, surprisingly small for a tetratomic molecule. Lin and Bauer[400] have pointed out that a chain mechanism could be involved, in which case k^{∞} may be a composite quantity which has a weaker dependence on pressure than the rate coefficient for the unimolecular initiating step. The chain mechanism can account for the observed activation energy of 17 kcal. mole^{-1} providing it is assumed that the initiating step is

$$F_2O_2 + M \rightarrow F + FO_2 + M \quad (\Delta H^0_{300} \sim 23 \text{ kcal.mole}^{-1})^{402}$$

rather than the considerably more endothermic step

$$F_2O_2 \rightarrow 2\,FO + M \quad (\Delta H^0_{300} \sim 58 \text{ kcal.mole}^{-1})^{400,\,402}$$

6.2 CHLORINE OXIDES

The chlorine oxides provide the kineticist with a chemistry as fascinating and as challenging as that of the nitrogen oxides. Four stable oxides are known: Cl_2O, ClO_2, Cl_2O_6 (ClO_3 in the vapour phase) and Cl_2O_7; a fifth compound, probably Cl_2O_3[403], has been discovered recently. In addition, ClO[404] and the peroxy radical ClOO[405] have been detected spectroscopically; both of these species are short-lived, but are important intermediates in many of the reactions of chlorine oxides. The complexity of the decomposition kinetics of the two simplest oxides, Cl_2O and ClO_2, revealed in early studies[406,407] perhaps discouraged further efforts to elucidate the mechanisms of those reactions. Hopefully, the recent rate coefficient determinations[408–410] for a number of elementary reactions of atoms and radicals with chlorine oxides will generate a renewed interest in the decomposition mechanisms.

6.2.1 Decomposition of Cl_2O

Thermal decomposition of Cl_2O proceeds at a measurable rate at temperatures in the region of 100 °C[406,411,412]. The slow self-acceleration of the reaction rate to a maximum is an indication that the reaction proceeds by a degenerately branched-chain mechanism[406]. The mild explosion which occurs towards the end of a decomposition when the reaction rate has declined to a value well below the maximum, is perhaps the result of self-ignition of the metastable intermediate responsible for branching. This intermediate has not been positively identified however, so that mechanisms proposed for the thermal decomposition are purely speculative. Szabo *et al.*[413] have shown that the experimental concentration–time curves are adequately described by a formal, four-stage mechanism, consisting of an initiation step second-order in Cl_2O, propagation and branching steps first-order in both Cl_2O and radical concentration, and a termination step second-order in radical concentration. By assuming that the reaction of two ClO radicals was the termination step, he derived Arrhenius expressions for the remaining steps; since it is by no means certain that the reaction of two ClO radicals is faster than possible linear terminations at the low radical concentrations appropriate to a thermal decomposition, it is felt that Szabo's rate coefficient expressions have no mechanistic significance. Bodenstein and Szabo[411] showed that in the very early stages of decomposition the kinetics of Cl_2O disappearance followed the rate law

$$-\frac{d[Cl_2O]}{dt} = k[Cl_2O]^2$$

very closely, with $k = 10^{8.7}\,T^{0.5}\exp-(21{,}230/\boldsymbol{R}T)$ l.mole^{-1}.sec^{-1} for the range 356–413 °K. The Arrhenius parameters associated with k were, by chance, very reasonable for a homogeneous bimolecular reaction between two Cl_2O molecules; however, comparison of k with values deduced from the results of other studies[406, 412] clearly showed that the initial reaction had an important heterogeneous component. There were also indications of possible heterogeneous contributions to the chain branching and termination reactions.

The mechanism of the photochemical decomposition of Cl_2O is better established[414, 415] than that of the thermal decomposition. Undoubtedly some of the elementary processes occurring during photolysis are also involved in the pyrolysis. In both cases the products of reaction are Cl_2, O_2 and minor amounts of ClO_2. Traces of a higher oxide of chlorine, possibly Cl_2O_3, are produced during photolysis[414]. This latter compound could perhaps be the branching agent responsible for the degenerate explosions in Cl_2O, as has been suggested[416] for explosions of ClO_2.

The flash photolysis study of Norrish *et al.*[415] has confirmed the suggestions of earlier workers that ClO is an important reaction intermediate. It reaches its maximum concentration during the photolytic flash and is clearly generated in

$$Cl_2O + h\nu(\lambda < 9130\ \text{A}) \rightarrow Cl + ClO \tag{6}$$

followed by

$$Cl + Cl_2O \rightarrow Cl_2 + ClO \tag{7}$$

or in

$$Cl_2O + h\nu(\lambda < 3020\ \text{A}) \rightarrow 2\,Cl + O \tag{6a}$$

followed by the rapid[410, 415, 417] reactions (7) and

$$O + Cl_2O \rightarrow 2\,ClO \tag{8}$$

The decay of ClO subsequent to the flash is bimolecular, and the rate coefficient for

$$2\,ClO \rightarrow Cl_2 + O_2 \tag{9}$$

calculated[415] from $-d[ClO]/dt = 2k_9[ClO]^2$ agrees well with other determinations (Table 28). The rapid decline of [ClO] is accompanied by a very slow removal of Cl_2O and an even slower formation of ClO_2. Norrish *et al.*[415] suggested reactions

$$ClO + Cl_2O \rightarrow ClO_2 + Cl_2 \quad (10)$$

and

$$ClO + Cl_2O \rightarrow Cl + O_2 + Cl_2 \quad (11)$$

followed by (7), to account for these facts, and were able to estimate the corresponding rate coefficients (Table 28) from the time dependence of $[ClO_2]$ and $[Cl_2O]$. Reaction (10) makes a negligible contribution to the total rate of loss of

TABLE 28

REACTION RATE COEFFICIENTS OF CHLORINE OXIDES

Reaction	*Rate coefficient* (*l.mole*$^{-1}$*.sec*$^{-1}$)	*Temp.* (°*K*)	*Ref.*
$Cl + Cl_2O \xrightarrow{7} Cl_2 + ClO$	$k_7 > 4 \times 10^8$	300	415
$O + Cl_2O \xrightarrow{8} 2\,ClO$	$k_8 = 8.2 \times 10^9$	300	417
$2\,ClO \xrightarrow{9} Cl_2 + O_2$	$k_9 = (4 \pm 1) \times 10^8 \exp-(2{,}500 \pm 300)/RT^{a,b}$	294–495	410
	$k_9 = 3.8 \times 10^6\ ^{a,b}$	298	421
	$k_9 = 1.0 \times 10^7\ ^{b,c}$	300	418
	$k_9 = (1.2 \pm 0.2) \times 10^7\ ^{b,c}$	300	415
	$k_9 = 4.6 \times 10^7 \exp(0 \pm 650)/RT^{b,c}$	293–433	420, 410
$ClO + Cl_2O \xrightarrow{10} ClO_2 + Cl_2$	$k_{10} \sim 10^5$	300	415
$ClO + Cl_2O \xrightarrow{11} Cl + O_2 + Cl_2$	$k_{11} = 5.3 \times 10^4$	300	415
$Cl + ClO_2 \xrightarrow{14} 2\,ClO$	$k_{14} > 10^{8.7}$	298	410
$\xrightarrow{15} 2\,ClO$	$k_{15} < 0.1\,k_{14}$	298	410
$O + ClO_2 \xrightarrow{17} ClO + O_2$	$k_{17} > 2 \times 10^{10}$	298	409
$O + ClO \xrightarrow{18} Cl + O_2$	$k_{18} > 8 \times 10^9$	300	417, 422
	$k_{18} = 6 \times 10^9$	300	419
	$0.45 > k_{18}/k_{17} > 0.15$	298	409

[a] These are genuine low-pressure results with little[410] or no[421] contribution from reactions operative at high pressures.
[b] k_9 is defined here in the conventional manner, *i.e.* $-d[ClO]/dt = 2k_9[ClO]^2$. Some of the quoted rate coefficients have been defined in the original literature by $-d[ClO]/dt = k_9[ClO]^2$ and are therefore twice as large as the values in the table.
[c] These values should be regarded with some caution since the interpretation of the flash photolysis results is complex; there is also good reason to believe[410, 421] that the recombination mechanisms at high and low pressures may differ.

ClO radicals during the first few milliseconds after the photoflash. A delay in the appearance of ClO_2 is perhaps associated with a shift to the left in the equilibrium

$$ClO+ClO_2 \underset{-12}{\overset{12}{\rightleftharpoons}} Cl_2O_3 \qquad (12, -12)$$

as [ClO] decays. The long time taken for $[ClO_2]$ to reach its maximum suggests that ClO_2 is removed in a molecular reaction, possibly

$$ClO_2+Cl_2O \rightarrow Cl_2+O_2+ClO \qquad (13)$$

For the photolysis at normal intensities Schumacher *et al.*[414] obtained $\Phi_{-Cl_2O} \sim 3.5$ at 10 °C, independent of Cl_2O and added Cl_2 pressures and of wavelength in the region 4360–3130 A. (An earlier study[423] gave $\Phi_{-Cl_2O} = 2$.) The quantum yield increased with increasing temperature. Since the primary process (6a) cannot occur in this wavelength region, they concluded that the large quantum yield was the result of a short chain consisting of (11) and (7). However, calculation of the rate of (9) for Schumacher's light intensities shows that very slow linear termination reactions, such as (10) with the value k_{10} suggested by Edgecombe *et al.*[415], may be more effective in removing ClO than the quadratic termination. The scheme of reactions (6), (7) and (10) certainly accounts for a quantum yield of 4 even without the inclusion of (11), although it suggests that ClO_2 should be an important product; if (11) takes place the predicted quantum yield is > 4. A scheme in which important contributions to the destruction of ClO are made both by reactions first-order in ClO and by reactions second-order in ClO cannot account for a Φ_{-Cl_2O} which is greater than two and which is at the same time independent of $[Cl_2O]$ and I_a. A mechanism consisting of steps (6) and (7) together with a predominantly second-order removal of ClO radicals can, however, account for most of the experimental findings provided that the reaction between two ClO's results at least partly in Cl atom formation. Recent studies do in fact suggest that Cl may be a product of reaction (9), at least at low pressures (see later in this section)[408–410].

The larger quantum yield of 4.5 observed[424] for wavelengths between 2350 and 2750 A is possibly an indication of a contribution from (6a) to the primary process in this region, as is also suggested by the absorption spectrum[414, 425]. The extent to which the rather unlikely[414] dissociation process

$$Cl_2O+h\nu(\lambda < 7610 \text{ A}) \rightarrow Cl_2+O \qquad (6b)$$

occurs has not been established.

A quantum yield $\Phi_{-Cl_2O} \sim 1.8$ has been reported[426] for the photolysis at 4358 A of Cl_2O in CCl_4 solution. ClO_2 is an important product of the photolysis in solution ($\Phi_{ClO_2} \sim 0.3$) whose intense absorption of light relative to that of Cl_2O complicates the interpretation of results even at low conversions.

6.2.2 *Decomposition of* ClO_2

ClO_2 is thermally unstable at temperatures above 30 °C. The decomposition kinetics are those characteristic of a degenerate chain-branching reaction[407, 416]. At temperatures in excess of 45 °C, self-acceleration of the decomposition rate culminates in violent explosion for $p_{ClO_2} > 0.2$ torr[416]. A recent study[416] of the dependence of the induction periods prior to self-ignition on various experimental parameters has provided a tentative reaction mechanism

$$2\,ClO_2 \xrightarrow{\text{wall}} ClO + ClO_3 \qquad \text{Initiation}$$

$$\left.\begin{aligned} ClO + ClO_2 &\rightleftharpoons Cl_2O_3 \\ ClO + Cl_2O_3 &\rightarrow ClOOCl + ClO_2 \\ ClOOCl &\rightarrow 2\,Cl + O_2 \end{aligned}\right\} \text{Propagation}$$

$$Cl + ClO_2 \rightarrow 2\,ClO \qquad \text{Branching} \qquad (14)$$

$$\left.\begin{aligned} Cl + ClO_2 &\rightarrow Cl_2 + O_2 \\ \text{Cl, ClO diffuse to wall} \end{aligned}\right\} \text{Termination} \qquad (15)$$

The most significant result of this study[416] has been the identification of the unstable intermediate, Cl_2O_3, which is responsible for delayed branching. By applying to this scheme the criterion for isothermal explosion, *viz.* that the rate of the branching step exceed the rate of termination, McHale and von Elbe[416] were able to account for the observed variations of induction period τ with pressures of ClO_2 and inert additives and with surface area. The temperature dependence of τ was determined by the energy barrier (~ 11 kcal.mole^{-1}) to the activated adsorption step of the initiation reaction. Because of dominant contributions from heterogeneous reactions to the overall rate of decomposition, it is not possible to extract information on the rates of any of the suggested propagation and branching steps from the observed kinetics. The initiation reaction is presumably the source of the $ClO_3(Cl_2O_6)$ which accompanies the principal products Cl_2 and O_2 in small amounts. Added ClO_3 has an inhibiting effect, perhaps exerted *via*

$$ClO_3 + ClO \rightarrow 2\,ClO_2$$

ClO_3 is therefore not the intermediate responsible for branching as has been suggested earlier[407]. Cl_2O has a slight inhibiting effect[407] and Cl_2O_7 has no effect on the reaction[416].

ClO_2 commences to absorb light below 5100 A. The onset of diffuseness in the bands at $\lambda \sim 3750$ A is almost certainly due to the predissociation[2]

$$ClO_2 + h\nu \rightarrow ClO + O(^3P) \tag{16}$$

although this process is possible, according to currently accepted heats of formation[1, 402], for $\lambda < 4900$ A. Dissociation to $ClO + O(^1D_2)$ can only occur with $\lambda < 2760$ A. Photodecomposition takes place at wavelengths longer than the observed threshold for predissociation[427]. The products of photolysis are Cl_2, O_2 and ClO_3, the latter increasing in importance with decreasing temperature and decreasing light intensity. Cl_2O and Cl_2O_7 are minor products of the photolysis in solution[428].

Flash photolysis studies[418, 419] have provided much insight into the mechanism of photolysis. The production of ClO and vibrationally excited $O_2(^3\Sigma_g^-)$ during the photolytic flash is attributed to the sequence of reactions (16) and

$$O + ClO_2 \rightarrow ClO + O_2^*(v \leqq 15) \tag{17}$$

Subsequent to a photolytic flash of low energy, the decay of [ClO] follows the second-order rate law expected for removal by (9). At high flash energies the reaction

$$O + ClO \rightarrow Cl + O_2 \tag{18}$$

contributes noticeably to the decline of [ClO], and an estimate of k_{18} has been made[419] from deviations from linearity of second-order plots of $[ClO]^{-1}$ *versus* time. Investigations[409] of reaction (17) in a fast-flow system have confirmed that both (17) and (18) are rapid reactions. ClO_3 is a minor product of the flash photolysis; evidently under conditions of high light intensity reaction (18) and the destruction of ClO_3 by photons and atoms ensure a low yield of trioxide. Surprisingly, the amount of ClO_3 which is formed during a flash survives the period of rapid [ClO] decay and, at constant flash energy, is independent of the pressure of inert additives. This suggests that the removal of ClO_3 by ClO is rather slow at room temperature and that the association

$$O + ClO_2 \rightarrow ClO_3 \tag{19}$$

is second-order for pressures of the order of 100 torr.

While reactions (16)–(19) and (9) are adequate to explain the small quantum yields for the photodecomposition in solution ($\Phi_{-ClO_2} \rightarrow 2$ in CCl_4 for $\lambda < 4100$ A[428, 429]), clearly additional steps are required to account for the $\Phi_{-ClO_2} \sim 12$ observed[427] for the photolysis of dry gaseous ClO_2 at 3650 A. Possibly Cl_2O_3

is responsible for the establishment of a chain reaction, as in the thermal decomposition. There appears to be some evidence for its formation during the flash photolysis[418].

6.2.3 *Reactions of* ClO

Evidence has recently been adduced by Clyne and Coxon[410] that, at pressures of the order of a few torr, reaction (9) proceeds through the intermediate formation of the chlorine peroxy radical

$$2\,ClO \underset{-a}{\overset{a}{\rightleftharpoons}} ClOO + Cl$$

$$Cl + ClOO \overset{b}{\rightarrow} Cl_2 + O_2$$

$$M + ClOO \overset{c}{\rightarrow} M + Cl + O_2$$

For stationary concentrations of ClOO and Cl,

$$-\frac{d[ClO]}{dt} = \frac{2\,k_a k_b}{k_{-a} + k_b}[ClO]^2 \cong 2\,k_a[ClO]$$

since $k_b \gg k_{-a}$[421, 430]. Measurements of the rate of [ClO] decay thus provided a value for k_a. The presence of Cl atoms in ClO undergoing decomposition was inferred from the rapid consumption of added ClO_2, O_3 or H_2 in chain reactions whose rates were proportional to the rate of (a). For example, the rate of disappearance of ClO_2 was dependent only on [ClO], *viz.*

$$-\frac{d[ClO_2]}{dt} = k_a[ClO]^2$$

Clearly, this discovery has an important bearing on the kinetics of the decompositions of chlorine oxides and of the halogen-sensitized decompositions of oxides, *e.g.* O_3 and N_2O. Johnston *et al.*[421] have recently shown that an additional, pressure-dependent reaction occurs between two ClO radicals, *viz*

$$2\,ClO + M \underset{-d}{\overset{d}{\rightleftharpoons}} Cl_2O_2 + M \quad (K_d = k_d/k_{-d})$$

$$Cl_2O_2 + M \overset{e}{\rightarrow} Cl_2 + O_2 + M$$

which was unimportant at the pressures employed by Clyne and Coxon, but which undoubtedly made a major contribution to reaction under the conditions

TABLE 29

REACTIONS OF THE CHLORINE PEROXY RADICAL ClOO

Reaction	*Rate coefficient*[α]	*Temp.(°K)*	*Ref.*
$Cl+O_2+M \xrightarrow{-c} ClOO+M$	$k_{-c}^{Ar} = 2\times10^8$	300	410
	$k_{-c}^{N_2} = 6.2\pm1.1\times10^8$	293	430
$M+ClOO \xrightarrow{c} M+Cl+O_2$	$k_c^{N_2} = 9\times10^{7\beta}$	293	430
	$k_c^{N_2} = 2.8\times10^8$	298	430, 421
$2\,ClO \xrightarrow{a} ClOO+Cl$	$k_a = (4\pm1)\times10^8 \exp-(2{,}500\pm300/RT)^{\gamma}$	294–495	410
	$k_a = 3.8\times10^6$	298	421
$ClOO+Cl \xrightarrow{-a} 2\,ClO^{\delta}$	$k_{-a} = 2\times10^{9\beta}$	294–495	410
	$k_{-a} = 8.7\times10^8$	298	421
$ClOO+Cl \xrightarrow{b} Cl_2+O_2$	$k_b = 9.4\times10^{10}$	298	421
	$k_b/k_{-a} = 15$	293	430
$2\,ClO+M \overset{K_d}{\rightleftharpoons} Cl_2O_2+M$	$k_eK_d = 1.8\times10^{10}$ $(M = O_2)$	298	421
$Cl_2O_2 \xrightarrow{k_e} Cl_2+O_2+M$	$= 1.2\times10^{10}$ $(M = Ar)$	298	421

[α] Units: $l.mole^{-1}.sec^{-1}$ (for $k_c^{N_2}$, k_a, k_{-a}, k_b); $l^2.mole^{-2}.sec^{-1}$ (for k_{-c}, k_eK_d); non-dimensional (k_b/k_{-a}).
[β] Calculated using $\Delta H_{f300}^0(ClOO) = 22$ kcal.mole^{-1}, $S_{300}^0(ClOO) = 63.0$ cal.deg^{-1}, $S_{300}^0(ClO) = 52.9$ cal.deg^{-1} and other data[1, 402].
[γ] k_a is defined here by $-d[ClO]/dt = 2\,k_a[ClO]^2$. In the original reference[410] this coefficient is defined by $-d[ClO]/dt = k_a[ClO]^2$ and is therefore twice as large.
[δ] k_{-a} defined by $-d[Cl]/dt = k_{-a}[Cl][ClOO]$.

employed in flash-photolysis studies. Using their method of molecular modulation kinetic spectrometry they were able[421] to derive a value for the effective rate coefficient K_dk_e of the overall reaction, as well as values for k_{-a} and k_b. The good agreement between their value of k_{-a} and that calculated from Clyne and Coxon's k_a (Table 29) encourages confidence in the proposed mechanism. The observation[421] of a linear dependence of reaction rate on [M] at $p > 50$ torr conflicts with the lack of dependence on [M] reported by Porter and Wright[420]. The true pressure dependence may well have been obscured in the latter investigation, however, by unsuspected variations of the adiabatic temperature rise following flash photolysis. Table 29 summarizes rate coefficient values determined for reactions in the Cl, ClO, ClOO systems as well as values calculated with the aid of subsidiary data[402, 420, 430].

If (14) and (15) are included in the peroxy radical scheme above, the resulting mechanism can account for Clyne and Coxon's observations on the rate of decomposition of ClO_2 in the presence of ClO. The addition of initiating and terminating steps results in a branched chain mechanism for the thermal decomposition of ClO_2 which is an attractive alternative to the degenerately branched chain proposed by McHale and von Elbe[416]. A branched chain scheme is not necessarily inconsistent with the observation of long induction periods: if (a) controls the rate of branching (*i.e.* quadratic branching), the induction period will be strongly dependent on the rate of initiation and may well be quite long. Reactions (a)–(c) and (7) constitute a straight chain scheme for the decomposition of Cl_2O; auto-acceleration of the rate of thermal decomposition may result from the accumulation of ClO_2 and the steadily increasing rate of (14), provided the rate of ClO_2 formation in (10) is not too fast relative to the rate of chain initiation.

There is good reason to believe[410, 431] that the regeneration of Cl atoms from ClO *via* reactions (a) and (c) is responsible for the propagation of chains in the chlorine-sensitized decomposition of O_3 and N_2O. The decomposition of ozone photosensitized by chlorine is a complex and poorly understood reaction[432]. Quantum yields as high as 60 have been reported[433]. Photodissociation of Cl_2 is undoubtedly followed by the fast[410] reaction ($k > 4 \times 10^8$ l.mole^{-1}.sec^{-1})

$$Cl + O_3 \rightarrow ClO + O_2$$

but the subsequent fate of ClO is not certain. The direct reaction with O_3 is slow[410], but continuation of the chain may alternatively occur *via* (a) and (c). Clyne and Coxon's studies[410] of the reaction of Br and O_3 indicate that the Br_2-sensitized O_3 decomposition may proceed by an analogous mechanism.

The thermal decomposition of N_2O catalyzed by the halogens, Cl_2, Br_2, I_2, proceeds according to the rate law[434]

$$-\frac{d[N_2O]}{dt} = k[N_2O][X]_{eq}$$

in the temperature range 800–1000 °K, where $[X]_{eq}$ is the equilibrium halogen atom concentration resulting from dissociation of the halogen molecule, *e.g.*

$$Cl_2 + M \rightleftharpoons 2\,Cl + M$$

Values for k are summarized in Table 30. Benson and Buss[431] argue convincingly that the reaction catalyzed by Cl_2 involves a Cl atom chain which consists either of

TABLE 30

RATE COEFFICIENTS FOR HALOGEN-CATALYZED N_2O DECOMPOSITION[434]

Halogen	*Rate coefficient k (l.mole^{-1}.sec^{-1}) for $-d(N_2O)/dt = k[N_2O][X]_{eq}$*[a]	*Temp (°K)*
Cl_2	$1.3 \times 10^{11} \exp-(33{,}500/RT)$	930–1030
Br_2	$2.0 \times 10^{11} \exp-(37{,}000/RT)$	880–980
I_2	$2.8 \times 10^{11} \exp-(38{,}000/RT)$	880–980

[a] $[X]_{eq}$ is the equilibrium concentration of halogen atoms according to $X_2 \rightleftharpoons 2X$.

$$Cl + N_2O \rightarrow ClO + N_2 \quad (20)$$

followed by (a) and (c), so that $k = k_{20}$, or of (20) followed by

$$ClO + N_2O \rightarrow N_2 + ClOO$$

and (c), so that $k = 2\,k_{20}$. Similar schemes presumably apply to the Br_2- and I_2-catalyzed reactions.

6.2.4 *Decomposition of* Cl_2O_7

The thermal decomposition of this compound has been studied[435] in the gas phase from 100–120 °C by following the pressure change accompanying reaction. The kinetics of the decomposition in the pressure range 1–80 torr were those of a homogeneous, unimolecular reaction close to its high-pressure limit. Extrapolation of plots of inverse rate coefficient against inverse pressure gave

$$k^{\infty} = 4.5 \times 10^{15} \exp-(32{,}900/RT) \text{ sec}^{-1}$$

as the high-pressure limit of the first-order rate coefficient for the disappearance of Cl_2O_7. Fission of a central Cl–O bond

$$Cl_2O_7 \rightarrow ClO_3 + ClO_4 \quad (21)$$

which, according to a recent estimate[436], requires 30 ± 4 kcal.mole^{-1}, is thought to be the rate-determining step in the decomposition[435]. The secondary processes which lead to the end products, Cl_2 and O_2, have not been established. Indirect evidence for the occurrence of (21) comes from observations[435] made on the decomposition of Cl_2O_7 containing F_2. Fisher[436], however, in a mass spectrometric study of the pyrolysis, failed to detect ClO_3 and ClO_4, but did observe ClO_2 and ClO as reaction intermediates.

6.2.5 *Decomposition of* Cl_2O_6

This compound exists as the monomer (ClO_3) in the gas phase and almos exclusively as the dimer in the liquid phase or solution. Paramagnetism of the liquid is probably not due to ClO_3, as was once thought, but rather to traces of ClO_2[437]. Qualitative studies have been made of the thermal and photochemical decomposition of ClO_3[438, 439].

6.3 BROMINE AND IODINE OXIDES

Iodine oxides and bromine oxides are solid compounds which are beyond the scope of this article and will therefore not be discussed in any detail. Clyne and Coxon[410] have found that BrO decays in a similar manner to ClO, with the second order rate coefficient of the order of 2×10^8 l.mole^{-1}.sec^{-1}. In view of the important role of ClO in the decomposition of chlorine oxides, it is conceivable that BrO may play a similar role in the decomposition of bromine oxides. However, no kinetic information on the decomposition of bromine oxides in the gas phase appears to be available at the present time.

6.4 OTHER HALOGEN OXIDES

The decomposition of a number of inorganic compounds which contain both halogen and oxygen atoms, such as $COCl_2$, NOCl, $SOCl_2$, ClO_2F, are discussed in Chapter 3.

6.5 OXYACIDS

6.5.1 Decomposition of $HClO_4$

Studies of the decomposition of this substance have been carried out in order to further understanding of the mechanism of the deflagration of ammonium perchlorate[440] and of the reactions occurring in perchloric acid fuel flames[441].

For temperatures in excess of 320 °C the gas-phase decomposition of anhydrous $HOClO_3$ (or of its dihydrate) is a homogeneous, unimolecular reaction close to its first-order limit at pressures in excess of 20 torr[440]. Levy[440] obtained a value

$$k^{\infty} = 5.8 \times 10^{13} \exp-(45{,}100/\boldsymbol{RT}) \text{ sec}^{-1}$$

for the first-order rate coefficient for the disappearance of $HOClO_3$ in the temperature range 350–440 °C. The close agreement between the activation energy and the measured[442] value of D(HO–ClO) = 46 kcal.mole^{-1} indicates that the reaction rate is controlled by

$$HOClO_3 \rightarrow HO + ClO_3$$

The subsequent reaction steps which lead to the overall stoichiometry

$$HOClO_3 = 0.5\ H_2O + 0.5\ Cl_2 + 1.75\ O_2$$

have not been established with certainty, but mass spectrometric observations[441] have clearly indicated the importance of secondary reactions of ClO and ClO_2.

REFERENCES

1 *JANAF Thermochemical Tables*, D. R. Stull (Ed.), Dow Chemical Co., Midland, Mich., 1965.
2 G. Herzberg, *Electronic Spectra of Polyatomic Molecules*, Van Nostrand, Princeton, 1966.
3 H. B. Palmer and T. J. Hirt, *J. Am. Chem. Soc.*, 84 (1962) 113;
H. B. Palmer and W. D. Cross, *Carbon*, 3 (1966) 475.
4 K. Bayes, *J. Am. Chem. Soc.*, 83 (1961) 3712; 84 (1962) 4077; 85 (1963) 1730;
C. Willis and K. Bayes, *J. Am. Chem. Soc.*, 88 (1966) 3203;
D. Williamson and K. Bayes, *J. Am. Chem. Soc.*, 90 (1968) 1957.
5 R. N. Smith, R. A. Smith and D. A. Young, *Inorg. Chem.*, 5 (1966) 145.
6 E. Tschuikow-Roux and S. Kodama, *J. Chem. Phys.*, 50 (1969) 5297.
7 C. Devillers, *Compt. Rend.*, 262c (1966) 1485.
8 T. Morrow and W. D. McGrath, *Trans. Faraday Soc.*, 62 (1967) 3142.
9 R. B. Cundall, A. S. Davies and T. F. Palmer, *J. Phys. Chem.*, 70 (1966) 2503.
10 A. Forchioni and C. Willis, *J. Phys. Chem.*, 72 (1968) 3105.
11 L. J. Stief and V. J. DeCarlo, *J. Am. Chem. Soc.*, 91 (1969) 839.
12 W. Braun, A. M. Bass, D. D. Davis and J. D. Simmons, *Proc. Roy. Soc. (London), Ser. A*, 312 (1969) 417.
13 F. F. Martinotti, M. I. Welch and A. P. Wolf, *Chem. Commun.*, (1968) 115.
14 A. R. Fairbairn, *Proc. Roy. Soc. (London), Ser. A*, 312 (1969) 207; *J. Chem. Phys.*, 48 (1968) 515.
15 W. O. Davies, *Quart. Rept. No.* 8, *IITRI*-T200-8, IIT Research Inst., Chicago, 1964.
16 L. L. Presley, C. Chackerian and R. Watson, *AIAA Paper No.* 66-518, 1966; *Accession No. A*66-33660.
17 S. A. Pursley, R. A. Matula and O. W. Witzell, *J. Phys. Chem.*, 70 (1966) 3768.
18 W. Groth, W. Pessara and H. J. Rommel, *Z. physik. Chem. (Frankfurt)*, 32 (1966) 192;
H. J. Rommel, *Chem. Abstr.*, 69 (1968) 48231c.
19 P. Harteck, R. R. Reeves and B. A. Thompson, *Z. Naturforsch.* 19a (1964) 2.
20 G. Liuti, S. Dondes and P. Harteck, *J. Chem. Phys.*, 44 (1966) 4051.
21 T. G. Slanger and G. Black, *J. Chem. Phys.*, 51 (1969) 4534.
22 R. J. Donovan and D. Husain, *Trans. Faraday Soc.*, 63 (1967) 2879.
23 A. R. Anderson, J. F. V. Best and M. J. Willett, *Trans. Faraday Soc.*, 62 (1966) 595.
24 S. Dondes, P. Harteck and H. von Weyssenhoff, *Z. Naturforsch.*, 19a (1964) 13.
25 H. W. Buschmann and W. Groth, *Z. Naturforsch.*, 22a (1967) 954.
26 H. W. Buschmann and W. Groth, *Chem. Phys. Letters*, 2 (1968) 245.
27 J. P. Briggs and P. G. Clay, *Nature*, 218 (1968) 355.
28 J. P. Briggs and P. G. Clay, *Nature*, 217 (1968) 947.
29 C. Willis and C. Devillers, *Chem. Phys. Letters*, 2 (1968) 51.
30 G. M. Meaburn and D. Penner, *Nature*, 212 (1966) 1042.
31 K. W. Michel, H. A. Olschewski, H. Richtering and H. Gg. Wagner, *Z. Physik. Chem. (Frankfurt)*, 44 (1965) 160.
32 T. A. Brabbs, F. E. Belles and S. A. Zlatarich, *J. Chem. Phys.*, 38 (1963) 1939.
33 W. O. Davies, *J. Chem. Phys.*, 41 (1964) 1846.
34 W. O. Davies, *J. Chem. Phys.*, 43 (1965) 2809.
35 E. S. Fishburne, K. R. Bilwakesh and R. Edse, *J. Chem. Phys.*, 45 (1966) 160.
36 S. A. Losev, N. A. Generalov and V. A. Maksimenko, *Dokl. Akad. Nauk. SSSR*, 150 (1963) 839;
S. A. Losev and L. B. Terebenina, *Zh. Priklad. Mehkan i Tekhn. Fiz.*, (1966) 133.
37 M. Steinberg, *N.A.S.A. Contractors Rept.* 166, Feb., 1965.
38 H. A. Olschewski, J. Troe and H. Gg. Wagner, *Ber. Bunsenges. Physik. Chem.*, 70 (1966) 1060.

39 J. Troe and H. Gg. Wagner, *Ber. Bunsenges. Physik. Chem.*, 71 (1967) 937.
40 R. S. Brokaw, *11th Intern. Symp. Combustion, Berkeley, Calif.*, (1966), p. 1063; K. G. P. Sulzmann, B. F. Myers and E. R. Bartle, *J. Chem. Phys.*, 42 (1965) 3969.
41 T. C. Clark, S. H. Garnett and G. B. Kistiakowski, *J. Chem. Phys.*, 51 (1969) 2885.
42 E. R. Bartle and B. F. Myers, *Abstr. No.* 152, Physical Chemistry Section 157th Am. Chem. Soc. Annual Meeting, Minneapolis, Minn., April, 1969.
43 S. H. Garnett, G. B. Kistiakowsky and B. V. O'Grady, *J. Chem. Phys.*, 51 (1969) 84.
44 B. H. Mahan, *J. Chem. Phys.*, 33 (1960) 959.
45 R. A. Young and A. Y. M. Ung, *J. Chem. Phys.*, 44 (1966) 3038; 47 (1967) 1566.
46 A. Y. M. Ung and H. I. Schiff, *Can. J. Chem.*, 44 (1966) 1981.
47 P. Warneck, *Discussions Faraday Soc.*, 37 (1964) 57; *J. Chem. Phys.*, 41 (1964) 3435.
48 T. Slanger, *J. Chem. Phys.*, 45 (1966) 4127.
49 R. R. Reeves, Jr., P. Harteck, B. A. Thompson and R. W. Waldron, *J. Phys. Chem.*, 70 (1966) 1637.
50 D. S. Sethi and H. A. Taylor, *J. Chem. Phys.*, 49 (1968) 3669.
51 J. Y. Yang and F. M. Servedio, *Can. J. Chem.*, 46 (1968) 338.
52 M. Clerc and F. Barat, *J. Chem. Phys.*, 46 (1967) 107; *J. Chim. Phys.*, 63 (1966) 1525.
53 W. Groth, *Z. Physik. Chem.*, B37 (1937) 307.
54 J. F. Noxon, *J. Chem. Phys.*, 52 (1970) 1852.
55 L. M. Quick and R. J. Cvetanović, unpublished results.
56 G. Paraskevopoulos and R. J. Cvetanović, *J. Am. Chem. Soc.*, 91 (1969) 7572.
57 M. Clerc and A. Reiffsteck, *J. Chem. Phys.*, 48 (1968) 2799.
58 M. Clerc and F. Barat, *J. Chim. Phys.*, 65 (1968) 832.
59 M. Arvis, *J. Chim. Phys.*, 66 (1969) 517.
60 D. L. Baulch and W. H. Breckenridge, *Trans. Faraday Soc.*, 62 (1966) 2768.
61 K. F. Preston and R. J. Cvetanović, *J. Chem. Phys.*, 45 (1966) 2888.
62 D. Katakis and H. Taube, *J. Chem. Phys.*, 36 (1962) 416.
63 O. F. Raper and W. B. DeMore, *J. Chem. Phys.*, 40 (1964) 1053.
64a R. A. Young, G. Black and T. S. Slanger, *J. Chem. Phys.*, 49 (1968) 4758;
b T. G. Slanger and G. Black, private communication.
65 N. G. Moll, D. R. Clutter and W. E. Thompson, *J. Chem. Phys.*, 45 (1966) 4469.
66 E. Weissberger, W. H. Breckenridge and H. Taube, *J. Chem. Phys.*, 47 (1967) 1764.
67 Y. Mori, *Bull. Chem. Soc. Japan*, 34 (1961) 1128.
68 C. M. Wolff and R. Pertel, *J. Phys. Chem.*, 69 (1965) 4047.
69 O. P. Strausz and H. E. Gunning, *Can. J. Chem.*, 39 (1961) 2244.
70 L. J. Stief, V. J. DeCarlo and W. A. Payne, *J. Chem. Phys.*, 51 (1969) 3336.
71 M. H. J. Wijnen, *J. Chem. Phys.*, 24 (1956) 851.
72 L. J. Stief, V. J. DeCarlo, W. A. Payne, R. Gorden Jr., and P. Ausloos, *J. Chem. Phys.*, 53 (1970) 475.
73 A. R. Anderson and D. A. Dominey, *Radiation Res. Rev.*, 1 (1968) 269.
74 M. Anbar and P. Perlstein, *Trans. Faraday Soc.*, 62 (1966) 1803.
75 D. A. Dominey and T. F. Palmer, *Discussions Faraday Soc.*, 36 (1963) 35.
76 A. R. Anderson and J. V. F. Best, *Trans. Faraday Soc.*, 62 (1966) 610.
77 A. R. Anderson and J. V. F. Best, *Advances in Chemistry, Series No.* 82, The American Chem. Soc., 1968, p. 231.
78 F. S. Dainton, *Discussions Faraday Soc.*, 36 (1963) 237.
79 D. L. Baulch, F. S. Dainton and R. L. S. Willix, *Trans. Faraday Soc.*, 61 (1965) 1146.
80 A. G. Gaydon, G. H. Kimball and H. B. Palmer, *Proc. Roy. Soc.* (*London*), *Ser. A*, 279 (1964) 313.
81 H. A. Olschewski, J. Troe and H. G. Wagner, *Z. Phys. Chem.* (*Frankfurt*), 45 (1965) 329.
82 S. J. Arnold, W. G. Brownlee and G. H. Kimball, *J. Phys. Chem.*, 72 (1968) 4344.
83 J. Troe, *Ber. Bunsenges.*, 72 (1968) 908; J. Troe and H. Gg. Wagner, *Ber. Bunsenges. Physik. Chem.*, 71 (1967) 937.
84 A. Kondratjev and A. Yakovleva, *J. Phys. Chem.* (*U.S.S.R.*), 14 (1940) 853.
85 A. B. Callear, *Proc. Roy. Soc.* (*London*), *Ser. A*, 276 (1963) 401.
86 M. DeSorgo, A. J. Yarwood, O. P. Strausz and H. E. Gunning, *Can. J. Chem.*, 43 (1965) 1886.

87 J. Heicklen, *J. Am. Chem. Soc.*, 85 (1963) 3562.
88 F. J. Wright, *J. Phys. Chem.*, 64 (1960) 1648.
89 N. Basco and A. E. Pearson, *Trans. Faraday Soc.*, 63 (1967) 2684.
90 J. R. Partington and H. H. Neville, *J. Chem. Soc.*, (1951) 1230.
91 H. G. Schecker, quoted in ref. 83.
92 A. J. Hay and R. L. Belford, *J. Chem. Phys.*, 47 (1967) 3944.
93 H. E. Gunning and O. P. Strausz, *Advan. Photochem.*, 4 (1966) 143.
94 K. S. Sidhu, I. G. Csizmadia, O. P. Strausz and H. E. Gunning, *J. Am. Chem. Soc.*, 88 (1966) 2412.
95 P. Fowles, M. DeSorgo, A. J. Yarwood, O. P. Strausz and H. E. Gunning, *J. Am. Chem. Soc.*, 89 (1967) 1352.
96 R. J. Donovan, *Trans. Faraday Soc.*, 65 (1969) 1419;
R. J. Donovan, L. J. Kirsch and D. Husain, *Nature* 222 (1969) 1164.
97 W. D. McGrath, J. I. McGarvey and D. N. Dempster, *Can. J. Chem.*, 45 (1967) 2454.
98 A. B. Callear and W. J. R. Tyerman, *Trans. Faraday Soc.*, 61 (1965) 2395; 62 (1966) 371.
99 I. R. Beattie, *Progr. Inorg. Chem.*, 5 (1963) 1.
100 A. G. Vosper, *J. Chem. Soc.*, (1970) 625, and references quoted therein.
101 C. J. Hochanadel, J. A. Ghormley and P. J. Ogren, *J. Chem. Phys.*, 50 (1969) 3075.
102 T. C. Hall and F. E. Blacet, *J. Chem. Phys.*, 20 (1952) 1745.
103 L. G. Wayne and D. M. Yost, *J. Chem. Phys.*, 19 (1951) 41.
104 I. C. Hisatsune, *J. Phys. Chem.*, 72 (1968) 269.
105 H. S. Johnston, *J. Chem. Phys.*, 19 (1951) 663.
106 E. S. Fishburne and R. Edse, *J. Chem. Phys.*, 41 (1964) 1297.
107 D. Gutman, R. L. Belford, A. J. Hay and R. Pancirov, *J. Phys. Chem.*, 70 (1966) 1793.
108 W. Jost, K. W. Michel, J. Troe and H. G. Wagner, *Z. Naturforsch.*, 19a (1964) 59.
109 H. A. Olschewski, J. Troe and H. G. Wagner, *Ber. Bunsenges., Physik. Chem.*, 70 (1966) 450.
110 J. Troe and H. G. Wagner, *7th AGARD Colloquium*, Oslo, 1966, p. 22.
111 E. K. Gill and K. J. Laidler, *Can. J. Chem.*, 36 (1958) 1570.
112 E. E. Nikitin, *Doklad. Akad. Nauk SSSR*, 129 (1959) 157.
113 G. M. Wieder and R. A. Marcus, *J. Chem. Phys.*, 37 (1962) 1835.
114 H. S. Johnston, *J. Chem. Phys.*, 20 (1952) 1103.
115 L. S. Kassel, *J. Chem. Phys.*, 21 (1953) 1093.
116 H. S. Johnston and J. R. White, *J. Chem. Phys.*, 22 (1954) 1969.
117 J. N. Bradley and G. B. Kistiakowsky, *J. Chem. Phys.*, 35 (1961) 256.
118 L. J. Drummond and S. W. Hiscock, *Aust. J. Chem.*, 20 (1967) 815.
119 A. P. Modica, *J. Phys. Chem.*, 69 (1965) 2111.
120 F. J. Lindars and C. Hinshelwood, *Proc. Roy. Soc. (London), Ser. A*, 231 (1955) 162, 178.
121 B. G. Reuben and J. W. Linnett, *Trans. Faraday Soc.*, 55 (1959) 1543.
122 L. Friedman and J. Bigeleisen, *J. Am. Chem. Soc.*, 75 (1953) 2215.
123 F. Kaufman, N. J. Gerri and R. E. Bowman, *J. Chem. Phys.*, 25 (1956) 106.
124 F. Kaufman and J. R. Kelso, *J. Chem. Phys.*, 23 (1955) 603.
125 C. P. Fenimore and G. W. Jones, *8th Intern. Symp. Combustion, Pasadena, Calif.*, (1960), p. 127; *J. Phys. Chem.*, 62 (1958) 178.
126 C. P. Fenimore, *J. Chem. Phys.*, 35 (1961) 2243.
127 E. S. Fishburne and R. Edse, *J. Chem. Phys.*, 44 (1966) 515.
128 A. Martinego, J. Troe and H. Gg. Wagner, *Z. Physik. Chem. (Frankfurt)*, 51 (1966) 104.
129 F. Kaufman, *Progr. Reaction Kinetics*, 1 (1961) 28.
130 D. L. Bunker, *J. Chem. Phys.*, 40 (1964) 1946; 37 (1962) 393.
131 S. C. Lind, *Radiation Chemistry of Gases*, ACS Monograph No. 151, Reinhold, New York, 1961, p. 243.
132 J. W. T. Spinks and R. J. Wood, *An Introduction to Radiation Chemistry*, Wiley, New York, 1964, p. 206.
133 P. Harteck and S. Dondes, *Nucleonics*, 14 (1956) 66.
134 B. P. Burtt and J. F. Kircher, *Radiation Res.*, 9 (1958) 1.
135 R. Gorden, Jr. and P. Ausloos, *J. Res. Natl. Bur. Std., A.*, 69 (1965) 79.

136 A. R. Anderson, in *Fundamental Processes in Radiation Chemistry*, P. Ausloos, (Ed.), Interscience, 1968, p. 281.
137 H. Sponer and L. G. Bonner, *J. Chem. Phys.*, 8 (1940) 33.
138a M. Zelikoff, K. Watanabe and E. Y. C. Inn, *J. Chem. Phys.*, 21 (1953) 1643;
b A. B. F. Duncan, *J. Chem. Phys.*, 4 (1936) 638.
139 H. Yamazaki and R. J. Cvetanović, *J. Chem. Phys.*, 41 (1964) 3703;
G. Paraskevopoulos and R. J. Cvetanović, *J. Am. Chem. Soc.*, 91 (1969) 7572.
140 G. A. Castellion and W. A. Noyes, Jr., *J. Am. Chem. Soc.*, 79 (1957) 290.
141 M. Zelikoff and L. M. Aschenbrand, *J. Chem. Phys.*, 22 (1954) 1680, 1685.
142 W. A. Noyes, Jr., *J. Chem. Phys.*, 5 (1937) 807.
143 N. R. Greiner, *J. Chem. Phys.*, 47 (1967) 4373.
144 H. Yamazaki and R. J. Cvetanović, *J. Chem. Phys.*, 39 (1963) 1902;
H. Yamazaki and R. J. Cvetanović, *J. Chem. Phys.*, 40 (1964) 582.
145 K. F. Preston and R. J. Cvetanović, *J. Chem. Phys.*, 45 (1966) 2888.
146 J. W. Zabor and W. A. Noyes, Jr., *J. Am. Chem. Soc.*, 62 (1940) 1975.
147 J. P. Doering and B. H. Mahan, *J. Chem. Phys.*, 34 (1961) 1617; 36 (1962) 1682.
148 F. S. Dainton and P. Fowles, *Proc. Roy. Soc. (London), Ser. A*, 287 (1965) 295.
149 J. Y. Yang and F. M. Servedio, *J. Chem. Phys.*, 47 (1967) 4817.
150 W. E. Groth and H. Schierholz, *Planetary Space Sci.*, 1 (1959) 333.
151 R. A. Young, G. Black and T. G. Slanger, *J. Chem. Phys.*, 49 (1968) 4769.
152 H. Okabe, *J. Chem. Phys.*, 47 (1967) 101;
K. H. Welge, *J. Chem. Phys.*, 45 (1966) 166;
N. Ya. Dodonova, *Opt. Spectry. (USSR)*, 20 (1966) 271;
K. H. Becker and K. H. Welge, *Z. Naturforsch.*, 20a (1965) 442;
D. K. Sen Gupta, *Proc. Roy. Soc. (London), Ser. A*, 146 (1934) 824.
153 M. Zelikoff and L. M. Aschenbrand, *J. Chem. Phys.*, 27 (1957) 123.
154 R. A. Young, G. Black and T. G. Slanger, *J. Chem. Phys.*, 50 (1969) 303.
155 R. A. Young, G. Black and T. G. Slanger, *J. Chem. Phys.*, 50 (1969) 309.
156 G. B. Kistiakowsky and G. G. Volpi, *J. Chem. Phys.*, 27 (1957) 1141.
157 M. A. A. Clyne and B. A. Thrush, *Proc. Roy. Soc. (London), Ser. A*, 261 (1961) 259.
158 K. Schofield, *Planetary Space Sci.*, 15 (1967) 643.
159 I. M. Campbell and B. A. Thrush, *Trans. Faraday Soc.*, 62 (1966) 3366.
160 W. M. Manning and W. A. Noyes, Jr., *J. Am. Chem. Soc.*, 54 (1932) 3907.
161 R. J. Cvetanović, *J. Chem. Phys.*, 23 (1955) 1203.
162 R. J. Cvetanović, *J. Chem. Phys.*, 23 (1955) 1208.
163 H. E. Gunning, *Can. J. Chem.*, 36 (1958) 89.
164 J. Noxon, *J. Chem. Phys.*, 52 (1970) 1852.
165 S. V. Filseth, F. Stuhl and K. H. Welge, *J. Chem. Phys.*, 52 (1970) 239.
166 F. Kaufman and J. R. Kelso, *J. Chem. Phys.*, 23 (1955) 1702.
167 F. Kaufman and L. J. Decker, *7th Intern. Symp. Combustion*, Butterworths, London, 1958, p. 57.
168 E. Freedman and J. W. Daiber, *J. Chem. Phys.*, 34 (1961) 1271.
169 K. L. Wray and J. D. Teare, *J. Chem. Phys.*, 36 (1962) 2582.
170 C. P. Fenimore and G. W. Jones, *J. Phys. Chem.*, 61 (1957) 654.
171 J. Zel'dovich, *Acta Physicochem. U.R.S.S.*, 21 (1946) 577.
172 W. E. Wilson, *J. Chem. Phys.*, 46 (1967) 2017.
173 A. S. Vlastaras and C. A. Winkler, *Can. J. Chem.*, 45 (1967) 2837.
174 H. S. Glick, J. J. Klein and W. Squire, *J. Chem. Phys.*, 27 (1957) 850.
175 R. E. Duff and N. Davidson, *J. Chem. Phys.*, 31 (1959) 1018.
176 E. L. Yuan, J. I. Slaughter, W. E. Koerner and F. Daniels, *J. Phys. Chem.*, 63 (1959) 952.
177 H. Wise and M. F. Frech, *J. Chem. Phys.*, 20 (1952) 22, 1724;
S. Dushman, *J. Am. Chem. Soc.*, 43 (1921) 411.
178 T. P. Melia, *J. Inorg. Nucl. Chem.*, 27 (1965) 95.
179 J. Heicklen and N. Cohen, *Advan. Photochem.*, 5 (1968) 157.
180 R. J. Fallon, J. T. Vanderslice and E. A. Mason, *J. Phys. Chem.*, 63 (1959) 2082.

181 O. P. STRAUSZ AND H. E. GUNNING, *Can. J. Chem.*, 39 (1961) 2549.
182 M. H. HOFFMAN AND R. B. BERNSTEIN, *J. Phys. Chem.*, 64 (1960) 1769.
183 A. B. CALLEAR AND R. G. W. NORRISH, *Proc. Roy. Soc. (London), Ser. A*, 266 (1962) 299; A. B. CALLEAR AND G. J. WILLIAMS, *Trans. Faraday Soc.*, 60 (1964) 2158.
184 W. A. NOYES, JR., *J. Am. Chem. Soc.*, 53 (1931) 514.
185 R. J. CVETANOVIĆ, *Progr. Reaction Kinetics*, 2 (1964) 41.
186 G. KARL, P. KRUUS, J. C. POLANYI AND I. W. M. SMITH, *J. Chem. Phys.*, 46 (1967) 244.
187 G. E. MOORE, O. R. WULF AND R. M. BADGER, *J. Chem. Phys.*, 21 (1953) 2091.
188 N. BASCO, A. B. CALLEAR AND R. G. W. NORRISH, *Proc. Roy. Soc. (London), Ser. A*, 260 (1961) 459.
189 J. J. MCGEE AND J. HEICKLEN, *J. Chem. Phys.*, 41 (1964) 2974.
190 J. HEICKLEN, *J. Phys. Chem.*, 70 (1966) 2456.
191 A. B. CALLEAR AND I. W. M. SMITH, *Trans. Faraday Soc.*, 59 (1963) 1720.
192 K. WATANABE, M. ZELIKOFF AND E. C. Y. INN, *Air Force Cambridge Res. Center Tech. Rep.* 53–23, 1953.
193 J. Y. MACDONALD, *J. Chem. Soc.*, (1928) 1.
194 M. JEUNEHOMME AND A. B. F. DUNCAN, *J. Chem. Phys.*, 41 (1964) 1692.
195 P. J. FLORY AND H. L. JOHNSTON, *J. Am. Chem. Soc.*, 57 (1935) 2641; *J. Chem. Phys.*, 14 (1946) 212.
196 A. B. CALLEAR AND I. W. M. SMITH, *Discussions Faraday Soc.*, 37 (1964) 96; *Intern. Conf. Photochem., Tokyo*, 1965, preprints, p. 5; *Trans. Faraday Soc.*, 61 (1965) 2383.
197 R. A. YOUNG AND R. A. SHARPLESS, *Discussions Faraday Soc.*, 33 (1962) 228.
198a A. G. LEIGA AND H. A. TAYLOR, *J. Chem. Phys.*, 42 (1965) 2107;
b D. S. SETHI AND H. A. TAYLOR, *J. Chem. Phys.*, 48 (1968) 533.
199 K. WATANABE, *Advan. Geophys.*, 5 (1958) 153.
200 B. H. MAHAN, *J. Chem. Phys.*, 43 (1965) 1853.
201 R. C. GUNTON AND T. M. SHAW, *Phys. Rev.*, 140 (1965) A748.
202 R. C. GUNTON AND T. M. SHAW, *Phys. Rev.*, 140 (1965) A756.
203 B. H. MAHAN AND J. C. PERSON, *J. Chem. Phys.*, 40 (1964) 392.
204 R. P. STEIN, M. SCHIEBE, M. W. SYVERSON, T. M. SHAW AND R. C. GUNTON, *Phys. Fluids*, 7 (1964) 1641.
205 S. C. LIN AND J. D. TEARE, *Phys. Fluids*, 6 (1963) 355.
206 S. C. LIND, *Radiation Chemistry of Gases*, ACS Monograph 151, Reinhold, 1961, Chap. 13, p. 232.
207 W. MUND AND R. GILLEROT, *Bull. Soc. Chim. Belges*, 38 (1929) 343.
208 P. HARTECK AND S. DONDES, *J. Chem. Phys.*, 27 (1957) 546.
209 P. HARTECK AND S. DONDES, *J. Chem. Phys.*, 28 (1958) 975.
210 M. BODENSTEIN AND H. RAMSTETTER, *Z. Physik. Chem. (Leipzig)*, 100 (1922) 106.
211 W. A. ROSSER AND H. WISE, *J. Chem. Phys.*, 24 (1956) 493.
212 P. G. ASHMORE AND B. P. LEVITT, *Research (London)*, 9 (1956) S25.
213 P. G. ASHMORE AND M. G. BURNETT, *Trans. Faraday Soc.*, 58 (1962) 253.
214 E. W. GRAHAM, *Ph. D. Thesis*, Univ. of California, Berkeley, 1963.
215 J. D. RAY AND R. A. OGG, *J. Chem. Phys.*, 26 (1957) 984.
216 H. FORD, *Can. J. Chem.*, 38 (1960) 1780.
217 D. R. HIRSCHBACH, H. S. JOHNSTON, K. S. PITZER AND R. E. POWELL, *J. Chem. Phys.*, 25 (1956) 736.
218 S. W. BENSON, *The Foundations of Chemical Kinetics*, McGraw-Hill, New York, 1960, p. 416.
219 P. G. ASHMORE, M. G. BURNETT AND B. J. TYLER, *Trans. Faraday Soc.*, 58 (1962) 685.
220 R. A. OGG, *J. Chem. Phys.*, 21 (1953) 2079.
221 W. A. GUILLORY AND H. S. JOHNSTON, *J. Chem. Phys.*, 42 (1965) 2457.
222 R. E. HUFFMAN AND N. DAVIDSON, *J. Am. Chem. Soc.*, 81 (1959) 2311.
223 H. HIRAOKA AND R. HARDWICK, *J. Chem. Phys.*, 39 (1963) 2361.
224 E. S. FISHBURNE, D. M. BERGBAUER AND R. EDSE, *J. Chem. Phys.*, 43 (1965) 1847.
225 L. S. KASSEL, *The Kinetics of Homogeneous Gas Reactions*, Chemical Catalogue Co., New York, 1932.

226 G. Schott and N. Davidson, *J. Am. Chem. Soc.*, 80 (1958) 1841.
227 I. C. Hisatsune, B. Crawford and R. A. Ogg, *J. Am. Chem. Soc.*, 79 (1957) 4648.
228 H. Blend, *Chem. Abstr.*, 60 (1964) 4849e.
229 W. T. Richards, *Rev. Mod. Phys.*, 11 (1939) 36;
W. T. Richards and J. A. Reid, *J. Chem. Phys.*, 1 (1933) 114.
230 M. Cher, *J. Chem. Phys.*, 37 (1962) 2564.
231 T. Carrington and N. Davidson, *J. Phys. Chem.*, 57 (1953) 418.
232 P. D. Brass and R. C. Tolman, *J. Am. Chem. Soc.*, 54 (1932) 1003.
233 S. H. Bauer and M. R. Gustavson, *Discussions Faraday Soc.*, 17 (1954) 69;
S. H. Bauer, *J. Phys. Chem.*, 57 (1953) 424.
234 A. E. Douglas and K. P. Huber, *Can. J. Phys.*, 43 (1965) 74.
235 R. K. Ritchie, A. D. Walsh and P. A. Warsop, *Proc. Roy. Soc. (London) Ser. A*, 266 (1962) 257.
236 R. K. Ritchie and A. D. Walsh, *Proc. Roy. Soc. (London), Ser. A*, 267 (1962) 395.
237 P. A. Leighton, *Photochemistry of Air Pollution*, Academic Press, New York, 1961.
238 T. Nakayama, M. Y. Kitamura and K. Watanabe, *J. Chem. Phys.*, 30 (1959) 1180.
239 H. H. Holmes and F. Daniels, *J. Am. Chem. Soc.*, 56 (1934) 630.
240 R. G. W. Norrish, *J. Chem. Soc.*, (1929) 1158.
241 R. G. W. Norrish, *J. Chem. Soc.*, (1929) 1611.
242 D. Neuberger and A. B. F. Duncan, *J. Chem. Phys.*, 22 (1954) 1693.
243 S. Sato and R. J. Cvetanović, *Can. J. Chem.*, 36 (1958) 279.
244 J. N. Pitts, J. H. Sharp and S. I. Chan, *J. Chem. Phys.*, 40 (1964) 3655.
245 K. F. Preston and R. J. Cvetanović, *Can. J. Chem.*, 44 (1966) 2445.
246 H. W. Ford and S. Jaffe, *J. Chem. Phys.*, 38 (1963) 2935.
247 F. E. Blacet, T. C. Hall and P. A. Leighton, *J. Am. Chem. Soc.*, 84 (1962) 4011.
248 G. H. Myers, D. M. Silver and F. Kaufman, *J. Chem. Phys.*, 44 (1966) 718.
249 H. W. Ford, *Can. J. Chem.*, 38 (1960) 1780.
250 D. Husain and R. G. W. Norrish, *Proc. Roy. Soc. (London), Ser. A*, 273 (1963) 165.
251 J. T. Herron and F. S. Klein, *J. Chem. Phys.*, 40 (1964) 2731; 41 (1964) 1285.
252 M. A. A. Clyne and B. A. Thrush, *Trans. Faraday Soc.*, 58 (1962) 511; *J. Chem. Phys.*, 38 (1963) 1252.
253 H. W. Ford and N. Endow, *J. Chem. Phys.*, 27 (1957) 1156, 1277.
254 J. King, *NASA Rep. N*65–32454 *No. NASA-CR*-64605.
255 F. Kaufman and J. R. Kelso, *J. Chem. Phys.*, 46 (1967) 4541.
256 R. G. Dickinson and W. P. Baxter, *J. Am. Chem. Soc.*, 50 (1928) 774.
257 L. F. Phillips and H. I. Schiff, *J. Chem. Phys.*, 36 (1962) 1509.
258 P. Harteck and S. Dondes, *J. Chem. Phys.*, 22 (1954) 953.
259 M. T. Dmitriev and L. V. Saradzhev, *Zh. Fiz. Khim.*, 35 (1961) 727.
260 M. T. Dmitriev, S. A. Kamenetskaya and S. Ya. Pshezhetskii, *Khim. Vysok. Energ.*, 1 (1967) 205.
261 M. A. A. Clyne and B. A. Thrush, *Trans. Faraday Soc.*, 57 (1961) 69.
262 F. Daniels and E. H. Johnston, *J. Am. Chem. Soc.*, 43 (1921) 53.
263 E. C. White and R. C. Tolman, *J. Am. Chem. Soc.*, 47 (1925) 1240.
264 J. K. Hunt and F. Daniels, *J. Am. Chem. Soc.*, 47 (1925) 1602.
265 H. S. Hirst, *J. Chem. Soc.*, 127 (1925) 657.
266 F. O. Rice and D. Getz, *J. Phys. Chem.*, 31 (1927) 1572.
267 M. E. Nordberg, *Thesis*, California Institute of Technology, Pasadena, 1928.
268 H. C. Ramsperger, M. E. Nordberg and R. C. Tolman, *Proc. Natl. Acad. Sci. U.S.*, 15 (1929) 453.
269 H.-J. Schumacher and G. Sprenger, *Z. Physik. Chem.*, 140 (1929) 281.
270 H.-J. Schumacher and G. Sprenger, *Proc. Natl. Acad. Sci. U.S.*, 16 (1930) 129.
271 J. H. Hodges and E. F. Linhorst, *Proc. Natl. Acad. Sci. U.S.*, 17 (1931) 28.
272 H. S. Johnston and Y. Tao, *J. Am. Chem. Soc.*, 73 (1951) 2948.
273 E. F. Linhorst and J. H. Hodges, *J. Am. Chem. Soc.*, 56 (1934) 836.
274 H. C. Ramsperger and R. C. Tolman, *Proc. Natl. Acad. Sci. U.S.*, 16 (1930) 6.
275 R. A. Ogg, *J. Chem. Phys.*, 15 (1947) 337.

276 H. S. Johnston, *J. Am. Chem. Soc.*, 73 (1951) 4542.
277 R. A. Ogg, W. S. Richardson and M. K. Wilson, *J. Chem. Phys.*, 18 (1950) 573.
278 G. Schott and N. Davidson, *J. Am. Chem. Soc.*, 80 (1958) 1841.
279 T. Katan, *L.C. Card No. MIC*60-2395, Univ. Microfilms, Ann Arbor; *Dissertation Abstr.*, 21 (1960) 68.
280 J. H. Smith and F. Daniels, *J. Am. Chem. Soc.*, 69 (1947) 1735.
281 I. C. Hisatsune, B. Crawford and R. A. Ogg, *J. Am. Chem. Soc.*, 79 (1957) 4648.
282 R. L. Mills and H. S. Johnston, *J. Am. Chem. Soc.*, 73 (1951) 938.
283 H. S. Johnston and R. L. Perrine, *J. Am. Chem. Soc.*, 73 (1951) 4782.
284 H. S. Johnston, *J. Am. Chem. Soc.*, 75 (1953) 1567.
285 D. J. Wilson and H. S. Johnston, *J. Am. Chem. Soc.*, 75 (1953) 5763.
286 R. H. Lueck, *J. Am. Chem. Soc.*, 44 (1922) 757.
287 H. Eyring and F. Daniels, *J. Am. Chem. Soc.*, 52 (1930) 1472, 1486.
288 H. Martin and W. Meise, *Z. Elektrochem.*, 63 (1959) 162.
289 R. A. Ogg, *J. Chem. Phys.*, 15 (1947) 613.
290 H. J. Schumacher and G. Sprenger, *Z. Physik. Chem.*, B2 (1929) 267.
291 G. Sprenger, *Z. Elektrochem.*, 37 (1931) 674.
292 E. J. Jones and O. R. Wulf, *J. Chem. Phys.*, 5 (1937) 873.
293 F. Cramarossa and H. S. Johnston, *J. Chem. Phys.*, 43 (1965) 727.
294 H. S. Johnston and D. M. Yost, *J. Chem. Phys.*, 17 (1949) 386.
295 H. C. Urey, L. H. Dawsey and F. O. Rice, *J. Am. Chem. Soc.*, 51 (1929) 3190.
296 H. H. Holmes and F. Daniels, *J. Am. Chem. Soc.*, 56 (1934) 630.
297 T. C. Castorina and A. O. Allen, *J. Phys. Chem.*, 69 (1965) 3547.
298 H. S. Johnston, L. Foering, Y. S. Tao and G. H. Messerly, *J. Am. Chem. Soc.*, 73 (1951) 2319.
299 H. S. Johnston, L. Foering and R. J. Thompson, *J. Phys. Chem.*, 57 (1953) 390.
300 H. S. Johnston, L. Foering and J. R. White, *J. Am. Chem. Soc.*, 77 (1955) 4208.
301 C. Fréjacques, *Thèses*, Univ. Paris, 1953; *Compt. Rend.*, 232 (1951) 2206.
302 H. Harrison, *Dissertation Abstr.*, 21 (1960) 773.
303 J. H. Smith, *J. Am. Chem. Soc.*, 69 (1947) 1741.
304 M. G. Burnett, *Ph. D. Thesis*, Cambridge, 1961.
305 T. Bérces and S. Förgeteg, *Trans. Faraday Soc.*, 66 (1970) 633.
306 T. Bérces and S. Förgeteg, *Trans. Faraday Soc.*, 66 (1970) 640.
307 T. Bérces, S. Förgeteg and F. Marta, *Trans. Faraday Soc.*, 66 (1970) 648.
308 L. A. Dmitriev, S. A. Kamenetskaya and S. Ya. Pshezhetskii, *Khim. Vysok. Energ.*, 2 (1968) 465.
309 S. W. Benson and A. E. Axworthy, *J. Chem. Phys.*, 26 (1957) 1718; 42 (1965) 2614.
310 W. M. Jones and N. Davidson, *J. Am. Chem. Soc.*, 84 (1962) 2868.
311 J. A. Zaslowsky, H. B. Urbach, F. Leighton, R. J. Wnuk and J. A. Wojtowicz, *J. Am. Chem. Soc.*, 82 (1960) 2682.
312 E. I. Intezarova and V. N. Kondrat'ev, *Izv. Akad. Nauk. SSSR, Ser. Khim.*, (1967) 2440.
313 S. Ya. Pshezhetskii, M. M. Morozov, S. A. Kamenetskaya, V. N. Siryatskaya and E. I. Gribova, *Russ. J. Phys. Chem.*, 33 (1959) 402.
314 A. Glissman and H. J. Schumacher, *Z. Physik. Chem.*, 21B (1933) 323.
315 M. A. A. Clyne, J. C. McKenney and B. A. Thrush, *Trans. Faraday Soc.*, 61 (1965) 2701.
316 W. D. McGrath and R. G. W. Norrish, *Proc. Roy. Soc. (London), Ser. A*, 242 (1957) 265.
317 W. D. McGrath and R. G. W. Norrish, *Proc. Roy. Soc. (London), Ser. A*, 254 (1960) 317.
318 C. J. Hochanadel, J. A. Chormley and J. W. Boyle, *J. Chem. Phys.*, 48 (1968) 2416.
319 M. C. Sauer, *J. Phys. Chem.*, 71 (1967) 3311.
320 M. C. Sauer and L. M. Dorfmann, *J. Am. Chem. Soc.*, 87 (1965) 3801.
321 G. M. Meaburn, D. Penner, J. LeCalve and M. Bourene, *J. Phys. Chem.*, 72 (1968) 3920.
322 J. T. Herron and F. S. Klein, *J. Chem. Phys.*, 44 (1966) 3645.
323 M. F. R. Mulcahy and D. J. Williams, *Trans. Faraday Soc.*, 64 (1968) 59.
324 D. Garvin, *J. Am. Chem. Soc.*, 76 (1954) 1523.
325 A. Mathias and H. I. Schiff, *Discussions Faraday Soc.*, 37 (1964) 38.

326 L. F. Phillips and H. I. Schiff, *J. Chem. Phys.*, 37 (1962) 924.
327 M. A. A. Clyne, B. A. Thrush and R. P. Wayne, *Nature*, 199 (1963) 1057.
328 E. J. Bowen, E. A. Moelwyn-Hughes and C. N. Hinshelwood, *Proc. Roy. Soc.* (*London*), *Ser. A*, 134 (1932) 211.
329 E. Castellano and H. J. Schumacher, *Z. Physik. Chem.* (*Frankfurt*), 34 (1962) 198; *J. Chem. Phys.*, 36 (1962) 2238.
330 G. B. Kistiakowsky, *Z. Physik. Chem.*, 117 (1925) 337.
331 H. J. Schumacher, *J. Am. Chem. Soc.*, 52 (1930) 2377; *Z. Physik. Chem.*, *Abt.* B17 (1932) 405.
332 R. G. W. Norrish and R. P. Wayne, *Proc. Roy. Soc.*, (*London*), *Ser. A*, 288, (1965) 200.
333 R. V. Fitzsimmons and E. J. Bair, *J. Chem. Phys.*, 40 (1964) 451.
334 V. D. Baiamonte, D. R. Snelling and E. J. Bair, *J. Chem. Phys.*, 44 (1966) 673; D. R. Snelling, V. D. Baiamonte and E. J. Bair, *J. Chem. Phys.*, 44 (1966) 4137.
335 D. R. Snelling and E. J. Bair, *J. Chem. Phys.*, 47 (1967) 228; 48 (1968) 5737.
336 U. Beretta and H. J. Schumacher, *Z. Physik. Chem.*, *Abt.* B17 (1932) 417.
337 E. Castellano and H. J. Schumacher, *Z. Physik. Chem.*, 65 (1969) 62.
338 L. J. Heidt and G. S. Forbes, *J. Am. Chem. Soc.*, 56 (1934) 2365; L. J. Heidt, *J. Am. Chem. Soc.*, 57 (1935) 1710.
339 G. S. Forbes and L. J. Heidt, *J. Am. Chem. Soc.*, 56 (1934) 1671.
340 O. F. Raper and W. B. DeMore, *J. Chem. Phys.*, 37 (1962) 2048; *J. Chem. Phys.*, 40 (1964) 1053.
341 D. Katakis and H. Taube, *J. Chem. Phys.*, 36 (1962) 416.
342 W. B. DeMore and O. F. Raper, *J. Chem. Phys.*, 44 (1966) 1780.
343 S. W. Benson, *J. Chem. Phys.*, 26 (1957) 1351; 33 (1960) 939.
344 R. Engleman, *J. Am. Chem. Soc.*, 87 (1965) 4193.
345 H. Taube, *Trans. Faraday Soc.*, 53 (1957) 656.
346 R. P. Wayne and J. N. Pitts, *J. Chem. Phys.*, 50 (1969) 3644.
347 R. J. McNeal and G. R. Cook, *J. Chem. Phys.*, 47 (1967) 5385.
348 R. E. Marsh, S. G. Furnival and H. I. Schiff, *Photochem. Photobiol.*, 4 (1965) 971.
349 T. P. J. Izod and R. P. Wayne, *Proc. Roy. Soc.* (*London*), *Ser. A.*, 308 (1968) 81.
350 O. R. Lundell, R. D. Ketcheson and H. I. Schiff, quoted in ref. 351.
351 H. I. Schiff, *Can. J. Chem.*, 47 (1969) 1903.
352 B. Lewis, *J. Phys. Chem.*, 37 (1933) 533.
353 P. C. Capron and R. Cloetens, *Bull. Soc. Chim.*, *Belges*, 44 (1935) 441.
354 J. T. Sears and J. W. Sutherland, *J. Phys. Chem.*, 72 (1968) 1166.
355 M. Gauthier and D. R. Snelling, *Chem. Phys. Letters*, 5 (1970) 93.
356 D. J. Meschi and R. J. Myers, *J. Am. Chem. Soc.*, 78 (1956) 6220; *J. Mol. Spectr.*, 3 (1959) 405.
357 B. P. Levitt and D. B. Sheen, *Trans. Faraday Soc.*, 63 (1967) 2955.
358 H. A. Olschewski, J. Troe and H. Gg. Wagner, *Z. Phys. Chem. N.F.*, 44 (1965) 173.
359 A. G. Gaydon, G. H. Kimball and H. B. Palmer, *Proc. Roy. Soc.* (*London*), *Ser. A*, 276 (1963) 461.
360 B. P. Levitt and D. B. Sheen, *J. Chem. Phys.*, 41 (1964) 584; *Trans. Faraday Soc.*, 61 (1965) 2404; 63 (1967) 540; B. P. Levitt and S. R. Fletcher, *Trans. Faraday Soc.*, 65 (1969) 1544.
361 C. J. Halstead and B. A. Thrush, *Proc. Roy. Soc.* (*London*), *Ser. A*, 295 (1966) 363; *Photochem. Photobiol.*, 4 (1965) 1007.
362 N. Cohen and R. W. F. Gross, *J. Chem. Phys.*, 50 (1969) 3119.
363 T. R. Rolfes, R. R. Reeves and P. Harteck, *J. Phys. Chem.*, 69 (1965) 849.
364 A. Sharma, J. P. Padur and P. Warneck, *J. Phys. Chem.*, 71 (1967) 1602.
365 A. McKenzie and B. A. Thrush, *Proc. Roy. Soc.* (*London*), *Ser. A*, 308 (1968) 133.
366 D. J. Williams, *Combust. Flame*, 12 (1968) 165.
367 E. L. Merryman and A. Levy, *J. Air Pollution Control Assoc.*, 17 (1967) 800.
368 K. Hoyermann, H. Gg. Wagner and J. Wolfrum, *Ber. Bunsenges. Physik. Chem.*, 71 (1967) 603.
369 R. J. Donovan, D. Husain and P. T. Jackson, *Trans. Faraday Soc.*, 65 (1969) 2930.

370 J. O. SULLIVAN AND P. WARNECK, *Ber. Bunsenges. Physik. Chem.*, 69 (1965) 7.
371 M. F. R. MULCAHY, J. R. STEVEN AND J. C. WARD, *J. Phys. Chem.*, 71 (1967) 2124.
372 R. D. CADLE AND J. W. POWERS, *Tellus*, 18 (1966) 176;
E. R. ALLEN AND R. D. CADLE, *Photochem. Photobiol.*, 4 (1965) 979.
373 F. KAUFMAN, *Proc. Roy. Soc. (London), Ser. A*, 247 (1958) 123.
374 P. WEBSTER AND A. D. WALSH, *10th Intern. Symp. Combustion*, Combustion Institute, Pittsburgh 1965, p. 463.
375 S. JAFFE AND F. S. KLEIN, *Trans. Faraday Soc.*, 62 (1966) 2150.
376 A. LEVY AND E. L. MERRYMAN, *Combust. Flame*, 9 (1965) 229;
C. P. FENIMORE AND G. W. JONES, *J. Phys. Chem.*, 69 (1965) 3593.
377 W. G. ROTHSCHILD, *J. Chem. Phys.*, 45 (1966) 3594; *J. Am. Chem. Soc.*, 86 (1964) 1307.
378 T. NAVANEETH RAO, S. S. COLLIER AND J. G. CALVERT, *J. Am. Chem. Soc.*, 91 (1969) 1609, 1616.
379 K. F. GREENOUGH AND A. B. F. DUNCAN, *J. Am. Chem. Soc.*, 83 (1961) 555.
380 R. G. W. NORRISH AND G. A. OLDERSHAW, *Proc. Roy. Soc. (London), Ser. A*, 249 (1958) 498.
381 A. JAKOVLEVA AND V. KONDRATJEV, *Acta Phys. Chim. U.R.S.S.*, 13 (1940) 241.
382 S. S. COLLIER, A. MORIKAWA, D. H. SLATER, J. G. CALVERT, G. REINHARDT AND E. DAMON, *J. Am. Chem. Soc.*, 92 (1970) 217.
383 R. B. CATON AND A. B. F. DUNCAN, *J. Am. Chem. Soc.*, 90 (1968) 1945.
384 S. OKUDA, T. N. RAO, D. H. SLATER AND J. G. CALVERT, *J. Phys. Chem.*, 73 (1969) 4412.
385 P. WARNECK, F. F. MARMO AND J. O. SULLIVAN, *J. Chem. Phys.*, 40 (1964) 1132.
386 H. D. METTEE, *J. Phys. Chem.*, 73 (1969) 1071.
387 H. D. METTEE, *J. Chem. Phys.*, 49 (1968) 1784.
388 J. N. DRISCOLL AND P. WARNECK, *J. Phys. Chem.*, 72 (1968) 3736.
389 R. W. FAIR AND B. A. THRUSH, *Trans. Faraday Soc.*, 65 (1969) 1557.
390 M. BODENSTEIN AND F. KRANENDIECK, *Z. Phys. Chem.*, 80 (1912) 48.
391 K. S. PANKHURST AND M. C. STYLES, *B.C.U.R.A. Monthly Bull.*, 27 (1963) 497.
392 E. FAJANS AND C. F. GOODEVE, *Trans. Faraday Soc.*, 32 (1936) 511.
393 W. KOBLITZ AND H. J. SCHUMACHER, *Z. Physik. Chem. (Leipzig)*, B25 (1934) 283.
394 W. C. SOLOMON, J. A. BLAUER AND F. C. JAYE, *J. Phys. Chem.*, 72 (1968) 2311.
395 L. DAUERMAN, G. E. SALSER AND Y. A. TAJIMA, *J. Phys. Chem.*, 71 (1967) 3999; 73 (1969) 1621.
396 J. TROE, H. GG. WAGNER AND G. WEDEN, *Z. Physik. Chem. N.F.*, 56 (1967) 238.
397 J. A. BLAUER AND W. C. SOLOMON, *J. Phys. Chem.*, 72 (1968) 2307.
398 R. GATTI, E. STARICCO, J. E. SICRE AND H. J. SCHUMACHER, *Z. Physik. Chem. N.F.*, 35 (1962) 343; 36 (1963) 211.
399 I. J. SOLOMON, A. J. KACMAREK AND J. RANEY, *J. Phys. Chem.*, 72 (1968) 2262.
400 M. C. LIN AND S. H. BAUER, *J. Am. Chem. Soc.*, 91 (1969) 7737.
401 H. J. SCHUMACHER AND P. FRISCH, *Z. Physik. Chem.*, B37 (1937) 1.
402 S. W. BENSON, *Thermochemical Kinetics*, Wiley, New York, 1968.
403 E. T. MCHALE AND G. VON ELBE, *J. Am. Chem. Soc.*, 89 (1967) 2795.
404 G. PORTER AND F. J. WRIGHT, *Discussions Faraday Soc.*, 14 (1953) 23.
405 E. D. MORRIS AND H. S. JOHNSTON, *J. Am. Chem. Soc.*, 90 (1968) 1918.
406 J. J. BEAVER AND G. STIEGER, *Z. Physik. Chem. (Leipzig)*, B12 (1931) 93.
407 H. J. SCHUMACHER AND G. STIEGER, *Z. Physik. Chem. (Leipzig)*, B7 (1930) 363.
408 M. A. A. CLYNE AND J. A. COXON, *Trans. Faraday Soc.*, 62 (1966) 2175.
409 M. A. A. CLYNE AND J. A. COXON, *Trans. Faraday Soc.*, 62 (1966) 1175.
410 M. A. CLYNE AND J. A. COXON, *Proc. Roy. Soc. (London), Ser. A*, 303 (1968) 207.
411 M. BODENSTEIN AND Z. SZABO, *Z. Physik. Chem.*, 39 (1938) 44.
412 C. N. HINSHELWOOD AND C. R. PRICHARD, *J. Chem. Soc.*, 123 (1923) 2730;
C. N. HINSHELWOOD AND J. HUGHES, *J. Chem. Soc.*, 125 (1924) 1841.
413 Z. G. SZABO, P. HUHN AND F. MARTA, *Trans. Faraday Soc.*, 55 (1959) 1131.
414 W. FINKELNBURG, H. J. SCHUMACHER AND G. STIEGER, *Z. Physik. Chem.*, B15 (1932) 127.
415 F. H. C. EDGECOMBE, R. G. W. NORRISH AND B. A. THRUSH, *Proc. Roy. Soc. (London), Ser. A*, 243 (1958) 24.
416 E. T. MCHALE AND G. VON ELBE, *J. Phys. Chem.*, 72 (1968) 1849.

417 C. G. Freeman and L. F. Phillips, *J. Phys. Chem.*, 72 (1968) 3025.
418 F. J. Lipscomb, R. G. W. Norrish and B. A. Thrush, *Proc. Roy. Soc. (London), Ser. A*, 233 (1955) 455.
419 N. Basco and S. K. Dogra, *Chem. Commun.*, (1968) 1071.
420 G. Porter and F. J. Wright, *Discussions Faraday Soc.*, 14 (1953) 23.
421 H. S. Johnston, E. D. Morris and J. Van den Bogaerde, *J. Am. Chem. Soc.*, 91 (1969) 7712.
422 H. Niki and B. Weinstock, *J. Chem. Phys.*, 47 (1967) 3249.
423 M. Bodenstein and G. Kistiakowsky, *Z. Physik. Chem.*, 116 (1925) 371.
424 H. J. Schumacher and R. V. Townend, *Z. Physik. Chem.*, B20 (1933) 375.
425 C. F. Goodeve and J. I. Wallace, *Trans. Faraday Soc.*, 26 (1930) 254.
426 R. G. Dickinson and C. E. P. Jeffreys, *J. Am. Chem. Soc.*, 52 (1930) 4288.
427 J. W. T. Spinks and J. M. Porter, *J. Am. Chem. Soc.*, 56 (1934) 264.
428 J. W. T. Spinks and H. Taube, *Can. J. Res.*, B15 (1937) 499.
429 E. J. Bowen and W. M. Cheung, *J. Chem. Soc.*, (1932) 1200.
430 J. E. Nicholas and R. G. W. Norrish, *Proc. Roy. Soc. (London), Ser. A*, 307 (1968) 391.
431 S. W. Benson and J. H. Buss, *J. Chem. Phys.*, 27 (1957) 1382.
432 R. G. W. Norrish and G. H. J. Neville, *J. Chem. Soc.*, (1934) 1864.
433 G. K. Rollefson and M. Burton, *Photochemistry and the Mechanism of Chemical Reactions*, Prentice-Hall, New York, 1942.
434 F. Kaufman, N. J. Gerri and D. A. Pascale, *J. Chem. Phys.*, 24 (1956) 32.
435 E. Coloccia, R. V. Figini and H. J. Schumacher, *Angew. Chem.*, 68 (1956) 492; R. V. Figini, E. Coloccia and H. J. Schumacher, *Z. Physik. Chem. N.F.*, 14 (1958) 32.
436 I. P. Fisher, *Trans. Faraday Soc.*, 64 (1968) 1852.
437 V. N. Belevskii and L. T. Bugaenko, *Russ. J. Inorg. Chem.*, 12 (1967) 1203.
438 C. F. Goodeve and F. D. Richardson, *J. Chem. Soc.*, (1937) 294.
439 M. H. Kalina and J. W. T. Spinks, *Can. J. Chem.*, B16 (1938) 381.
440 J. B. Levy, *J. Phys. Chem.*, 66 (1962) 1092.
441 I. P. Fisher, *Trans. Faraday Soc.*, 63 (1967) 684.
442 G. A. Heath and J. R. Majer, *Trans. Faraday Soc.*, 60 (1964) 1783.

Chapter 3

Decomposition of halides and derivatives

D. A. ARMSTRONG AND J. L. HOLMES

1. The hydrogen halides

1.1 THE PHOTOCHEMISTRY OF HYDROGEN IODIDE

1.1.1 Absorption spectrum

Early work by Rollefson and Booher[1] in 1931 and by Goodeve and Taylor[2] in 1936 established that gaseous hydrogen iodide displays an absorption continuum from 2000 to 4000 A with a maximum at about 2200 A. It was recognised that at wavelengths below 3100 A excited iodine atoms ($^2P_{\frac{1}{2}}$) would be produced and that the relatively weak absorption above 3100 A must be due to a transition to a repulsive state producing only ground-state atoms I($^2P_{\frac{3}{2}}$). In 1948 a detailed study was made by Romand[3], and his results are shown as curve 1 in Fig. 1(a). He analysed his experimental curve, postulating that it is made up from curves 2 and 3 corresponding to the two transitions, $Q_0 \leftarrow N$ and $^3Q_1 \leftarrow N$ respectively. These processes correspond to the $N^1\Sigma_+ \rightarrow {}^3\Pi_{0+}$ and $^1\Sigma_+ \rightarrow {}^3\Pi_1$ transitions illustrated in Fig. 1(b) from the work of Mulliken[4]. Thus, curve 2 in Fig. 1(a) signifies production of an excited iodine atom ($^2P_{\frac{1}{2}}$) and curve 3, production of a ground-

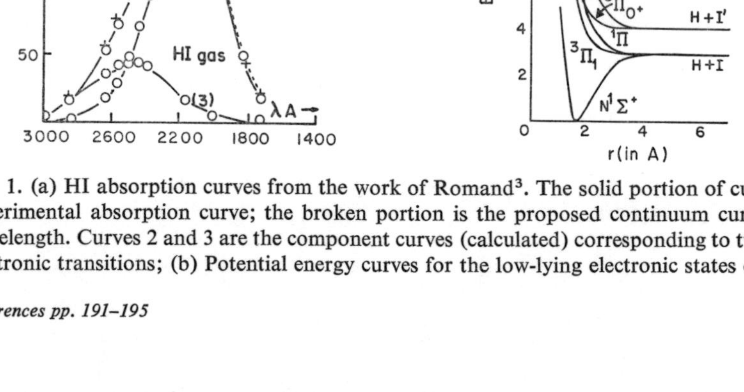

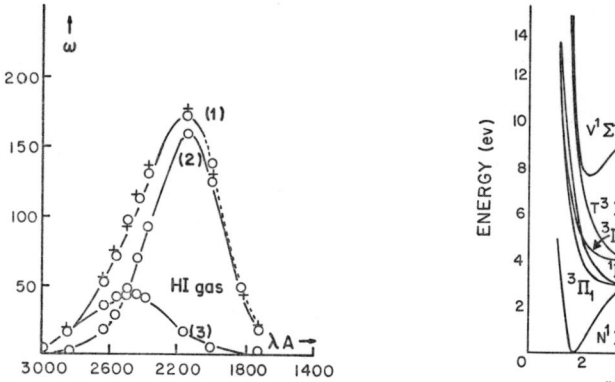

Fig. 1. (a) HI absorption curves from the work of Romand[3]. The solid portion of curve 1 is the experimental absorption curve; the broken portion is the proposed continuum curve at lower wavelength. Curves 2 and 3 are the component curves (calculated) corresponding to two different electronic transitions; (b) Potential energy curves for the low-lying electronic states of HI[4].

state ($^2P_{\frac{3}{2}}$) iodine atom. Since the potential energy curve for the $^1\Pi$ state lies above those of the $^3\Pi_{0+}$ and $^3\Pi_1$ on the vertical from the minimum in the ground state, then according to the Frank–Condon principle, absorption curve maxima should appear in the order $^1\Pi$, $^3\Pi_{0+}$, $^3\Pi_1$ with increasing wavelength. Production of ground-state iodine atoms should therefore be important at the extremes of the continuum while I($^2P_{\frac{1}{2}}$) production is important at the centre. Thus, for example, both 3130 A and 2537 A radiation will produce excited and ground-state iodine atoms while 3660 A will yield only ground-state atoms.

1.1.2 Photolysis of hydrogen iodide

(a) Energetics

The proportions of ground-state and ($^2P_{\frac{1}{2}}$) excited iodine atoms produced in a photolysis using monochromatic radiation can be approximately calculated from Fig. 1. To conserve momentum, essentially all the energy from the primary process in excess of that used in bond dissociation ΔH^0_{298} (HI $\rightarrow$ H+I) = (71.3 kcal.mole^{-1}) and electronic excitation, must appear as translational energy of the H atom. Thus, for example, the 2537 A photolysis can be represented by the two reactions

$$\text{HI}+h\nu(2537\text{ A}) \rightarrow \text{H}+\text{I}(^2P_{\frac{1}{2}})+19.6\text{ kcal}$$
$$\rightarrow \text{H}+\text{I}(^2P_{\frac{3}{2}})+41.4\text{ kcal}$$

and each process will be of roughly equal importance.

(b) Gas-phase photolysis

Early work (before 1930) by Warburg[5], Bodenstein and Lieneweg[6], Lewis[7] and Bonhoeffer and Farkas[8] showed that the quantum yield for the HI decomposition was close to 2 in the wavelength range 2000–2800 A and from 20 to 175 °C. This work was extended by Rollefson and Booher[1] in 1931 to 3660 A where the quantum yield remained close to two. It was therefore concluded that the mechanism of the photolysis was essentially the same for all effective wavelengths, even though the electronic state of the iodine atoms was wavelength dependent. An important observation by Bonhoeffer and Farkas[8] was that the quantum yield fell with increasing percentage decomposition and this effect was ascribed to the inhibition reaction (3) in the sequence

$$\text{HI}+h\nu \xrightarrow{I_a} \text{H}+\text{I}, \phi = 1 \quad (1)$$

$$\text{H}+\text{HI} \rightarrow \text{H}_2+\text{I} \quad (2)$$

$$\text{H}+\text{I}_2 \rightarrow \text{HI}+\text{I} \quad (3)$$

$$\text{I}+\text{I}+\text{M} \rightarrow \text{I}_2+\text{M} \quad (4)$$

A "steady-state" treatment of the kinetic scheme yields the differential equation

$$d[_2]dt = d[H_2]dt = I_{abs}/(1+k_3[I_2]/k_2[HI])$$

which on integration, and assuming k_3/k_2 constant, becomes

$$I_{abs}\times t-[I_2]_f = \frac{k_3}{k_2}\left\{\frac{[HI]_i}{4}\ln\frac{[HI]_i}{[HI]_i-2[I_2]_f}-\frac{[I_2]_f}{2}\right\}$$

where $[I_2]_f$ is the concentration of iodine at time t, $[HI]_i$ is the initial concentration of hydrogen iodide and $I_{abs}\times t$ is the total quantity of light absorbed, measured actinometrically.

Bonhoeffer and Farkas estimated $k_3/k_2 \sim 100$ and claimed that at 15 % decomposition the photolysis is completely self inhibited. More recent work by Ogg and Williams[9, 10] showed that for the photolysis with 2537 A radiation, k_3/k_2 is independent of HI pressure (50–150 torr), independent of temperature and has a value 3.5 ± 0.3. The effect of cyclohexane as an inert diluent[11] was to increase k_3/k_2 to 7.0 ± 0.4 at 155°, which value remained constant at high cyclohexane: hydrogen iodide ratios. This result was attributed to collisional thermalisation of the "hot" H atoms produced by 2537 A radiation and this limiting "high-pressure" value of $k_3/k_2 = (k_3/k_2)_\infty$ was considered to be that for thermally equilibrated H atoms.

These experiments were extended by Hamill *et al.*[12], who used helium, argon and hydrogen as inert diluents. Addition of these gases also produced an increase in k_3/k_2, which approached a limiting value, $(k_3/k_2)_\infty$, at high inert gas pressures.

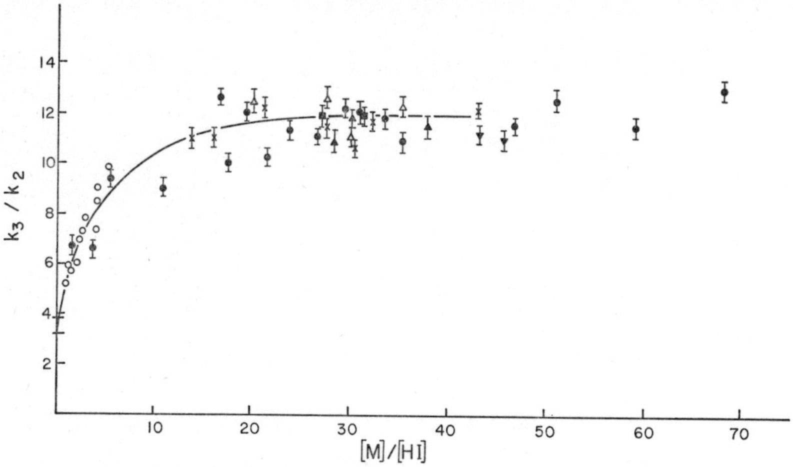

Fig. 2. HI photolyses in the presence of inert gases. Effect of inert gas on k_3/k_2. ▼, X, ●, ▲ Helium as inert gas, temperatures 53, 69, 102 and 200 °C respectively; △ argon, 79 °C; ■ hydrogen, 170 °C[15]; ○ from Hamill *et al.*[12], helium at 114 °C.

$(k_3/k_2)_\infty$ was independent of the inert gas and was temperature-dependent. At a given inert gas pressure k_3/k_2 was independent of $[I_2]/[HI]$. These results were explained simply by the "hot-atom" effect. Values of $(k_3/k_2)_\infty$ were obtained by extrapolation. A plot of log $(k_3/k_2)_\infty$ *vs.* $1/T$ for the temperature range 114–198° gave $E_3 - E_2 = -4.5$ kcal.mole^{-1}. Not only is this large value for E_2 difficult to understand, but Hamill's results are in poor agreement with those of Sullivan[13, 14], who has evaluated $(k_3/k_2)_\infty$ for the temperature range 360–530°, from careful studies of the thermal HI synthesis: in this work hot-atom effects were absent. His results yield $\log k_2 = (12.05 \pm 0.07) - (0 \pm 250)/4.58\,T$; $E_3 = 0$ and $(k_3/k_2)_\infty = 14 \pm 4$ at 394°, 12 ± 2 at 527°. (Hamill's results extrapolated to these temperatures give $(k_3/k_2)_\infty = 1.2$ and 0.66 respectively.) The photolytic experiments have recently been repeated by Holmes and Rodgers[15], also using He, Ar and H_2 as inert gases, in the temperature range 50–200°. These results are consistent both with the low value for $E_3 - E_2$ and with the experimental data of Hamill. Figure 2 shows Holmes' results and Hamill's; it can easily be seen how the latter's extrapolation to high inert gas pressures could be in error. Holmes' value for $E_3 - E_2$ is 0 ± 350 cal.mole^{-1}. Table 1 lists all the reported values for $(k_3/k_2)_\infty$. Within experimental error there is no temperature dependence in the range 53–527 °C and $(k_3/k_2)_\infty = 11 \pm 3$.

TABLE 1

PHOTOLYSIS OF HYDROGEN IODIDE

Values of $(k_3/k_2)_\infty$ (*i.e.* k_3/k_2 for thermal H atoms)

$(k_3/k_2)_\infty$	T (°C)	*Ref.*
11.2	53	15
11.7	69	15
11.9	102	15
13.5	114	12
9.01	154	12
7.0	155	11
4.69	198	12
11.4	200	15
10.2	393.6	13
14±4	394	14
8.85	437.1	13
8.20	464.7	13
12±2	527	14

The vacuum ultraviolet photolysis of HI has been studied by Martin and Willard[16] using the 1849 A mercury lines as the exciting radiation*. They estimated the molar extinction coefficient at 1849 A to be between 110 and 150 in fair agreement with Romand's[3] experimental value of 125. The hydrogen atoms produced have sufficient energy to cause the reaction

$$H^* + HI \rightarrow H + H + I$$

* See footnote on p. 147.

which would increase the overall quantum yield for hydrogen to greater than unity; (H^* will be 84 or 62 kcal.mole^{-1} "hot" depending on the electronic state of the iodine atom). However, their results at $-78°$ and $30°$ showed $\phi_{H_2} = 1.04 \pm 0.04$ and 1.07 ± 0.04 respectively.

The flash photolysis of HI has been investigated by Donovan and Husain[17] but with the purpose of studying energy transfer in the decay of $^2P_{\frac{1}{2}}$ iodine atoms to their ground state. An additional process which they observed was

$$H+HI \rightarrow H_2+I(^2P_{\frac{1}{2}})$$

the exothermicity being large enough to yield the electronically excited I atom. The same effect is observed with HBr[18].

1.2 THERMAL DECOMPOSITION OF HYDROGEN IODIDE

It is not possible to discuss the thermal decomposition of HI without at the same time discussing the HI synthesis. Until very recently it was common, particularly in Physical Chemistry textbooks, to find the reactions

$$H_2+I_2 = 2\,HI \tag{5}$$

$$2\,HI = H_2+I_2 \tag{6}$$

used as good examples of simple bimolecular processes involving no mechanistic subtleties. It is probably because of the above assumption that these are among the most thoroughly investigated gas-phase thermal reactions[19]. That (5) and (6) might not represent the *mechanism* of these reactions had been realised in 1955 by Benson and Srinivasan[20], who proposed that the participation of a simple chain reaction might be important in the synthesis above 325 °C. This proposal was thoroughly investigated in the experiments of Sullivan, who between 1959 and 1963 performed very careful studies of the HI synthesis. He concluded that the following elementary reactions were of importance in the temperature range 360–527°

$$H_2+I_2 \rightleftarrows 2\,HI \qquad (5)(6)$$

$$I+H_2 \rightleftarrows HI+H \qquad (7)(2)$$

$$H+I_2 \rightleftarrows HI+I \qquad (3)(8)$$

$$M+I_2 \rightleftarrows 2\,I+M \qquad (9)(4)$$

---

* The maximum for the $^1\Sigma_+ \rightarrow {}^1\Pi$ transition appears to lie well below 2000 *A*[16], but this transition probably contributes to the absorption at 1849A together with $^1\Sigma_+ \rightarrow {}^3\Pi_{0+}$

TABLE 2

ARRHENIUS PARAMETERS FOR THE H_2–I_2–HI SYSTEM

Reaction	*$\log_{10} A$ (l.mole^{-1}.sec^{-1})*	*E (kcal.mole^{-1})*	*Ref.*
$H+HI \rightarrow H_2+I$(2)	10.70±0.07	0.75±0.25	14
$H+I_2 \rightarrow HI+I$(3)		0.0	14
(3)/(2)	$k_3/k_2 = 12\pm2$		14
$H_2+I_2 \rightarrow 2\,HI$(5)	11.41±0.07	41.42±0.25	13
$HI+HI \rightarrow H_2+I_2$(6)	10.60±0.07	44.39±0.25	13
$H_2+I \rightarrow HI+H$(7)	11.20±0.07	33.53±0.25	14
$HI+I \rightarrow I_2+H$(8)	11.37±0.2	36.48±0.3	13
$D_2+I_2 \rightarrow 2\,DI$(5 D)	10.64	41.51	21
$D_2+I \rightarrow DI+D$(7D)	11.05	34.49	21
(3D)/(2D)	$k_{3D}/k_{2D} = 14$		21

and he evaluated the Arrhenius parameters shown in Table 2.

For the early stages of the synthesis, where reaction (6) may be neglected the overall rate is given by the expression

$$\frac{d[HI]}{dt} = k_5[H_2][I_2]\left\{1+\frac{2\,k_7K_{9,4}^{\frac{1}{2}}}{k_5[I_2]^{\frac{1}{2}}}\right\}$$

From this equation the relative importance of the molecular and free radical processes can be calculated. At 575 °C the contribution of (5) to the overall reaction is small. For the thermal decomposition (6) alone will be important in the early stages of the reaction.

In 1958 Semenov[22] pointed out that an alternative mechanism existed for reaction (5) which would give rise to the same kinetic order as found experimentally, namely

$$M+I_2 \rightleftarrows 2\,I+M \qquad (9)(4)$$

$$H_2+2\,I \rightleftarrows 2\,HI \qquad (5a)(5b)$$

It follows that

$$\frac{d[HI]}{dt} = 2\,k_{5a}[I]^2[H_2]$$
$$= 2\,k_{5a}K_{9,4}[I_2][H_2]$$

Since ΔH^0_{298} $[I_2(g) \rightarrow 2\,I(g)] = 36.2$ kcal.mole^{-1}, then the apparent activation energy, $E_5 = 41.4$ kcal.mole^{-1} is sufficient to allow dissociation of iodine molecules with 5.2 kcal.mole^{-1} left for an activation energy E_{5a}.

This possibility was explored by Sullivan[23] in 1967. His experimental test was photochemically to produce a known quantity of iodine atoms in excess of their

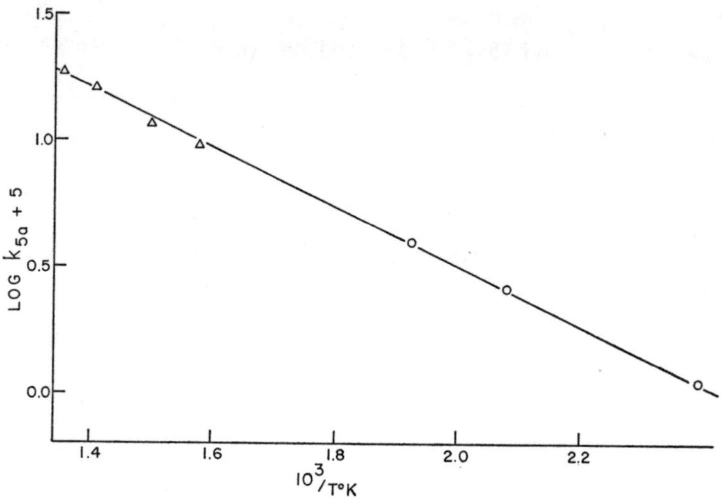

Fig. 3. Arrhenius plot for reaction (5a). ○ From photochemical experiments[23]; △ based on thermal reaction data.

thermal equilibrium concentration and observe any effect on the overall kinetics at low temperature where the chain mode of the synthesis is of small importance. His Arrhenius plot for the reaction is shown in Fig. 3 and his results indicate that the bimolecular reaction (5) does not occur. The rate of the third-order reaction was given by $k_{5a} = (6.66 \pm 0.5) \times 10^7 \exp[-(-5310 \pm 85)/RT]$ $l^2.mole^{-2}.sec^{-1}$. It follows that the thermal decomposition of hydrogen iodide must occur by reaction (5b). The details of reaction (5a) are still uncertain but it is probably made up of two bimolecular steps, *viz.*

$$I + H_2 \rightleftarrows H_2I \qquad (10)(11)$$

$$H_2I + I \rightarrow 2\,HI \qquad (12)$$

which, provided (12) is rate determining, will yield the observed order of three. At very high iodine atom concentrations, such as might be produced in a flash photolysis experiment, the synthesis reaction would be second order if (10) became rate determining. This effect would be observed provided that the intermediate H_2I is a sufficiently long-lived species. If H_2I has a finite existence, then hydrogen should be a good third body in the recombination of iodine atoms; Engleman and Davidson[24] showed that hydrogen was an appreciably better third body than helium, but was much less efficient than benzene. The latter is believed to act as a chaperone *via* a charge transfer complex. More experimental work is required to clarify the detailed mechanism of reaction (5a).

Noyes[25] has proposed that application of absolute-reaction-rate theory to the elementary reactions (5) and (5a) yields the paradox that Sullivan's most recent observations[23] could have been predicted *regardless* of the relative importance of mechanisms (5) and (5a). The paradox arises from the assumption that the transition states for both (5) and (5a) are one and the same. While symmetrical activated complexes of the type

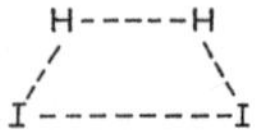

could arise from both H_2+I_2 and $H_2+2\,I$, the latter reactants may form an activated complex of lower energy which could not be formed from interaction of the two molecules, *e.g.*

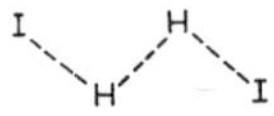

This has been discussed recently by Laidler and Murrell[26].

1.3 THE PHOTOLYSIS OF HYDROGEN BROMIDE

Less work has been done on this compound than on hydrogen iodide.

The ultraviolet absorption spectrum of HBr was examined by Goodeve and Taylor[27], Price[28] and by Romand[3]. The absorption continuum begins at 2850 A, rises to a maximum at 1800 A and becomes banded in the vacuum ultraviolet[27]. The primary process is believed to involve the same electronic states as for HI with production of ground-state ($^2P_{\frac{3}{2}}$) bromine atoms predominating at the ends of the continuum and a contribution from excited ($^2P_{\frac{1}{2}}$) bromine atoms near the middle of the band. Martin and Willard[29] used the 1849 A (mercury line) photolyses of HBr and DBr as a source of hot hydrogen and deuterium atoms to study their reactions with D_2, CD_4, C_2D_6 and H_2, CH_4, C_2H_6 respectively. They determined ε (1849 A) = 470, in good agreement with Romand[3], and from the work of Mulliken[30] concluded that the primary process should be represented by the reaction

$$HBr \xrightarrow{1849\text{ A}} H^* + 0.98\ Br(^2P_{\frac{3}{2}}) + 0.02\ Br(^2P_{\frac{1}{2}})$$

Donovan and Husain[18] used flash photolysis to produce $Br(^2P_{\frac{1}{2}})$ atoms in order to study their electronic relaxation in the presence of various added gases. Their experiments were complicated by the participation of

$$H^* + HBr \rightarrow H_2 + Br(^2P_{\frac{1}{2}})$$

as an additional source of excited bromine atoms. Hamill *et al.*[12] in their hot hydrogen atom experiments (described earlier, p. 146) were unable successfully to measure $(k_{14}/k_{13})_\infty$ for thermal hydrogen atoms.

$$H+HBr \rightarrow H_2+Br \tag{13}$$

$$H+Br_2 \rightarrow HBr+Br \tag{14}$$

The 2537 A mercury line was used to photolyse HBr in the presence of hydrogen as "inert gas". They quoted an approximate value of 9.5 ± 1 at 30° in fair agreement with Bodenstein and Jung's[31] earlier result of 8.6 evaluated in a study of the photobromination of hydrogen. For hot hydrogen atoms produced by the 2537 A photolysis in the absence of moderator they obtained an average value $k_{14}/k_{13} = 0.66 \pm 0.1$. That k_{13} should be greater than k_{14} is a surprising result and this system seems worthy of further investigation.

Britton and Cole[32], in a shock-tube investigation of the H_2–Br_2 reaction evaluated $(k_{14}/k_{13})_\infty = 8.3 \pm 0.7$ and 10.1 ± 1.7, with no temperature coefficient over the temperature range 1000–1400 °C. They also concluded that $E_{13} \approx E_{14} \approx 2$ kcal.mole^{-1}.

1.4 THERMAL DECOMPOSITION OF HYDROGEN BROMIDE

This reaction and the synthesis of HBr have also received much less attention than the corresponding reactions of the HI system. The problem of the mechanism of the H_2+Br_2 reaction which Bodenstein and Lind[33] found to be complex was later solved independently by Christiansen[34], Herzfeld[35] and Polanyi[36]. The well-known mechanism and the kinetic equation resulting from the stationary-state solution are given below

$$Br_2+M \rightleftarrows 2\,Br+M \tag{15)(16}$$

$$Br+H_2 \rightleftarrows HBr+H \tag{17)(13}$$

$$H+Br_2 \rightarrow HBr+Br \tag{14}$$

$$\frac{d[HBr]}{dt} = \frac{2\,k_{13}K_{15,16}^{\frac{1}{2}}[H_2][Br_2]^{\frac{1}{2}}}{1+k_{13}[HBr]/k_{14}[Br_2]}$$

No evidence has been found for the participation of the molecular reactions

$$H_2+Br_2 \rightleftarrows 2\,HBr$$

Arrhenius parameters for reactions in the hydrogen–bromine system are shown in Table 3.

The results shown in Table 3 have been discussed in some detail by Fettis and Knox[42]. They consider that the early work[33, 38, 39, 40] yields rather low values

TABLE 3

ARRHENIUS PARAMETERS FOR THE H_2–Br_2–HBr SYSTEM

Reaction	*log A (l.mole⁻¹.sec⁻¹)*		*E (kcal.mole⁻¹)*	*Ref.*
$Br+H_2 \rightarrow HBr+H$ (17)	11.25	T*	19.2	32
	11.2[a]	T	18.5[a]	33
	11.0	T	18.0	37
	10.8	P**	17.6	38
	10.8[a]	T	17.7[a]	39
	10.6	P	17.5	40
$Br+D_2 \rightarrow DBr+D$ (18)	11.08	T	20.4	32
	10.9	T	19.9	39

* T = Thermal reaction

** P = Photochemical reaction.

[a] Recalculated by Pease[41].

for A_{17} and E_{17} and favour the more recent results of Levy[37] and Britton and Cole[32]. Their "best" values

$$\log k_{17} = (11.43 \pm 0.14) - (19{,}700 \pm 380)/4.58\,T$$

$$\log k_{18} = (11.29 \pm 0.14) - (21{,}400 \pm 380)/4.5\,T$$

were obtained by combining all the data and imposing the conditions

(a) that $E_{18} - E_{17}$ must equal the difference in zero point energies for D_2 and H_2 (= 1.77 kcal.mole^{-1}) and

(b) that the ratio of the frequency factors $A_{17}/A_{18} = 1.40$.

1.5 HYDROGEN CHLORIDE

The photochemical and thermal decompositions of this compound have received very little attention. Most kinetic data have come from studies of the synthesis and chlorine and hydrogen atom reactions.

The hydrogen–chlorine chain reaction has proved to be one of the most controversial systems yet studied. After thirty years of investigation Bodenstein[43] was able to say in 1931 that every worker on the photochemical synthesis of HCl had produced his own mechanism; even as late as 1940 little positive information had been obtained. However, the accumulated techniques and experience had firmly established the importance of atom chain reactions. The mechanism of photo-initiation and propagation is the same as for the hydrogen bromide photosynthesis, a non-branching chain reaction

$$h\nu + Cl_2 \rightarrow 2\,Cl \qquad (20)$$

$$Cl + H_2 \rightleftarrows HCl + H \qquad (21)(22)$$

$$H + Cl_2 \rightleftarrows HCl + Cl \qquad (23)(24)$$

The complications which arose in the early photochemical work were due to the presence of impurities in the reactants, notably oxygen, NCl_3 and water which aided chain initiation or termination. In thermal reactions wall effects were in evidence.

Ashmore and Chanmugam[44] were the first workers to determine accurate Arrhenius parameters for the propagation reaction (21) in their study of the effects of NO and NOCl as sensitiser of the H_2–Cl_2 chain reaction. They found k_{21} (250°) $= 4.8 \pm 0.4 \times 10^8$ l.mole^{-1}.sec^{-1} and $E_{21} = 5.50$ kcal.mole^{-1}.

Steiner and Rideal[45] determined the Arrhenius parameters for (22), $\log k_{22} = 10.86 - 5200/4.58\,T$, from a study of the HCl-catalysed ortho–para hydrogen reaction in the temperature range 600–770°. When allowance is made for the better data for the hydrogen dissociation available to Steiner and Rideal, their results agree well with earlier work by Rodebush and Klingelhoefer[46].

The results from the last three references cover a wide temperature range, 0–800 °C and the values for k_{21} give a good straight Arrhenius plot. A least squares treatment yields[42] $\log k_{21} = (10.92 \pm 0.3) - (5{,}480 \pm 150)/4.58\,T$. However, Clyne and Stedman[47] have discredited the linear dependence of $\log k_{21}$ on $1/T$. These authors studied the kinetics of reaction (22) using a low-pressure discharge-flow system from −80 to 100°. They determined $k_{22} = (3.5 \pm 1.5) \times 10^8\,T^{\frac{1}{2}} \exp[(-2900 \pm 300)/RT]$ l.mole^{-1}.sec^{-1} for their temperature range. Combining their results with the earlier work[44,45,46] they showed that all data for $\log k_{22}$ lay on a smooth curve when plotted against I/T and could be reasonably represented by the simple collision-theory equation

$$k_{22} = (6.2 \pm 3.5) \times 10^8\,T^{\frac{1}{2}} \exp[(-3100 \pm 400)/RT]\ \text{l.mole}^{-1}.\text{sec}^{-1}$$

They also discussed the reaction in terms of model activated complexes for a transition state theory treatment. Jones[48] in 1951 determined the relative rates of reactions (21) and (25).

$$Cl + HT \rightarrow (HCl + TCl) + (H + T) \qquad (25)$$

He found $k_{21}/k_{25} = (1.35 \pm 0.03) \exp\{(552 \pm 7)/RT\}$. Bigeleisen *et al.*[49] evaluated $k_{21/26} = (1.24 \pm 0.03) \exp\{(490 \pm 6)/RT\}$.

$$Cl + HD \rightarrow (HCl + DCl) + (H + D) \qquad (26)$$

Both values were obtained by photochlorination of H_2, HT and HD. Bigeleisen *et al.* also showed that $k_{22}/k_{23} < 0.05$ at 298° by searching for HD in the photochlorination of H_2–D_2 mixtures where half the "hydrogen" remained unreacted; HD was undetected (< 0.02 %) in the residual hydrogen.

The rate coefficient ratio k_{22}/k_{23} has been determined by Klein and Wolfsberg[50] from the photochlorination of hydrogen in the presence of tritium-labelled HCl. For the temperature range 0–62°

$$k_{22}/k_{23} = (0.143 \pm 0.033) \exp\{(-1540 \pm 130)/RT\}$$

which yields $k_{22}/k_{23} = 0.012$ at 298°, consistent with Bigeleisen's limit.

The homogeneous thermal decomposition of HCl has only been studied in shock tubes. Fishburne[51] investigated the shock pyrolysis of HCl diluted with Ar in the temperature region 3300–5400 °K and obtained Arrhenius parameters for

$$\text{Ar} + \text{HCl} \rightarrow \text{H} + \text{Cl} + \text{Ar} \tag{27}$$

$$k_{27} = 1.92 \times 10^8\, T^{\frac{1}{2}} \exp(-69{,}700/RT)\ \text{l.mole}^{-1}.\text{sec}^{-1}$$

The reason for the low activation energy was and remains not understood. Jacobs *et al.*[52] found a similar effect in their shock-tube experiments at 2800–4600 °K using the same diluent as Fishburne. For HCl and DCl they obtained

$$k_{27}(\text{HCl}) = 6.8 \times 10^{18}\, T^{-2} \exp(-102{,}170/RT)\ \text{l.mole}^{-1}.\text{sec}^{-1}$$

but in strict Arrhenius equation form their results gave a low activation energy for

$$\text{HCl} \rightarrow \text{H} + \text{Cl} \tag{28}$$

viz. $k_{28} = 6.6 \times 10^{12} \exp(-70{,}000/RT)\ \text{sec}^{-1}$.

1.6 HYDROGEN FLUORIDE

The kinetics of neither the photochemical nor the thermal decomposition of this compound have received much attention. Bodenstein *et al.*[53] in 1937 showed that the hydrogen and fluorine reaction could not be photosensitised by chlorine at room temperature.

The kinetics of the thermal hydrogen–fluorine reaction were studied at 110° by Levy and Copeland[54] using a magnesium flow reactor. The reaction was found to be first order in F_2 but the rate was independent of both H_2 concentration and surface/volume ratio of the reaction zone. They concluded that the reaction was initiated and terminated at the walls. The same authors[55] investigated the inhibition of the reaction by oxygen in the same apparatus over the temperature range 122–162°. The kinetics were complex but they nevertheless proposed an activation energy of 5–7 kcal.mole^{-1} for the reaction

$$F+H_2 \rightarrow HF+H^{\cdot}$$

In view of the kinetic complexities and the small temperature range, this result should be regarded with caution.

Jacobs *et al.*[56] and Blauer[57] have studied the thermal dissociation of HF behind incident shock waves in argon. Both workers used an infrared emission technique to follow the dissociation reaction. Kinetic parameters for the reactions

$$HF+M \rightarrow H^{\cdot}+F+M \quad (29)$$

$$HF+H^{\cdot} \rightarrow H_2+F \quad (30)$$

were evaluated by both authors as $k_{29} = 1.13\times10^{16}\,T^{-1}\exp(-134{,}100/RT)$ l.mole^{-1}.sec^{-1} (temperature range 3500–5000°)[56]
$k_{29} = 0.47\times10^{16}\,T^{-1}\exp(-134{,}100/RT)$ l.mole^{-1}.sec^{-1} (temperature range 3400–5800°)[57]
$k_{30} = 2\times10^{9}\exp(-35{,}000/RT)$ l.mole^{-1}.sec^{-1} (Jacobs *et al.*[56])
$k_{30} = 1\times10^{10}\exp(-35{,}000/RT)$ l.mole^{-1}.sec^{-1} (Blauer[57])

Both authors found that the data for reaction (29) better fitted the above expressions rather than the conventional Arrhenius equation. Blauer examined the effect of added F_2 and concluded that even in the presence of large amounts of fluorine, the reaction

$$HF+F \rightarrow H+F_2$$

made a negligible kinetic contribution.

1.7 THE RADIOLYSIS OF HYDROGEN HALIDES

The authors were unable to find any published reports on the radiolysis of hydrogen fluoride. However, there is much work on hydrogen chloride and significant studies of hydrogen bromide and hydrogen iodide have been reported. Since the majority of the reaction types which occur in HCl are also known to occur in HBr and HI, it will be convenient to discuss the gas-phase radiolysis of all three systems in the same section, with a division under the sub-headings: (i) Primary processes, reactions of positive ions and radiolytic yields; (ii) Hydrogen formation by electrons, negative ions and hydrogen atoms; and (iii) Combination reactions of ions and halogen atoms. The radiolysis of liquid HCl is considered in subsection (iv). Main reactions are summarised at the beginning of each subsection, and intermediates which initially possess excess kinetic energy have been underlined to distinguish them from the corresponding thermal species. The subscript (c) next to an ion indicates that it is clustered.

1.7.1 Primary processes, reactions of positive ions and radiolytic yields

$$HX \rightsquigarrow HX^* \rightarrow \underline{H} + X \tag{a}$$

$$HX \rightsquigarrow HX^+ + \underline{e} \tag{b}$$

$$HX + HX^+ \rightarrow H_2X^+ + X \tag{c}$$

$$H_2X^+ + n\,HX \rightarrow H_2X^+(c) \tag{d}$$

(a) Excited molecules

The excitation of HCl molecules to states lying below the ionisation potential is shown in the electron-impact spectrum in Fig. 4. This spectrum was obtained by Compton *et al.*[58], using their threshold excitation technique. The initial energies of the impacting electrons correspond exactly to the transition energies given on the abscissa, and lie in the range 5–13 eV. Earlier work by Frances[59] showed that similar transitions were induced by interactions with 400 eV electrons. It is evident from these electron impact spectra that the cross-sections for excitation are comparable to those for ionisation. Therefore many electronically excited molecules will be formed in gaseous HCl along the tracks of primary electrons and δ-rays, as well as by electrons of lower energy in the spurs and clusters.

The broad peak with a maximum at 7.5 eV in Fig. 4 corresponds to excitation to the $^1\Pi$ state, and probably also the $^3\Pi_1$ and $^3\Pi_{0+}$ states. These are analogous to the lower states of HI (Fig. 1), and will decompose to halogen atoms in the

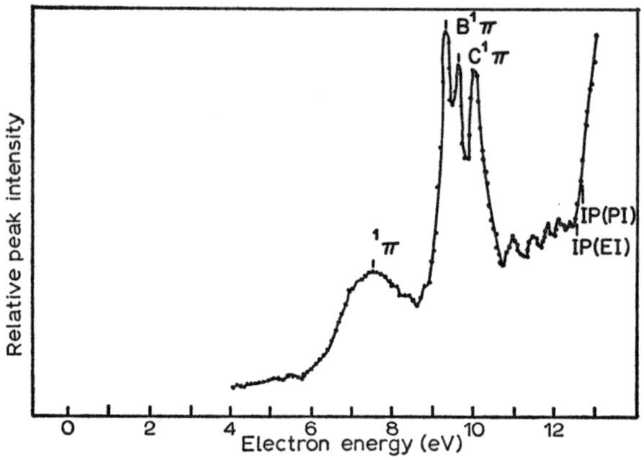

Fig. 4. Electron-impact threshold excitation spectrum of HCl. (Reproduced from ref. 58 by courtesy of *J. Chem. Phys.*)

$^2P_{\frac{3}{2}}$ or $^2P_{\frac{1}{2}}$ state and ground-state hydrogen atoms with excess kinetic energy. The well-defined peaks in the 9.3–10.5 eV energy range correlate well with transitions to the $B^1\Pi$ and $C^1\Pi$ states, which were observed spectroscopically by Price[28]. Photo-chemical studies of these states are at present lacking. However, detailed spectroscopic studies by Barrow and Stamper[60] have shown that they are strongly predissociated in the case of HBr. From an inspection of the energy-level diagrams for HCl[61] and HI, it seems quite likely that the $B^1\Pi$ and $C^1\Pi$ states in these two hydrogen halides will also be predissociated. Thus the dissociation of neutral excited states will play a major part in the radiolytic decomposition of hydrogen halides.

(b) Primary molecule ions and fragment ions

HX^+ ions are the predominant positive ions in the mass spectra of the hydrogen halides. The $X^2\Pi_i$ ground states and $A^2\Sigma^+$ excited states of these ions are well characterised[62, 63], and, with the exception of the $A^2\Sigma^+$ state of HI^+, are stable to dissociation. Ionisation efficiency curves indicate that both states are formed in interactions with fast electrons. In the case of HI the $^2\Pi_{\frac{3}{2}}$ and $^2\Pi_{\frac{1}{2}}$ spin orbital

TABLE 4

ELECTRON IMPACT IONISATION PROCESSES IN HCl

Ion	*Relative abundance at 140 eV (%)*	*Appearance potentials*[g] *(eV)*	*Probable ionisation process*	
HCl^+	81 ± 2	12.8[a], 12.9[b] 12.78[c], 12.56[d]	$HCl \rightarrow HCl^+ + \underline{e}$	(i)
		(14.2[d])†	$HCl \rightarrow HCl^+(A^2\Sigma^+) + \underline{e}$	(ii)
H^+, Cl^-	0.1–1	14.5[e]	$HCl \rightarrow \underline{H}^+ + Cl^-$	(iii)
Cl^+	12 ± 2	17.6[a], 17.2[b]	$HCl \rightarrow \underline{H} + Cl^+ + \underline{e}$	(iv)
		20.8[a], 21.3[b] 22.6[a]	$HCl \rightarrow \underline{H} + Cl^+ + \underline{e}$	(v)
H^+	3 (approx)	18.4[a], 18.6[b]	$HCl \rightarrow \underline{H}^+ + Cl + \underline{e}$	(vi)
		28.2[a], 28.4[b] 29.5[c]	$HCl \rightarrow \underline{H}^+ + Cl^* + \underline{e}$	(vii)
HCl^{2+}	0.5–1.5	36.0[a], 35.7[b] 35.5[f]	$HCl \rightarrow HCl^{2+} + 2\,\underline{e}$	(viii)
Cl^{2+}	2.5–4.0	45.7[b]		

[a] Thorburn[67].
[b] Nier and Hansen[65].
[c] Morrison and Nicholson[66].
[d] Fox[68].
† does not agree with spectroscopic data[62, 63] which require 16.2.
[e] Fox[70].
[f] Dorman and Morrison[73].
* Probably in the 4 ^4P or ^2P excited state.
[g] All values subject to an uncertainty of about ±0.2 eV. For exact quoted error limits see the original publications.

components of the ground state have also been resolved[64]. The 70 eV mass spectra show that H^+, X^+, HX^{2+} and X^{2+} ions are produced, their relative abundance increasing with the atomic number of the halogen.

The ionisation of HCl by electron impact has been investigated by Nier and Hanson[65], Morrison and Nicholson[66], Thorburn[67], Fox[68], and others. Their observations are summarised in Table 4. The ionisation processes in the last column are those postulated by Nier and Hanson[65], and Hanson[69]. Unless otherwise specified, the product ions and free radicals are taken to be in their ground electronic states. The percentage relative abundance given in the second column of Table 4 for ion-pair formation, process (iii), was estimated from Fox's results[70] by the method described by Armstrong *et al.*[71]. Obviously the formation of H^+ and Cl^- in this process is quite negligible, and electrons are the only primary negative ions which need be considered in the radiation chemistry of HCl. The percentage relative abundances of the predominant positive ions in Table 4 were calculated for an electron impact energy of 130–150 eV from the results of Nier and Hanson[65] and Johnson and Arnold[72]. The work of Hanson[69] has shown that the H^+ ions are initially hot. This means that their collection efficiency by the mass spectrometer is likely to be different from that of the heavier ions and their relative abundances in Table 4 are subject to considerable uncertainty. It is also evident that the states from which H^+ arises are too short-lived to permit bimolecular reactions and deactivation in gases at normal pressure. Similar considerations apply in the case of the electronic states involved in process (v), and also in process (iv) above its calculated threshold energy of 17.44 eV[63]. Mass spectral data for other hydrogen halides are incomplete. However, for an electron impact energy of 70–85 eV the ion abundances in HBr are 100 : 46 : 7 : 10 for $HBr^+ : Br^+ : HBr^{2+} : Br^{2+}$ respectively[72,74].

(*c*) *Ion–molecule reactions*

Several of the primary ionisation processes listed in Table 4 yield H and Cl atoms as well as ions, and similar processes are observed with HBr and HI. Atoms may also be formed in secondary ion–molecule reactions which are discussed below. As shown in Table 5, the ionisation potentials of the halogen atoms and their hydrides are such that the charge transfer reaction

$$X^+ + HX \rightarrow X + HX^+ \qquad \text{(e)}$$

is in each case exothermic; (e) is likely to be rapid, since the differences in the ionisation potentials are small. Some evidence for its occurrence in HCl has been reported[82]. The reactions of H^+, Cl^{2+} and HCl^{2+} are clearly of much less importance. However, the following points are of interest. Because the ionisation potential of hydrogen (13.5 eV[76]) exceeds those of HCl, HBr and HI, reaction (f)

$$H^+ + HX \rightarrow H + HX^+ \qquad \text{(f)}$$

TABLE 5

IONISATION DATA FOR HYDROGEN HALIDES

	X = Cl	X = Br	X = I
Ionisation potentials (eV)			
(a) $HX \rightarrow HX^+ + e$	$12.7_4 \pm 0.01$[a]	$11.6_7 \pm 0.0_5$[a]	$10.3_9 \pm 0.0_2$[a]
(b) $X \rightarrow X^+ + e$	12.9_7[b]	11.8_1[b]	10.4_5[b]
Second ionisation			
Potentials (eV)			
(a) $HX^+ \rightarrow HX^{2+} + e$	23.0 ± 0.5[c]	22.0 ± 0.5[c]	20.0 ± 0.5[c]
(b) $X^+ \rightarrow X^{2+} + e$	23.8_0[d]	19.2[d]	19.0_1[d]
Rate coefficient of the reaction ($l.mole^{-1}.sec^{-1}$)			
$HX^+ + HX \rightarrow H_2X^+ + X$	4×10^{10} [e]	1×10^{11} [f]	
Values of W (eV)			
(a) for fast electrons	24.8 ± 0.4[g]	24.4 ± 0.4[g]	
	25.3 ± 0.5[h]		
(b) for α-particles	27.0 ± 0.5[i]	27.0 ± 0.5[i]	

[a] Mean values computed from the spectroscopic data[62,63].
[b] From ref. 75; rounded off to second significant figure, error < ±0.01 eV.
[c] Calculated from the data of refs.[67,65,73].
[d] From ref. 76; the value for Br is approximate only.
[e] From ref. 77.
[f] From ref. 78.
[g] From ref. 79 for fast electrons generated by ^{60}Co γ-rays.
[h] From ref. 80 for fast electrons generated by ^{60}Co γ-rays.
[i] Calculated from the data of ref. 81 using $W_\alpha = 35.0$ eV for air.

may be expected in these systems. This process is favoured over the reaction

$$H^+ + HX \rightarrow H_2X^{+*} \tag{g}$$

since the H_2X^{+*} ions would require collisional stabilisation and because a large fraction of the H^+ fragment ions are initially "hot". Reactions of doubly charged HX^{2+} and X^{2+} ions cannot be predicted with certainty, but the second ionisation potentials of HCl and Cl (see Table 5) are such that the energy released in electron capture by either of these ions is about 23 eV. Processes (v) and (vi) were observed in the electron impact experiments with HCl at this energy, and thus the following reactions are probable

$$HCl^{2+} + HCl \rightarrow HCl^+ + H + Cl^+$$

$$HCl^{2+} + HCl \rightarrow HCl^+ + H^+ + Cl$$

$$Cl^{2+} + HCl \rightarrow Cl^+ + H + Cl^+$$

$$Cl^{2+} + HCl \rightarrow Cl^+ + H^+ + Cl$$

Similar reactions are likely to occur with the other hydrogen halides.

The above charge-transfer reactions should be complete within about 10^{-9} sec. Their net result is conversion of all fragment ions to HX^+ with the simultaneous formation of additional hydrogen and halogen atoms. If these are the only important reactions of the fragment and multiple charged ions, the overall stoichiometry of the ionisation process becomes

$$HX(1+x) \rightsquigarrow HX^+ + xX + xH + e \tag{h}$$

The mass spectral data of Nier and Hanson show that the value of x is insensitive to changes in electron beam energy over the range 25–400 eV, and that it lies between 0.17 and 0.19, depending on the fraction of charge-transfer reactions of HX^{2+} and X^{2+} which result in dissociation. In view of the uncertainty in the relative abundance of H^+, the value of x may be taken as 0.20 ± 0.02. Similarly, the value of x for HBr is 0.40 ± 0.04.

Reaction (c) has been shown to occur

$$HX^+ + HX \rightarrow H_2X^+ + X \tag{c}$$

for HCl and HBr and the rate coefficients are given in Table 5. Their magnitudes are such that the lifetimes of the HX^+ ions will be of the order of 10^{-10} sec at normal pressures. From studies of ion mobilities in gaseous HCl[83], and from experiments with other polar gases (*cf.* ref. 84) it is apparent that H_2X^+ ions and any other stable ions will be clustered. In liquid hydrogen halides the lifetime of the parent ions would be of the order of 10^{-13} sec, and solvation of the H_2X^+ ions would be expected to occur in about 10^{-11} sec.

In summary, the predominant primary processes may be adequately represented by reactions (a)–(d), and (h).

(*d*) *Radiolytic yields of products and primary species*

Since the radiolytic product yields are determined by the yields of primary species, it is convenient to discuss them together. Ion pair yields for the decomposition of gaseous hydrogen halides are summarised in Table 6. The early experiments with radon α-particles were conducted in accordance with the procedures of Mund and Lind[85]. These require complex corrections for energy lost in the cell walls. Gunther and Leichter[95] based their dosimetry for the determination of the ion pair yield of the X-radiolysis of gaseous hydrogen iodide on an involved calculation of the energy absorbed from the X-ray beam. Later studies[87–89] of the X- and γ-radiolysis of hydrogen bromide and hydrogen chloride have been performed in pyrex irradiation cells with graphite collecting electrodes. Ion pair yields were determined directly by dividing the rate of product formation by the saturation ionisation currents obtained under identical radiolysis conditions. The yields were unaffected by the graphite coating. The major difficulty lay in the attainment of satisfactory saturation ionisation currents. However, these can be achieved readily in HCl and HBr provided the dose rates are not excessively large,

TABLE 6

SUMMARY OF ION PAIR YIELDS FOR THE RADIOLYSIS OF GASEOUS HCl, HBr AND HI

	Initial pressure of HX (*torr*)	M_{-HX}/N	*Radiation type*	*Ref.*
HCl				
	320–705	4.10 ± 0.20^a	Radon α-particles	86
	204–619	3.92 ± 0.30	120 kvp X-rays	87
	614–641	3.90 ± 0.14	^{60}Co γ-rays	88
	268–772	4.12 ± 0.10	^{60}Co γ-rays	89
Overall weighed mean value of $M_{-HCl}/N = 4.01\pm0.14$ for X- and γ-radiolysis				
Overall weighed mean value of $M_{H_2}/N = 2.00\pm0.07$ for X- and γ-radiolysis				
Overall weighed mean value of $G(H_2) = 8.00\pm0.30$ for X- and γ-radiolysis				
HBr				
	760	4.3	Radon α-particles	90
	169–172	4.26^b	Radon α-particles	91
	223–406	4.7 ± 0.1	^{60}Co γ-rays	89
	203–304	4.8 ± 0.1	120 kvp X-rays	92
		$4.0–5.2^c$	48 kvp X-rays	93
Overall mean value of M_{-HBr}/N for X- and γ-radiolysis $= 4.75\pm0.10$				
Overall mean value of M_{H_2}/N for X- and γ-radiolysis $= 2.38\pm0.05$				
Overall mean value of $G(H_2)$ for X- and γ-radiolysis $= 9.80\pm0.20$				
HI				
	383–806	6.2 ± 0.6	Radon α-particles	94
	170–684	8.2 ± 1.0	Radon α-particles	86
	228–370	6.5 ± 0.6	X-ray	95

[a] This value was obtained after corrections for back reactions involving chlorine (see text).
[b] Calculated from the initial data of experiments 13 and 14 of ref. 19 on the basis of ionisation in HBr only. All values except b are the means of several determinations with initial pressures lying in the range indicated in column one.
[c] Range of values reported for the radiolysis sensitised by Ar, Kr, and Xe.

and cells of appropriate geometry are employed[92]. The ion pair yields obtained by the ionisation chamber method should therefore be the most reliable.

Product yields in irradiated hydrogen halides are subject to the effects of back reactions, particularly those involving the halogen molecules. These effects are most serious in the case of HCl, where a 20 % reduction in yield is observed at chlorine concentrations as low as 0.30 mole %[86, 88]. In the more recent studies of the X- and γ-radiolysis of HCl and HBr percentage conversions were kept below 1×10^{-2} mole %. Under these conditions product yields were dose independent, showing that back reactions were negligible.

Lee *et al.*[88] used the results of their investigation of the effects of chlorine on the radiolysis of HCl to correct Vandame's α-particle data[86]. As shown in Table 6, his corrected yields agree reasonably well with those for the X- and γ-radiolysis. However, the true initial α-particle yields may still be slightly higher than those given in Table 6, since no allowance was made for the effects of reaction (21)

$$Cl+H_2 \rightarrow HCl+H \tag{21}$$

which is also dose-dependent and would occur at very large doses, such as those used by Vandame. A correction for this would require an exact knowledge of the rates of alternative reactions of chlorine atoms, *viz.*

$$Cl+\text{wall} \rightarrow \tfrac{1}{2}\, Cl_2 \tag{31a}$$

$$Cl+Cl+M \rightarrow Cl_2+M \tag{31b}$$

and these are not well known. With one or two exceptions the ion pair yields for the α-particle radiolysis of HBr were determined at percentage conversions corresponding to bromine concentrations greater than 2 mole %. Thus they are also susceptible to the effects of back reactions, and must be regarded as minimum yields.

In the room-temperature radiolysis of HI, back reactions can be neglected owing to the very low vapour pressure of I_2 (0.3 torr at 25°). The most reliable determination of (M_{-HI}/N) appears to be that obtained in Lind's laboratory by Brattain[94]. The results quoted in Table 6 are confined to determinations in which less than 30 % of the initial HI was decomposed. Vandame's yields are surprisingly high and Gunther and Leichter[95] suggest that this may be due to inadequate correction for thermal reactions.

From Table 6 we may note that the decomposition yields for HCl and HBr are, within experimental error, independent of pressure over a reasonably broad range. Also there is no significant difference between the yields for ^{60}Co γ- and 120 kVp X-rays.

Table 5 lists the values of *W*, the mean energy expended in creating an ion pair, for both α-particles and fast electrons in gaseous hydrogen halides. In the case of HCl the weighed mean of the two determinations for fast electrons is 25.0±0.5 eV. This corresponds to an ion yield ($G_{\text{ionisation}} = G_e$) of 4.00±0.07 per 100 eV, and it follows from this and the ion pair yield of hydrogen that $G(H_2)$ for the X- or γ-radiolysis is 8.0±0.3. Similarly, $G(H_2)$ for the X- or γ-radiolysis of HBr is 9.8±0.2 molecules per 100 eV.

For reasons which will become apparent in section 1.8.2, it is assumed that each electron–ion pair eventually produces one molecule of hydrogen. In an earlier section (p. 160) it was seen that in HCl an additional 0.20±0.2 hydrogen atoms per ion pair should come from dissociative ionisation and charge transfer processes. In the absence of hydrogen atom scavengers they will undergo the reaction

$$H+HCl \rightarrow H_2+Cl \tag{22}$$

Consequently the yield of hydrogen resulting from primary ionisation events will be close to 1.2 molecules per ion pair. The remainder of the total yield M_{H_2}/N=

2.0 is probably due to dissociations of neutral excited molecules, each of which yields a hydrogen atom and subsequently, by reaction (22) or (32),

$$\underline{H}+HCl \rightarrow H_2+Cl \tag{32}$$

a single molecule of hydrogen. Thus the observed yield requires the formation of about 0.8 neutral excited molecules per ion pair.

For HBr $M_{H_2}/N = 2.38$. From the value of x estimated in the preceding section it can be seen that about 0.4 hydrogen atoms per ion pair may arise from dissociative ionisation and charge transfer. From arguments similar to those used for HCl, the ion pair yield of dissociative excited molecules must be close to unity. The results of an investigation by Zubler *et al.*[93] of the inert gas-sensitised radiolysis of HBr, are particularly interesting. Their observation that M_{H_2}/N remains essentially unchanged at 2.3 ± 0.3 implies that the transfer of excitation energy from the inert gas atoms to HBr molecules is highly efficient.

The importance of neutral excited molecules in the radiolysis of HBr is contrary to the conclusions of Eyring *et al.*[96], who proposed the reactions

$$e+HBr \rightarrow H+Br^- \tag{33}$$

$$Br^-+HBr^+ \rightarrow 2\,Br+H \tag{34}$$

as the main sources of hydrogen atoms. Presently accepted thermochemical data show that reaction (33) is endothermic and cannot be of major importance (see section 1.7.2) below. Neither of the above reactions is compatible with current experimental work[92], and the Eyring mechanism requires extensive revision. The first serious suggestion that excited molecules were involved in hydrogen halide radiolyses was made by Brattain[94] in 1938. Her ion pair yield of 6.2 for the α-particle decomposition of HI corresponds to $M_{H_2}/N = 3.1$. It was attributed to the concomitant formation of two "activated" molecules per ion pair. This ratio is probably too high, since dissociative ionisation and charge transfer reactions would make a larger contribution in HI than in HBr. Following the trend observed for HCl and HBr, x for HI might have a value of about 0.6. The remainder of the total hydrogen yield of 3.1 molecules per ion pair could then be explained by the production of 1.5 dissociative excited molecules per ion pair, since HI has lower excitation potentials than HBr and transitions to the low-lying triplet states will become more likely with increase of atomic number. However, this represents an extremely efficient use of the radiation energy, and further investigations of the HI system should be made.

1.7.2 Hydrogen formation by electrons, negative ions and hydrogen atoms

$$\underline{e}+\mathrm{HX} \rightarrow \mathrm{H}+\mathrm{X}^- \qquad \text{(i)}$$

$$\underline{e}+\mathrm{HX} \rightarrow e+\mathrm{HX}\ \text{(with excess energy)} \qquad \text{(j)}$$

$$e+m\,\mathrm{HX} \rightarrow \mathrm{H}+\mathrm{X{-}H{-}X}^-_{(c)} \qquad \text{(k)}$$

$$e+m\,\mathrm{HX} \rightarrow (\mathrm{HX})^-_{(c)} \qquad \text{(l)}$$

$$\underline{\mathrm{H}}+\mathrm{HX} \rightarrow \mathrm{H}+\mathrm{HX}\ \text{(with excess energy)} \qquad \text{(m)}$$

$$\mathrm{H}+\mathrm{HX} \rightarrow \mathrm{H_2}+\mathrm{X} \qquad \text{(n)}$$

$$\mathrm{H}+\mathrm{HX} \rightarrow \mathrm{H_2}+\mathrm{X} \qquad \text{(o)}$$

$$\mathrm{H}+\mathrm{X_2} \rightarrow \mathrm{HX}+\mathrm{X} \qquad \text{(p)}$$

(*a*) *Resonance capture*

Electrons of initial energy greater than the lowest electronic transition are brought below this within a few collisions. Those with energy in the range 0.1–5 eV will be moderated to thermal energies (reaction (j)) within about 10^{-10} sec[97], primarily by exciting vibrational and rotational transitions. During thermalisation, a certain fraction, *0*, will be lost in resonance capture.

$$e+\mathrm{HX} \rightarrow \mathrm{H}+\mathrm{X}^- \qquad \text{(i)}$$

The available data on this reaction have been summarised by Christophorou *et al.*[98].

The value of θ is given by the equation

$$\ln(1-\theta) = \int_{E_i}^{E_{th}} \frac{\sigma_c(E)}{\sigma(E)\cdot[f(E)\cdot E]}\cdot \mathrm{d}E$$

which is based on expression (4) of Magee and Burton[99]. E_i and E_{th} are the initial and thermal electron energies; $\sigma_c(E)$ and $\sigma(E)$ are the capture and total collision cross-sections at energy E; and $f(E)$ is the average fractional energy loss per collision at this energy. The integrated capture cross-section (*i.e.* $\int_0^\infty \sigma_c(E).\mathrm{d}E$) for HCl is reported to be 7.4×10^{-18} cm^2.eV[98]. From this and the available data for $f(E)$ and $\sigma(E)$ it can be shown (*cf.* ref. 71) that θ should be about 0.08. Hence, reaction (i) is clearly of minor importance in HCl. There are no determinations of $\sigma(E)$ and $f(E)$ for HBr and HI. However, marked changes in these quantities are unlikely, and it is evident from the trend in the integrated cross-sections in Table 7 that the importance of reaction (i) will increase in the sequence HCl < HBr < HI. This is consistent with the fact that G_e, the yield of electrons which

TABLE 7

BASIC DATA FOR THE RESONANCE CAPTURE REACTION $e+HX \rightarrow H+X^-$

	X = Cl	X = Br	X = 1
Bond dissociation energy of HX (eV)[a]	4.43	3.75	3.06
Electron affinity of X (eV)[b]	3.61	3.36	3.06
ΔH at 0 °K (eV)	0.82	0.39	~0
Cross section at maximum (cm²)[c]	2×10^{-17}	27×10^{-17}	2.3×10^{-14}
Integrated capture cross section (cm².eV)[c]	7.4×10^{-18}	7.4×10^{-17}	3.7×10^{-16}

[a] From reference 101.
[b] From reference 102.
[c] From reference 98.

are scavenged by SF_6 (see below), is only 71±5 % of the total of electron yield in the case of HBr[92, 100], whereas in the case of HCl it is 95±5 %[97].

Thermalised electrons may also undergo resonance capture. The minimum activation energies for HCl, HBr and HI can be obtained from the difference between D_{H-X} and the electron affinity of the halogen atom. From Table 7, the values are 0.82 and 0.39 eV (or 18.9 and 9.0 kcal.mole^{-1}) for HCl and HBr respectively. Thus this reaction can be ruled out as a competitive process for thermal electrons in these two gases. On the other hand, in HI resonance capture has a threshold energy of zero[98]; here it should be the predominant mode of electron loss and no distinction between the capture processes for thermal and faster electrons is necessary. In pure hydrogen halides reaction (i) is followed by (o) so that each capture yields a single hydrogen molecule.

Although reaction (i) cannot be important in gaseous HCl there are clear indications that electrons do not remain free in the gas for long. Loeb[83] found that the mobility of the long-lived negative ions was low and similar to that of the positive ions. This observation has been confirmed in more recent work by McDaniel and McDowell[103]. It implies that the negative ion is of high mass and probably a clustered Cl^- ion. More recent electron swarm experiments[98] have shown that low-energy electrons are captured in nitrogen gas containing small concentrations of HCl or HBr. There is evidence that the capture process reaches its peak at zero energy. Furthermore, it appears to be second order in the hydrogen halide pressure, implying that the capture reaction involves activated complexes of the formula $(HX)_2^-$. Further information about the electron attachment processes has come from studies of the effects of chlorine, bromine and sulphur hexafluoride on the hydrogen yields. The last of these cannot compete with the hydrogen halides for hydrogen atoms (see section on sulphur halides), and its effects are due to the scavenging of electrons in the reaction[71]

$$e+SF_6 \rightarrow SF_6^- \quad (35)$$

(*b*) *Experiments with scavengers in* HCl

As stated above, the total yield of hydrogen, $G(H_2)$, from irradiated HCl is 8.0 ± 0.3 molecules per 100 eV. On the addition of SF_6 this falls rapidly, reaching a plateau at a $(SF_6)/(HCl)$ ratio of 1.5×10^{-2} (see Fig. 5(a)). The reduction in $G(H_2)$ is then within experimental error equal to $G_{\text{ionisation}}$. This indicates that each electron normally reacts with HCl to form a single molecule of hydrogen. Since the value of k_{35} is known[104], it was possible to determine the mean lifetime, τ_e, of the electrons from competition kinetic experiments at lower $(SF_6)/(HCl)$ ratios. The number of electrons undergoing reaction (35) was assumed to be equal to the decrement, $\Delta G(H_2)$, in the hydrogen yield, and the data were analysed in terms of the equation

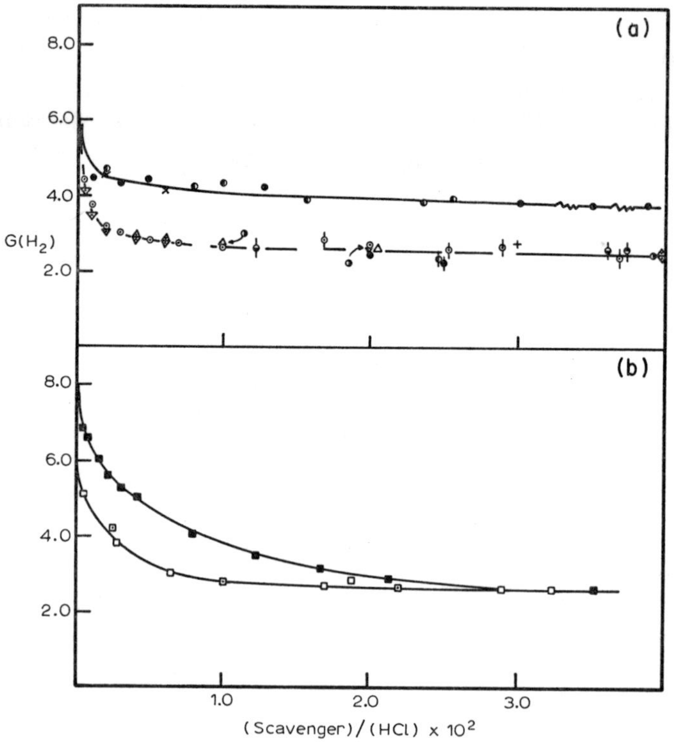

Fig. 5. Effects of scavengers on the hydrogen yields from irradiated gaseous HCl at different temperatures, pressures and dose rates, and in cells of different surface-to-volume ratios. (a) SF_6 and Br_2 in HCl at 23 ± 2°C. SF_6 in 580 torr HCl in 7-cm diameter cell: X, 5×10^{13} eV ml^{-1} min; ◐, 2×10^{14} eV ml^{-1} min^{-1}. SF_6 in 390 torr HCl in 18-cm diameter cell: ●, 2×10^{12} eV ml^{-1} min^{-1}. Br_2 in 500 torr HCl in 7-cm diameter cell; +, 5×10^{12} eV ml^{-1} min^{-1}. Br_2 in 600 torr HCl in 7-cm diameter graphite-coated cell; ◑, 5×10eV ml^{-1} min^{-1}. Br_2 in HCl in 18-cm diameter cell: ▽, 190 torr HCl and 1×10^{12} eV ml^{-1} min^{-1}; △, 910 torr HCl and 4×10^{12} eV ml^{-1} min^{-1}; ○, 1150 torr HCl and 1×10^{13}eV ml^{-1} min^{-1}. (b) Cl_2 in HCl at 550–640 torr in 7-cm diameter cell and 2×10^{14} eV ml^{-1} min^{-1}: ■, at 1 ± 1°C, □, at -77 ± 2°C. (Data taken from refs. 88 and 97)

$$\frac{1}{\Delta G(\mathrm{H_2})} = \frac{1}{G_e}\left\{1+\frac{1}{k_{35}(\mathrm{SF_6})\cdot\tau_e}\right\}$$

The yield G_e, of scavengeable electrons was 3.8 ± 0.2, in agreement with $G_{\text{ionisation}}$ ($= 4.00 \pm 0.07$). τ_e was found to be 1×10^{-9} sec in HCl at 560 torr. This is several orders of magnitude smaller than the lifetimes if there were homogeneous ion combination at conventional dose rates, *viz.*

$$e + \mathrm{H_2Cl^+_{(c)}} \rightarrow \mathrm{H} + m\,\mathrm{HCl}$$

or wall reactions[105], and these clearly cannot be major electron-loss processes.

Davidow and Armstrong[97] proposed that the capture of thermal electrons by HCl resulted from the two reactions

$$e + 2\,\mathrm{HCl} \rightarrow \mathrm{H} + \mathrm{Cl{-}H{-}Cl^-}$$

and

$$e + 2\,\mathrm{HCl} \rightarrow \mathrm{HCl^-} + \mathrm{HCl}$$

However, further competition experiments with SF_6 have shown that the capture is approximately third order in HCl over the pressure range 1–4 atm. To explain these results Johnson and Redpath[106] suggested that the electron might react with hydrogen-bonded agglomerates of HCl, as was previously proposed for the liquid phase[107] *viz.*

$$m\,\mathrm{HCl} \rightleftharpoons (\mathrm{HCl})_m$$

$$e + (\mathrm{HCl})_m \rightarrow \mathrm{H} + \mathrm{Cl{-}HCl^-_{(c)}}$$

The overall reaction is

$$e + m\,\mathrm{HCl} \rightarrow \mathrm{H} + \mathrm{Cl{-}HCl^-_{(c)}} \quad (36)$$

An alternative was the formation of an electron–HCl cluster[106]

$$e + m\,\mathrm{HCl} \rightarrow (\mathrm{HCl})^-_{(c)} \quad (37)$$

which subsequently reacted to produce hydrogen molecules or atoms.

Lee *et al.*[88] and Davidow and Armstrong[97] investigated the effects of chlorine and bromine on the hydrogen yields from HCl. Although the ground states of the molecular negative ions, Cl_2^- and Br_2^-, are stable to dissociation[108, 109] the positions of the energy levels of the X_2 and X_2^- systems are such that the capture of a free electron by an isolated chlorine or bromine molecule is a dissociative

References pp. 191–195

process. The threshold energy of the resonance capture reaction

$$e+Br_2 \rightarrow Br+Br^- \tag{38}$$

is zero[110] and it is evident that bromine will compete for thermal electrons as does SF_6. In HCl this gives rise to the marked drop in $G(H_2)$ as the ratio $[Br_2]/[HCl]$ increases from 0 to 10^{-3} (see Fig. 5(a)). For larger bromine concentrations there is a continued but more gradual decrease. This has been attributed to the scavenging of hydrogen atoms, which do not have electrons as precursors[71]. Finally, at $[Br_2]/[HCl] = 10^{-2}$ $G(H_2)$ reaches a plateau, indicating that there is a residual hydrogen yield, $G_{H''}$, which cannot arise from reactions of electrons or thermal hydrogen atoms. Since the magnitude of $G_{H''}$ is independent of dose rate, irradiation cell volume and surface coating (see data of Fig. 5(a)), it cannot be attributed to heterogeneous processes. It has been ascribed[71] to the formation of hydrogen in reaction (32). This is discussed later.

High concentrations of chlorine give the same value of $G_{H''}$, within experimental error, as is observed with bromine[97]. However, there is an important difference between the two halogens. Although it is exothermic by 26 kcal.mole^{-1}, the resonance capture process

$$e+Cl_2 \rightarrow Cl+Cl^-$$

has a threshold energy of 1.6 eV[110]. Thus, unlike bromine and SF_6, chlorine cannot scavenge free thermal electrons. This difference between the two halogens explains why the hydrogen yields from HCl at 23 °C are suppressed far more efficiently by bromine than by chlorine[88]. It might appear that chlorine reacts only with hydrogen atoms; however at −77 °C low concentrations of chlorine have a strong effect on $G(H_2)$ (see Fig. 5(b)), which cannot be attributed to competition between reactions (22) and (23). This is also true, but less obvious, for the results at 1 °C. From experiments at −77 °C with HCl containing both SF_6 and chlorine it was concluded[97] that only 1.5±0.2 thermal electrons per 100 eV form hydrogen atoms. The remaining 2.5±0.2 produce a clustered electron–HCl complex, $(HCl)^-_{(c)}$, similar to that proposed by Johnson and Redpath[106] and Lee *et al.*[88]. This either reacts with HCl, *viz.*

$$(HCl)^-_{(c)}+HCl \rightarrow H+Cl\text{–}H\text{–}Cl^-_{(c)} \text{ or } H_2+Cl^-_{2(c)} \tag{39}$$

or is scavenged by chlorine, which has an electron affinity[109] of about 50 kcal. mole^{-1}, *viz.*

$$(HCl)^-_{(c)}+Cl_2 \rightarrow Cl^-_{2(c)} \tag{40}$$

The rate coefficient ratio k_{39}/k_{40} was found to be 8.0×10^{-6} at -77 °C and 1.2×10^{-3} at 1 °C, indicating that $E_{39}-E_{40}$ is about 5.2 kcal.mole^{-1} and very different from $E_{22}-E_{23}$ (= 1.54 kcal.mole^{-1}).

Much work will be required before the details of these electron reactions and the structure of $(HCl)^-_{(c)}$ are fully understood. However, Raff and Pohl[111] have estimated that the binding energy of an electron in an HCl dimer complex (*i.e.* ClH–*e*–HCl) would be 22 kcal.mole^{-1}. Since the electron affinity of HCl cannot be more[112] than a few kcal.mole^{-1}, a complex of this type would be expected to form a more stable nucleus for $(HCl)^-_{(c)}$ than localisation of the excess electron on a single HCl molecule in the centre of the cluster. Nevertheless, there may be a stage in the formation of $(HCl)^-_{(c)}$ at which the electron is associated with only one molecule.

The scavenging of $(HCl)^-_{(c)}$ by trace concentrations of chlorine explains the abrupt initial drop in the yield of hydrogen from HCl at -77 °C (see Fig. 5(b)). The more gradual reduction of $G(H_2)$ over the $[Cl_2]/[HCl]$ range from 10^{-4} to 10^{-2} has been attributed to a competition between HCl and chlorine for hydrogen atoms, *viz.*

$$H + Cl_2 \rightarrow HCl + Cl \tag{22}$$

$$H + HCl \rightarrow H_2 + Cl \tag{23}$$

the yield of these being 3.3 per 100 eV. From a kinetic analysis of the -77 °C data k_{22}/k_{23} was found to be 2.6×10^{-3}, in good agreement with 2.5×10^{-3} calculated for this temperature from Klein and Wolfsberg's results[50].

Since reaction (36) can only account for 1.5 per 100 eV of the scavengeable hydrogen atoms observed in HCl, the remaining 1.8 per 100 eV must arise from excitation and dissociative ionisation. Apart from those produced in process (iv) (see Table 4) near its threshold energy, all of these would initially be hot. However, a significant fraction of them may be moderated to thermal energies before reacting[29]. This fraction would be determined by the relative cross-section for reaction (32) and collisional energy loss, (41), *viz.*

$$\underline{H}(K.E. = E) + HCl \rightarrow H_2 + Cl + \text{Energy} \tag{32}$$

$$\underline{H}(K.E. = E) + HCl \rightarrow \underline{H}(K.E. = E - \delta E) + HCl^{\dagger} \tag{41}$$

It should be unaffected by the presence of Cl_2 and Br_2, which could not compete effectively with these reactions at the halogen: HCl ratios shown in Fig. 5(a) and (b). Thus hot atom reactions can account for the residual (or unscavengeable) hydrogen, $G_{H''} = 2.2$, as well as contribute to the yield of scavengeable hydrogen atoms.

† With excess internal and kinetic energy = δE.

(*c*) *Experiments with scavengers in HBr*

Chen and Armstrong[92] have performed competition experiments with SF_6 in HBr at 200 and 558 torr. Their results indicated an inverse dependence of the electron lifetime on the square of the HBr pressure. They fitted the equation

$$\frac{1}{\Delta G(H_2)} = \frac{1}{G_e}\left\{1+\frac{k_{42}(HBr)^2}{k_{35}(SF_6)}\right\}$$

which is based on the reaction

$$e+2\,HBr \rightarrow \text{hydrogen}+\text{negative ion} \tag{42}$$

The ratio k_{42}/k_{35} had a value of 3.9×10^{-1} l.mole^{-1}. Taking $k_{35} = 1.9\times10^{14}$ l.mole^{-1}.sec^{-1}, it follows that k_{42} is 7.3×10^{13} l^2.mole^{-2}.sec^{-1}. The details of this reaction await investigation. However, the magnitude of the third-order rate coefficient indicates that, whatever the mechanism, a relatively long-lived complex is involved. This may be $(HBr)_2$ or vibrationally excited HBr^- ions[92]. It may also be noted that the hydrogen bond dissociation energy, $D_{Br^- - HBr}$[113], equals or exceeds 12.8 kcal.mole^{-1}. Thus reaction

$$e+2\,HBr \rightarrow [(HBr)_2^-] \rightarrow H+Br\text{–}H\text{–}Br^- \tag{43}$$

is exothermic by at least 3.8 kcal.mole^{-1} and therefore a feasible process. Its order is also consistent with the order implied by the swarm experiments mentioned above. However, contrary to statements by Chen and Armstrong[92], the beam experiments of Christophorou *et al.*[98] cannot be taken as unequivocal evidence for its occurrence.

Bromine is as effective as SF_6 in suppressing the hydrogen yield from HBr and it is evident[100] that, at low $[Br_2]:[HBr]$ ratios, the reaction

$$e+Br_2 \rightarrow Br+Br^- \tag{38}$$

is the main process responsible for this. However, reactions (13) and (14)

$$H+HBr \rightarrow H_2+Br \tag{13}$$

$$H+Br_2 \rightarrow HBr+Br \tag{14}$$

must also occur in irradiated HBr. Indeed it was on the basis of a competition between them that Eyring *et al.*[96] explained the results of Lind and Livingston. Assuming a value of $k_{14}/k_{13} = 8$ and reasonable ionisation parameters, they calculated decomposition yields which agreed well with experiment over a wide range of $HBr:Br_2$ ratios. They made no allowance for the capture of thermal electrons by bromine, but the effects of this would not have been seen in any case, since reaction (42) would have been almost completely suppressed at all of the bromine concentrations present in Lind and Livingston's experiments[100]. A more serious problem is that most of the hydrogen atoms involved in the competition would have been hot. In fact the striking success of their mechanism implies that k_{45}/k_{44} differs little from k_{14}/k_{13}.

$$\underline{\mathrm{H}}+\mathrm{HBr} \rightarrow \mathrm{H_2}+\mathrm{Br} \tag{44}$$

$$\underline{\mathrm{H}}+\mathrm{Br_2} \rightarrow \mathrm{HBr}+\mathrm{Br} \tag{45}$$

Since the ratio of the pre-exponential factors of reactions (14) and (13) is the same as k_{14}/k_{13}, this is not inherently unreasonable. However, $k_{45}/k_{44} = 8$ does not agree with the observations of Hamill *et al.*[12], who reported a value of 0.66 from their photochemical experiments. This problem obviously requires further investigation.

Competition kinetic studies with irradiated HI are as yet lacking and in any case would be very difficult because of its high reactivity.

1.7.3. Combination reactions of ions and halogen atoms

$$\mathrm{H_2X^+_{(c)}}+\mathrm{X{-}H{-}X^-_{(c)}} \rightarrow \text{neutral HX molecules} \tag{q}$$

$$\mathrm{H_2X^+_{(c)}}+\mathrm{X^-_{2(c)}} \rightarrow \text{neutral atoms and HX molecules} \tag{r}$$

$$\mathrm{X}+\mathrm{X}+\mathrm{M} \rightarrow \mathrm{X_2}+\mathrm{M} \tag{s}$$

$$\mathrm{X} \rightarrow \mathrm{X}\ (\text{on wall}) \rightarrow \tfrac{1}{2}\mathrm{X_2} \tag{t}$$

Armstrong and Back[105] have employed an ionisation chamber with a pulsed collecting field to determine the mean ion lifetimes in irradiated gaseous HCl. For typical radiolysis conditions these lay between 6 and 30 millisec, varying inversely with the square root of the dose rate in the range 8.6×10^{11}–5.5×10^{10} eV.cm^{-3}.sec^{-1}. Wall combination was unimportant, and the addition of 1.9 mole % of SF_6 had no effect on the lifetimes. The homogeneous ion combination coefficient, α(HCl), increased with pressure, attaining a constant value of $3.1\pm0.3\times10^{-6}$ cm^3.ion-pair^{-1}.sec^{-1} over the range 250–660 torr. The observed inverse dependence of α(HCl) on temperature agrees with the predictions of the Thompson theory of ion combination.

It appears that the ion-combination reactions make little or no contribution to the formation of hydrogen[100, 114]. This is reasonable, since clustered H_2X^+ and X–H–X^- ions would be expected to reform HX, *viz.*

$$\mathrm{H_2X^+_{(c)}}+\mathrm{X{-}H{-}X^-_{(c)}} \rightarrow \text{neutral HX molecules} \tag{q}$$

as do the solvated ions in liquid hydrogen halides[115]. The products of combination of H_2X^+ and X_2^- are unknown, but these ions are unlikely to form hydrogen atoms or molecules. A possible reaction is

$$H_2X^+_{(c)}+X^-_{2(c)} \rightarrow y\,HX+X$$

In all three systems the halogen atoms have been assumed to undergo recombination, either in the presence of a third body (s) or on the wall of the reaction cell (t). In HCl there is also evidence for the reaction

$$Cl+H_2 \rightarrow HCl+H \qquad (21)$$

but this is not significant below 0 °C and at low conversions[71].

1.7.4 The radiolysis of liquid HCl

The γ-radiolysis of liquid HCl has also received attention[107]. The yield of hydrogen at -79 °C is 6.5 ± 0.1 molecules per 100 eV. From the known concentration of HCl molecules and the absolute value of k_{22} ($= 3.5\pm0.4\times10^6$ l. mole^{-1}.sec^{-1} at this temperature) the lifetime of thermal hydrogen atoms with respect to reaction (22) is estimated as 10^{-8} sec. This is greater than the period of track expansion in irradiated liquids[116] and a significant fraction of the thermal hydrogen atoms will undergo track combination; dissociative ionisations will be fewer and these two effects can explain why the liquid-phase yield is 1.5 molecules per 100 eV less than that for the gas.

Two mole per cent sulphur hexafluoride caused only a slight reduction in $G(H_2)$. By contrast, in gaseous HCl 2×10^{-2} mole % sufficed to halve the yield of hydrogen from electron reactions. The difference is readily understood since the HCl concentration in the liquid is such that the reaction

$$e+m\,HCl \rightarrow H+Cl\text{–}H\text{–}Cl^-_{(\text{solvated})} \qquad (36)$$

will occur within one or two encounters after the electrons are thermalised. It is unlikely that HCl^- or any solvated-electron type of structure can exist for more than a few vibrations and there is certainly no evidence for it.

The effects of chlorine, bromine and ethylene on the hydrogen yields were interpreted in terms of competitions between

$$H+HCl \rightarrow H_2+Cl \qquad (22)$$

and the reactions

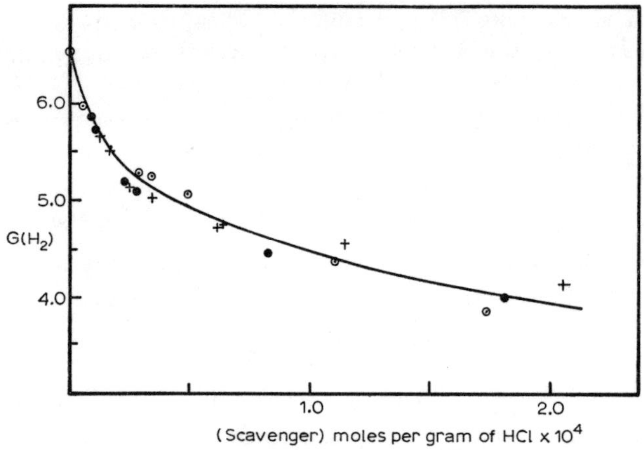

Fig. 6. Effects of scavengers on the hydrogen yields from γ-irradiated liquid HCl. $+Br_2$, ● Cl_2, ⊙ C_2H_4. Concentrations of Cl_2 and C_2H_4 have been multiplied by k_{H+Cl_2}/k_{H+Br_2} and $k_{H+C_2H_4}/k_{H+Br_2}$ respectively (taken from the data of refs. 107, 117).

$$H+Cl_2 \rightarrow HCl+Cl \tag{23}$$

$$H+Br_2 \rightarrow HBr+Br \tag{14}$$

$$H+C_2H_4 \rightarrow C_2H_5 \tag{46}$$

A kinetic analysis of the data gave $k_{22}/k_{23} = 1.7\times10^{-3}$, $k_{22}/k_{14} = 7.9\times10^{-4}$ and $k_{22}/k_{46} = 7.9\times10^{-3}$. The dependence of $G(H_2)$ on the solute concentrations is shown in Fig. 6, where the chlorine and ethylene concentrations have been normalised by multiplying them by the values of k_{23}/k_{14} and k_{46}/k_{14} respectively. These were calculated from the experimental ratios given above. They are summarised in Table 8.

In other experiments varying concentrations of bromine were added to liquid HCl, which contained a fixed amount of HBr. At -79 °C the rate coefficients of reactions (13) and (14) are both much larger than k_{22}. Thus kinetic analysis of the hydrogen yields from these solutions leads to a determination of k_{13}/k_{14}. The

TABLE 8

RATIOS OF RATE COEFFICIENTS FOR REACTIONS OF HYDROGEN ATOMS IN LIQUID HCl AT -79 °C

S	k_{H+S}/k_{H+Br_2}
Cl_2	0.46
C_2H_4	0.10
HBr[a]	0.12
HBr[b]	0.11

[a] Obtained from a 4.4 mole % solution of HBr in HCl.
[b] Obtained from a 7.6 mole % solution of HBr in HCl.

results are also given in column two of Table 8. The experimental value of k_{22}/k_{23} in liquid HCl is in reasonable agreement with the gas-phase value at this temperature.

The yield of hydrogen from thermal hydrogen atoms was found to be 2.2 ± 0.3. This leaves 4.3 ± 0.3 molecules per 100 eV to be accounted for, and most of these are probably formed from excited molecules. The work of Armstrong[117] shows that high chlorine concentrations can suppress the formation of hydrogen by these species. The scavenger may react with hot hydrogen atoms or absorb energy which normally becomes localised on individual HCl molecules to cause their dissociation.

2. Hydrogen cyanide

Pure hydrogen cyanide is stable at ordinary temperatures, but it tends readily to undergo polymerisation. This property, together with its poisonous character, have been sufficient to limit studies of its photolysis or thermal decomposition. The only investigations are those which deal specifically with the polymerisation reaction, and even on this there are few quantitative kinetic data. However, it has been shown that the thermal polymerisation of liquid hydrogen cyanide is catalysed by bases, particularly hydroxide and cyanide ions[118]. Polymerisation may also be induced by the action of ^{60}Co γ-rays, and Hummel and Janssen[119] have studied the kinetics of this process. With carefully dried hydrogen cyanide they observed a linear dependence of polymer yield on dose, and there was no post-irradiation effect. They observed dose rate exponents which lay between 0.5 and 1.0, more frequently being closer to the lower value. The main product was a cross-linked azulmine with a formula which corresponded closely to $(HCN)_x$. At dose rates in the range 5×10^4–3×10^5 roentgens per hour x had a value of about sixteen. Smaller amounts of diaminomaleonitrile and triazine were also formed, and the overall polymerisation yield corresponded to $G_{(-HCN)} = 34$. The yields of nitrogen and hydrogen were at least an order of magnitude smaller. Ammonia was only observed from samples which contained moisture.

Hummel and Janssen suggested the following mechanism to explain the formation of the azulmine polymer

$$HCN \rightsquigarrow HCN^+ + e$$

$$HCN + e \rightarrow HCN^- \qquad (47)$$

$$HCN + e \rightarrow H + CN^- \qquad (48)$$

Reaction (48) is about 50 kcal endothermic, and certainly could not take place without solvation of the CN^- ion. On the other hand, there is spectroscopic

$$2\ HCN \rightleftarrows \underset{\underset{C\equiv N}{|}}{\overset{\overset{N-H}{\|}}{C}}-H$$

$$CN^- + \underset{\underset{C\equiv N}{|}}{\overset{\overset{N-H}{\|}}{C}}-H \longrightarrow N\equiv C-\underset{\underset{C\equiv N}{|}}{\overset{\overset{NH_2}{|}}{C^-}}$$

$$N\equiv C-\underset{\underset{C\equiv N}{|}}{\overset{\overset{NH_2}{|}}{C^-}} + \underset{\underset{C\equiv N}{|}}{\overset{\overset{N-H}{\|}}{C}}-H \longrightarrow N\equiv C-\underset{\underset{C\equiv N}{|}}{\overset{\overset{NH_2}{|}}{C}}-\underset{\underset{C\equiv N}{|}}{\overset{\overset{NH_2}{|}}{C^-}}$$

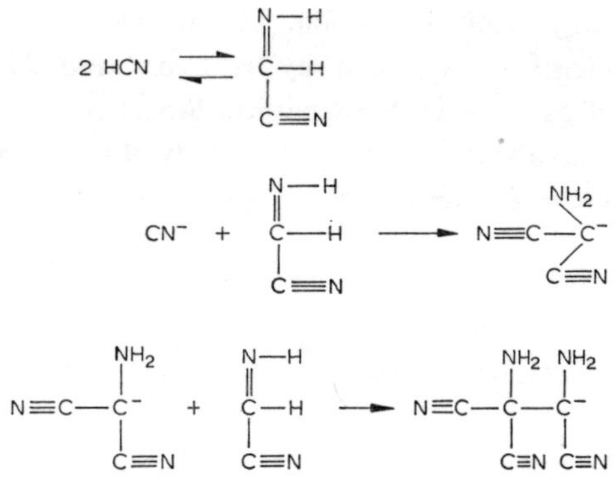

evidence[120] for HCN⁻, which could be formed in reaction (47). This may itself be an initiating ion[119] or undergo solvolysis to H and CN⁻.

The anionic mechanism is similar to that postulated to explain the thermal polymerisation. The apparent general similarity of the polymers produced by the two methods of initiation justifies this. Cross-linking involves additions to the free CN groups, and regular networks with two or more interconnected polymer chains are possible. Thus the structure of the azulmine is highly complex. Termination must occur by reactions of the growing anions with H_2CN^+ ions formed in the reaction

$$HCN^+ + HCN \rightarrow H_2CN^+ + CN$$

This reaction has been shown to be very rapid[77]. Sulphuric and acetic acids suppress the polymerisation. Evidently their anions are ineffective as initiators, and the enhanced proton concentration provided by them must reduce the chain lifetime. The slight retarding effect of oxygen could be due to electron scavenging. However, the authors suggest that there may be a small free radical component of the chain reaction, which is inhibited in the presence of oxygen.

When the HCN contained 1–2 % mole of water the yields tended to increase exponentially with dose. This was due to the occurrence of a superimposed thermal reaction, which became faster as the radiation products accumulated and persisted after irradiation ceased. Hummel and Janssen attributed it to the hydrolysis of imino and cyano groups on the polymer chain. They suggested that the ammonia generated in this process reacted with HCN to form cyanide ions, which then initiated the polymerisation.

Lind et al.[121] observed the polymerisation of gaseous hydrogen cyanide under the action of radon α-particles. The product was a reddish black solid which settled to the bottom of the reaction cell. The ion pair yield of about eleven was the same whether the reaction was induced by energy absorbed in HCN itself or in nitrogen, helium, argon, krypton or xenon[122]. Catalysis of the reaction by the last of these is interesting for positive charge transfer from xenon ions to hydrogen cyanide molecules is energetically forbidden. This problem is not encountered with the anionic mechanism suggested by Hummel and Janssen. Evidence for the polymerisation of gaseous TCN by tritium β-particles has been cited briefly by Staats *et al.*[123] and Takahashi[124] has made a study of the decomposition of hydrogen cyanide in high frequency electrical discharges (see ref. 124 for previous work).

3. Carbonyl halides

3.1 PHOSGENE

The thermal and photochemical syntheses of phosgene have been very thoroughly investigated. Apart from the initiation process the mechanisms of the two reactions are essentially the same. A long controvervy between Bodenstein *et al.*[125] and Rollefson and Lenher[126] was finally settled in the favour of the former and their mechanism is shown below.

$$Cl_2 + h\nu \rightarrow 2\,Cl \qquad (20)$$

$$(\text{or } M + Cl_2 \rightleftarrows M + 2\,Cl, \text{ thermal})$$

$$Cl + CO \rightleftarrows COCl \qquad (49), (50)$$

$$COCl + Cl_2 \rightleftarrows COCl_2 + Cl \qquad (51), (52)$$

$$COCl + Cl \rightarrow CO + Cl_2 \qquad (53)$$

The Rollefson mechanism had

$$Cl_2 + CO + Cl \rightarrow COCl + Cl_2$$

in place of reaction (49) but no evidence has been found to support the participation of Cl_2 as a special third body for this reaction.

The observed rate laws are

(a) In the region of room temperature (0–40 °C) and at reactant pressures above about 40 torr

$$\frac{d[COCl_2]}{dt} = K\,I_{abs}^{\frac{1}{2}}[CO]^{\frac{1}{2}}[Cl_2]$$

This requires

$$k_{51}[COCl][Cl_2] \ll k_{50}[COCl]$$

i.e. that the equilibrium (49, 50) is established with long chains

$$k_{51}[Cl_2] \ll k_{53}[Cl]$$

K in the above equation $= k_{51}(k_{20}/k_{53}K_{49,\,50})^{\frac{1}{2}}$

(b) At higher temperatures ($\sim$ 300°) and/or at low pressures

$$\frac{d[COCl_2]}{dt} = K'I_{abs}^{\frac{1}{2}}[CO][Cl_2]$$

The change of rate equation has been attributed to wall effects which were shown by Fowler and Beaver[127], to be important under these conditions.

The most recent values for Arrhenius parameters are those of Dainton and Burns[128], who performed a very careful study of the photochemical formation of phosgene, investigating the effects of light intensity, temperature and concentration and determined radical lifetimes by the rotating sector technique.

Their Arrhenius parameters for reactions 51 and 53 are listed below, together with $K_{49,\,50}$.

$$\log k_{51} = 9.4 - \frac{2960}{2.3\,RT}\ (\text{l.mole}^{-1}.\text{sec}^{-1})$$

$$\log k_{53} = 11.6 - \frac{830}{2.3\,RT}\ (\text{l.mole}^{-1}.\text{sec}^{-1})$$

$$\log K_{49,\,50} = -2.806 + \frac{6310}{2.3\,RT}\ (\text{l.mole}^{-1})$$

The latter is in good agreement with $\Delta H_{49} = -6.3$ kcal.mole^{-1} from combined photochemical and thermal data[129]; from thermodynamic data[130] and Dainton's results, Goldfinger[131] calculates $\log k_{52} = 11.4 - 19.9/2.3\,RT$ l.mole^{-1}.sec^{-1}. For the thermal reactions the Arrhenius parameters[129] are

$$\log k_{\text{formation}} = 8.023 - \frac{26{,}300}{4.58\,T}\ (\text{l.mole}^{-1}.\text{sec}^{-1})$$

$$\log k_{\text{decomposition}} = 13.48 - \frac{52{,}400}{4.58\,T}\ (\text{sec}^{-1})$$

The carbon isotope effect in the photochemical and thermal exchange reaction

$$^{14}CO + COCl_2 \rightarrow {}^{14}COCl_2 + CO$$

was investigated by Stranks[132].

Bradley and Tuffnell[133] studied the decomposition of $COCl_2$ and $(COCl)_2$ in a combined shock tube and flash photolysis apparatus. The shock raised the reactant gas mixture to a steady high temperature and steady high velocity; the rapid flow was used continuously to remove photolysed gas from the illuminated region. The experiments yielded no new kinetic data but results indicated that $(COCl)_2^*$ is an intermediate in the photolyses of both compounds.

The thermodynamic parameters for COCl have been evaluated by Jacox and Milligan[134], who identified COCl by its IR spectrum in an argon or CO matrix at 14 °K.

3.2 CARBONYL BROMIDE

Lenher and Schumacher[135] first studied the kinetics of the thermal dissociation of $COBr_2$ in 1928. They showed that the reaction

$$COBr_2 = CO + Br_2$$

was first order, but was probably heterogeneous under their reaction conditions. Schumacher and Bergman[136] determined the quantum yield for the photolysis in 1931. They found $\phi_{CO} = 1$ with no temperature coefficient and concluded that the photolysis was simply

$$COBr_2 + h\nu \rightarrow CO + Br_2$$

3.3 CARBONYL FLUORIDE

The equilibrium

$$2\,COF_2 \rightleftharpoons CO_2 + CF_4$$

was studied as long ago as 1934 by Ruff and Shih-Chang Li[137]; the latter author repeated his earlier work[138] and evaluated ΔH for the equilibrium as −26 kcal. mole^{-1}.

4. Halides of nitrogen

4.1 DINITROGEN TETRAFLUORIDE

When compared with other halides of nitrogen, nitrogen trifluoride and dinitrogen tetrafluoride are comparatively stable and easy to handle. For this reason they are far more amenable to detailed kinetic investigations. The tetrafluoride is of particular interest to kineticists, since it is dissociated into NF_2 radicals at relatively low temperatures. The presence of these radicals has been confirmed by electron spin resonance[139], ultraviolet[140] and infrared spectroscopy[141], and by mass spectrometry[142].

Herron and Dibeler[142] and Kennedy and Colburn[143] have calculated the bond dissociation energies of N_2F_4 and NF_3. Their results are summarised in Table 9.

TABLE 9

ESTIMATED BOND DISSOCIATION ENTHALPIES OF NITROGEN–FLUORINE COMPOUNDS

Units are, kcal.mole^{-1} at 298 °K

D_{NF_2-F}	$= 57.1 \pm 2.5$[a]	Average D_{N-F} in N_2F_4	$= 71$[a]
$D_{NF_2-NF_2}$	$= 20.1 \pm 0.3$[b]	$D_{(FN=NF)cis}$	$= 106 \pm 10$[c]
D_{NF-F}, D_{N-F} mean	$= 71 \pm 2.0$[a]	$D_{(FN=NF)trans}$	$= 103 \pm 10$[c]

Method	*ΔH for the reaction* $N_2F_4 \rightleftharpoons 2\ NF_2$ *in the temperature range* 360–425 °K
UV spectroscopy	21.7 ± 2.0[d]
Pressure–volume–temperature measurements	19.85 ± 0.5[d]
ESR	19.3 ± 1.0[e]
Mass spectrometry	21.5 ± 1.6[c]
Weighted mean	20.1 ± 0.3

[a] Kennedy and Colburn[143].
[b] Taken as mean value of ΔH given below. The temperature dependence is assumed to be negligible in the range 298–425 °K[140].
[c] Herron and Dibeler[142].
[d] Johnson and Colburn[140].
[e] Johnson *et al.*[139].

In contrast to the situation in NH_3, the first N–F bond in NF_3 is broken more easily than the second and third. The average strength of the latter two bonds is 71 kcal.mole^{-1}. Within experimental error this is the same as the average N–F bond energy in N_2F_4. Since D_{NF_2-F} is only 57.1 ± 2.5 kcal, the reaction

$$NF_2 + N_2F_4 \rightarrow NF_3 + N_2F_3$$

is endothermic, as is the disproportionation

$$NF_2 + NF_2 \rightarrow NF_3 + NF$$

Consequently, NF_2 radicals in N_2F_4 are chemically unreactive, and the equilibrium

$$N_2F_4 \rightleftharpoons 2\ NF_2$$

is completely reversible under normal conditions. In this regard the resemblance to the dinitrogen tetroxide system is rather striking.

The heat of dissociation of N_2F_4 has been determined from equilibrium pressure measurements[140] at different temperatures as well as by spectroscopic techniques. As shown in Table 9, the agreement between the different methods is good. The overall mean of these measurements is 20.1 kcal.mole^{-1}. A statistical treatment[140] of the molecular properties of NF_2 and N_2F_4 has shown that at 298 °K the entropy change of the dissociation should lie in the range 40–45 cal. mole^{-1}.degree^{-1}. The experimental results of Johnson and Colburn are in accord with this. A review of the chemistry of the NF_2–N_2F_4 system has been given by Johnson[144].

More recently, the dissociation of dinitrogen tetrafluoride has been investigated by the shock-tube technique. Modica and Hornig[145] monitored the optical density due to the NF_2 absorption band at 2600 A. They were thus able to follow the instantaneous concentrations of these radicals in shock heated mixtures of N_2F_4 with argon and other gases. From the equilibrium concentrations determined under a variety of conditions they calculated equilibrium constants, which were in excellent agreement with the data given in Table 9. In kinetic experiments the initial rate of dissociation was found to be proportional to the tetrafluoride concentration. The exponent with respect to argon concentration over the range 0.03–0.11 mole.l^{-1} was 0.9, indicating that the reaction was close to, if not within, the second-order low-pressure region. Thus for these conditions the reaction may properly be written as

$$\mathrm{Ar+N_2F_4 \rightarrow Ar+2\,NF_2} \tag{54}$$

Brown and Darwent[146] have also studied the reaction in argon, using essentially the same technique and under the same conditions as those employed by Modica and Hornig. The combined data of the two research groups are shown in the form of an Arrhenius plot in Fig. 7. The second-order rate coefficients of Brown and Darwent tend to be slightly higher than those of Modica and Hornig. However, the agreement between the two sets of results is quite good, and a least squares fit gives

$$k_{54} = 3.6\times 10^{13} \exp\left(-\frac{15{,}290}{RT}\right) \text{ l.mole}^{-1}\text{.sec}^{-1}$$

for the temperature range 350–450 °K†. In helium the activation energy was the

† The higher activation energy of 17.1 kcal mole^{-1} given by Brown and Darwent[146] was calculated on the assumption that the rate was independent of argon concentration over the concentration range 5.4–6.4$\times 10^{-2}$ mole.l^{-1}. Modica and Hornig's work[145] shows that this is not so.

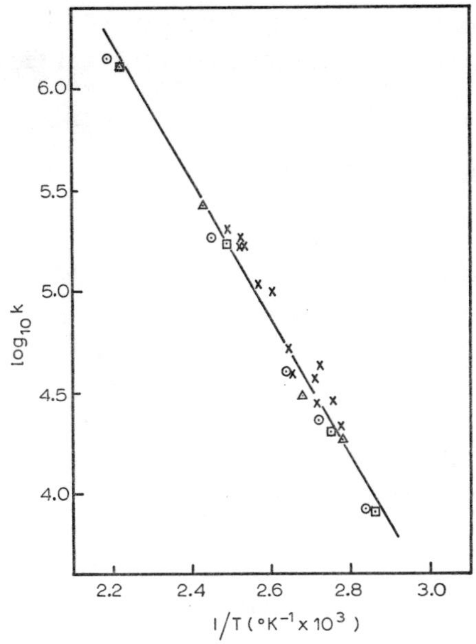

Fig. 7. Arrhenius activation energy plot for the reaction $Ar+N_2F_4 \rightarrow Ar+2\,NF_2$. × results of ref. 146; △ results of ref. 145, (Ar) = $3.6\pm0.5\times10^{-2}$ moles l^{-1}, and 1 : 1 N_2F_4 : Ar; ⊙ results of ref. 145, (Ar) = $5.6\pm0.2\times10^{-2}$ moles l^{-1}, and 1 : 1 N_2F_4: Ar; ⊡ results of ref. 145, (Ar) = $5.6\pm0.2\times10^{-2}$ moles l^{-1}, and 2 : 1 N_2F_4: Ar.

same within experimental error, but the pre-exponential factor was about 20 % larger[145]. This is due to the increased collision rate with helium. The efficiency of activation per collision for helium is about one half of that with argon.

The decomposition of dinitrogen tetrafluoride in shock-heated nitrogen has also been investigated[146]. At 400 °C and a total concentration of 7×10^{-2} mole. l^{-1} the apparent first-order rate coefficient, k_{55},

$$N_2F_4 \rightarrow 2\,NF_2 \tag{55}$$

was about 2.0×10^4 sec^{-1}. This is approximately twice as large as the corresponding rate coefficient in argon under identical conditions in the same apparatus. The difference cannot be explained solely on the basis of collision rates and nitrogen must be a more efficient activating molecule than argon. The nitrogen exponent was 0.6 at 0.02 mole.l^{-1}, and it approached zero as the nitrogen concentration was increased toward 0.2 mole.l^{-1}. Thus this pressure range must lie well above the region for pure second-order behaviour in nitrogen. From a linear plot of k_{55} *versus* the reciprocal of total concentration, Brown and Darwent obtained a value of $3.36\pm0.19\times10^4$ sec^{-1} for k_{55}^{∞}, the first-order rate coefficient in the high-pressure region at 400 °K. With total concentrations in the range of 0.06–0.08 mole.l^{-1}, k_{55} was $2.5\pm0.3\times10^4$ sec^{-1} at the same temperature. This is within 25 % of k_{55}^{∞}, and under these circumstances the activation energy should approximate to the critical energy of the reaction. The experimental data of Brown and Darwent for this concentration range fitted the Arrhenius expression

$$k_{55} = 10^{14.98\pm 0.42} \times \exp\left(\frac{-19{,}400\pm 700}{RT}\right) \text{ sec}^{-1}$$

and it may be noted that the activation energy is within experimental error equal to the value of $\Delta E_{55} = 19.3\pm 0.3$, obtained from the data of Table 9. The lower activation energy observed in the second-order pressure region with argon and helium has been noted as evidence for the participation of internal degrees of freedom in the activation process[145, 146]. From a classical treatment it was concluded that about four effective oscillators are involved, the same as in the decomposition of N_2O_4. In view of the similar properties of the two molecules and the dissociation of the N–N bond in both cases, this can hardly be coincidental. A more detailed quantum mechanical treatment of this aspect of the reaction would be of considerable interest. However, further experimental work on the pressure dependence of the reaction rate in argon and nitrogen is also desirable, for the difference in behaviour with these two gases is greater than would normally be expected[147].

At temperatures above 1400 °C the dissociation of dinitrogen tetrafluoride is essentially instantaneous, and the thermal decomposition of the resulting NF_2 radicals may be observed. Diesen[148] has employed a time-of-flight mass spectrometer, coupled to a shock tube, to follow their decay and the formation of reactive intermediates and products. The following mechanism is for the most part his, but Modica and Hornig[149] have also made important contributions. They employed the same spectroscopic technique for following the NF_2 concentration as in their work on the N_2F_4 decomposition.

$$NF_2 + M(\text{Ar or Ne}) \rightarrow NF + F + M \tag{56}$$

$$2\,NF \rightarrow N_2F_2^* \rightarrow N_2 + 2\,F \tag{57}$$

$$NF + NF_2 \rightarrow N_2F_2 + F \tag{58}$$

$$N_2F_2 + NF_2 \rightarrow N_2F + NF_3 \tag{59}$$

$$N_2F + NF_2 \rightarrow N_2 + NF_3 \tag{60}$$

$$NF_3 + M \rightarrow NF_2 + F + M \tag{61}$$

At temperatures in the range 1500–1800 °K, N_2, NF_3 and F are important products and there is a small peak at mass 47, which suggests the presence of N_2F_2 molecules, N_2F radicals or a combination of both[148]. There are measurable induction periods for both N_2 and NF_3. Both induction periods vary inversely with the initial NF_2 concentration, the one for NF_3 being the longer of the two. This compound is unstable at these temperatures; consequently its concentration passes through a maximum and then declines. Its rate of formation immediately after the induction period is twice the rate of production of nitrogen, and the two together account for the rate of NF_2 consumption. This can be explained by the foregoing mechanism if the fraction, θ, of NF radicals undergoing reaction (58) approaches unity.

A steady-state treatment of reactions (56) to (61) gives the approximate relationship

$$\theta = \frac{1}{1+\frac{1}{2}\left[\left\{1+\frac{8\,k_{56}k_{57}(\mathrm{M})}{k_{58}^2(\mathrm{NF_2})}\right\}^{\frac{1}{2}}-1\right]}$$

The rate expression for the disappearance of NF_2 then becomes

$$-\frac{\mathrm{d}[\mathrm{NF_2}]}{\mathrm{d}t} = k_{56}[\mathrm{M}][\mathrm{NF_2}][1+\theta] \qquad \text{(i)}$$

$$= k_{\mathrm{app}}[\mathrm{M}][\mathrm{NF_2}] \qquad \text{(ii)}$$

Using the above equation and the values of k_{57} and k_{58} (2.5×10^{10} and 2.0×10^{9} $\mathrm{l.mole^{-1}.sec^{-1}}$ respectively) given by Diesen, θ is found to vary between 0.76 and 0.86 as the argon: NF_2 ratio is reduced from 100 to 25 in the temperature range 1775–1800 °K. This corresponds to a variation of only 6 % in the $(1+\theta)$ term of expression (i). By the same token, changes in the argon: NF_2 ratio during the course of a single run will have only a minor effect on this term unless the experiment is followed to very high conversions. Consequently, the reaction should obey the simple rate law given by equation (ii), the apparent second-order rate coefficient being the product of k_{56} and $(1+\theta)$. The results of Modica and Hornig[149] are in accord with this prediction. They observed a first-order dependence of the reaction rate on NF_2 concentration, and a 0.7 ± 0.3 dependence on argon over the concentration range 0.9×10^{-2}–1.9×10^{-2} $\mathrm{mole.l^{-1}}$. True values of k_{56} may be obtained by dividing their values of k_{app} (actually reported as k_{56}) by $(1+\theta)$.

The radical reactions (57) and (58) are both highly exothermic and should have small or negligible activation energies. Any significant temperature dependence of θ must, therefore, arise from the temperature dependence of k_{56}. On this basis values of θ were calculated for the conditions employed by Modica and Hornig and used to derive the true values of k_{56}^{Ar} from their data by an iterative procedure. The temperature dependence of k_{56}^{Ar} over the range 1440–1975 °K was found to obey the expression

$$k_{56}^{\mathrm{Ar}} = 1.2\times10^{12}\exp(-51{,}800/\boldsymbol{R}T)\,\mathrm{l.mole^{-1}.sec^{-1}}, \qquad \text{(iii)}$$

which agrees well with that previously reported by Diesen[148]. The experimental activation energy is slightly larger than that found for k_{app}[149]. However, it is still significantly less than the mean energy of the two N–F bonds in NF_2. As stated by Modica and Hornig, this may reflect the participation of internal degrees of freedom in the energization process. However, it is also possible that $D_{(NF-F)}$ is less than the average bond energy of 71 kcal.mole^{-1}, and the difference between the critical energy and the experimental activation energy may not be as large as 20 kcal.mole^{-1}.

The calculated value of θ falls rapidly as the temperature rises above 1800 °K. This is in accord with the observed decrease in the ratio of the rates of formation of NF_3 and N_2. At temperatures in the range 2200–3000 °K and high dilutions with argon or neon, NF_2, NF, F and N_2 are the only species observed, and there is a good material balance[148]. After a short induction period the NF concentration rises to a maximum and then decays. These observations imply that (56) and (57) are now the only important reactions, and it follows that $2k_{57}[NF]^2 = k_{56}[NF_2][M]$ at the maximum NF concentration. From the experimental concentrations and the determined value of k_{56}, a semiquantitative, but absolute, value of k_{57} has been determined[148]. The result, 2.5×10^{10} l.mole^{-1}.sec^{-1}, is within a factor of two independent of temperature in the range 2300–3000 °K. It implies a steric factor of about 0.2 if a collision diameter of 3.3 A is assumed for the NF radical.

At temperatures above 1900 °K Modica and Hornig[149] observed a chemiluminescence, which lay in the wavelength range 2350–3000 A and interfered with the spectroscopic analysis for NF_2. They suggested that the emission arose from an excited $N_2F_2^*$ species, *viz.*

$$2\,NF \rightarrow N_2F_2^* \rightarrow h\nu + N_2F_2 \tag{62}$$

Since the N=N bond in N_2F_2 has a dissociation energy of about 104 kcal.mole^{-1}, the excess internal energy of the $N_2F_2^*$ species would be sufficient to account for the quantum energy of the emission. Reaction (62) is an alternative to (57). However, at these temperatures its kinetic effects would be indistinguishable since N_2F_2 would not be stable, but would decompose, *viz.*

$$N_2F_2 \rightarrow N_2 + 2\,F$$

For a temperature of 2550 °K and 7.17×10^{-4} mole.l^{-1} of argon the value of k_{56}^{Ar} calculated from equation (iii) is 3.7×10^7 l.mole^{-1}.sec^{-1}. This agrees reasonably well with the value 1.9×10^7 l.mole^{-1}.sec^{-1} derived from Diesen's results for these conditions. Taking $k_{56}^{Ar} = 2 \times 10^7$ l.mole^{-1}.sec^{-1} and k_{57} and k_{58} as previously given, θ is found to be 0.13. The dominance of reaction (57) under these conditions is thus confirmed, and thus the proposed mechanism agrees with the experimental observations presently available. More rigorous tests will require further detailed studies of the effects of temperature and argon: NF_2 ratio on the concentration–time curves of the intermediates and products.

Bumgardner and Lustig[150] have made a preliminary study of the 2537 A photolysis of NF_2 radicals in the presence of excess N_2F_4 at room temperature. Mass spectrometric analysis showed NF_3, N_2F_2 and N_2 to be major products. The following reactions were suggested

$$NF_2+h\nu \rightarrow NF+F$$

$$F+NF_2+M \rightarrow NF_3+M$$

$$2\,NF+M \rightarrow N_2F_2+M \qquad (63)$$

The occurrence of reaction (63) under these conditions is not surprising in view of the much lower temperature and that N_2F_4 molecules are probably better deactivators for excited $N_2F_2^*$ molecules than are argon atoms. However, attention must also be given to the possible occurrence of reaction (58). Spectroscopic investigations of the NF radical and its reactions in inert gas matrices at 4–20 °K have been conducted by Comeford and Mann[151]. They also reported the formation of N_2F_2, and were able to distinguish the *cis* and *trans* forms by means of their different infrared spectra.

4.2 NITROGEN TRICHLORIDE

Griffiths and Norrish[152] have studied the chlorine-photosensitized decomposition of NCl_3 by following the pressure increase from the overall reaction, *viz.*

$$2\,NCl_3 = N_2+3\,Cl_2$$

with a sensitive Bourdon gauge. There is an initial Budde effect. After this the reaction was zero order in NCl_3 until the very end, when a semi-explosion occurs. The quantum yield varied between 2 and 20, and the authors postulated a chain mechanism involving the following initiation and propagation steps

$$Cl_2+h\nu \rightarrow 2\,Cl$$

$$Cl+NCl_3 \rightarrow NCl_2+Cl_2 \qquad (64)$$

$$NCl_2+NCl_3 \rightarrow N_2+2\,Cl_2+Cl \qquad (65)$$

Although an "aging effect" was observed with some reaction cells, the rate of decomposition in the zero-order region was independent of the area of the illuminated surface and the surface : volume ratio. Thus it is reasonable to assume that the chain decomposition is homogeneous. The light intensity exponent was unity, and for light of wavelength 4358 or 3650 A the quantum yield varied with pressure in accordance with the equation

$$\Phi = (2.2 \pm 0.2) + 1/(0.0038\, P_{Cl_2} + k_x P_x)$$

P_{Cl_2} and P_x are the pressures in torr of chlorine and other gases, which were added to the system. The values of k_x are 0.0009_3 for He 0.001_6 for Ar, 0.001_7 for N_2, 0.002_5 for O_2, and 0.003_8 for CO_2. It was noted that these constants were roughly proportional to the respective third body and energy transfer efficiencies of the added gases. For this reason their effects were attributed to the collisional stabilization of an energy-rich species; Griffiths and Norrish suggested the formation of NCl_4, which subsequently reformed Cl and NCl_3, *viz.*

$$NCl_3 + Cl \underset{M}{\overset{M}{\rightleftharpoons}} NCl_4$$

or was destroyed catalytically on the surface of the reaction vessel, *viz.*

$$2\, NCl_4 + \text{wall} \rightarrow \text{products}$$

While claiming no finality for their mechanism, the authors pointed out that it accounted satisfactorily for the zero-order dependence on NCl_3 concentration, the intensity exponent of unity and many other characteristics of the reaction. However, a serious defect lies in the apparent absence of any effect of pressure on the rate of diffusion of NCl_4 to the reaction vessel surface. Such an effect would be expected under their conditions and it would yield a dependence of the reaction rate on total pressure opposite to that observed.

One further point deserves comment. Although reaction (65) is exothermic (see below) and probably has a low activation energy, it is inherently extremely complex. Indeed it would be surprising if the formation of the N≡N bond and the elimination of the chlorine atom and molecules occurred in a single step. Thus this reaction probably proceeds through a number of steps, *e.g.*

$$NCl_2 + NCl_3 \rightarrow N_2Cl_3 + Cl_2 \qquad (66)$$

$$N_2Cl_3 \rightarrow N_2Cl + Cl_2$$

$$N_2Cl \rightarrow N_2 + Cl$$

There is unfortunately no information on N–Cl bond energies in the chlorine analogues of hydrazine. However, reaction (66) is likely to be highly exothermic, and some of the excess energy may reside in the N_2Cl_3 species. Collisionally stabilized intermediates of this type should be unreactive toward chlorine or NCl_3. They would, however, react rapidly with Cl or NCl_2 radicals, and homogeneous termination reactions of the form

$$N_2Cl_3 + Cl \rightarrow N_2 + 2\ Cl_2$$

do lead to a rate expression which fits the experimental data of Griffiths and Norrish. There are, therefore, alternatives to the reactions of NCl_4, which involve collisional stabilisation but not surface termination.

In 1964 Briggs and Norrish[153] reported an investigation of the flash photolysis of Cl_2–NCl_3 mixtures. The trichloride was completely consumed 22 μsec after the flash, and two transients with absorption maxima at 2400 A and 2950 A were observed. The absorption at 2950 A was assigned to the NCl_2 radical. When the light from the flash lamp was passed through a pyrex filter the intensity from the 2400 A band was reduced. This and other observations lead to the assignment of the latter absorption to NCl radicals, formed by the photo-dissociation of NCl_2 at wavelengths below the pyrex cutoff. The absence of any real evidence for NCl_4 or N_2Cl_3 radicals is not surprising, for their concentrations would be rapidly reduced in the absence of excess NCl_3.

An intriguing investigation by Apin[154] has shown that nitrogen trichloride undergoes spontaneous thermal decomposition unless the total pressure of its environment exceeds a critical value. This critical pressure depends on the composition of the gaseous mixture and increases with temperature. The decomposition obviously involves the formation of free radicals, since the compound sensitizes the chain reaction between hydrogen and chlorine[155]. The attack of chlorine (and hydrogen) atoms on nitrogen trichloride is evidently very rapid, for the chlorine–hydrogen chain reaction does not proceed until essentially all of the added trichloride has been consumed.

Ashmore[155] has suggested that the reaction

$$NCl_3 \rightarrow NCl_2 + Cl \qquad (67)$$

is the primary process in the NCl_3 decomposition and this must be followed by reactions (64) and (65) as propagation steps. Again the effects of pressure imply that energy-rich intermediates are formed, and they may partly be explained on the same basis as the observations of Griffiths and Norrish. In addition, Ashmore has pointed out that the attack of NCl_2 on NCl_3 may lead to branching[155]. This probably involves the ejection of chlorine atoms from N_xCl_y intermediates of the type suggested above, and stabilizing collisions may thus have an important influence on the frequency of the overall branching process.

From the heat of formation of nitrogen trichloride[156] it may be shown[157] that the average N–Cl bond energy is 46 ± 3 kcal.mole^{-1}. The facile thermal decomposition of this compound and its ability to sensitize the hydrogen–chlorine reaction at ordinary temperatures imply that the first N–Cl bond is the weakest, as is the case for the N–F bonds in NF_3. Thus $D_{(NCl_2-Cl)}$ may reasonably be expected to lie in the range 25–35 kcal.mole^{-1}. On this basis the heat of formation of NCl_2 would be about 62 kcal.mole^{-1} and reactions (64) and (65) are all highly exothermic.

Shushunov and Pavlova[158] have investigated the thermal decomposition of nitrogen trichloride at a concentration of 0.7 mole.l^{-1} in liquid carbon tetrachloride. In the absence of light and air the reaction was first order up to at least 50 % decomposition, and the rate coefficients fitted the Arrhenius expression

$$k_{(\text{decomp.})} = 10^{16} \exp(-32{,}000/RT) \text{ sec}^{-1}$$

From the considerations of the preceding paragraphs it seems likely that the primary step is reaction (67). Some radicals may undergo cage reactions, *viz.*

$$NCl_2 + Cl \rightarrow NCl_3$$

$$NCl_2 + Cl \rightarrow NCl + Cl_2$$

and reactions of the NCl radicals with the solvent are also conceivable. However, a chain decomposition of the solute is unlikely. This follows from Apin's observations[154] of the effects of pressure on the gas phase decomposition of NCl_3 and the quantum yield of 2.0–2.5, which Griffiths and Norrish[152] observed for the chlorine photosensitized decomposition in the high-pressure limit. Furthermore, Bowen[159] found a quantum yield of 0.2–1.2 for the photolysis of NCl_3 in liquid carbon tetrachloride. Therefore the primary split of the NCl_2–Cl bond is probably the sole rate-controlling process, and it is interesting to note that the experimental activation energy is of the magnitude suggested above for the dissociation energy of this bond.

5. Halides of sulphur

Comparatively little kinetic work has been done on the halides of sulphur. As in the case of the nitrogen, the greatest attention has been given to the fluorides, and kinetic studies of SF_6 and S_2F_{10} are reported below. At the time of writing no other sulphur halides have been investigated in sufficient detail to warrant review.

5.1 SULPHUR HEXAFLUORIDE

Sulphur hexafluoride does not undergo hydrolysis as do the lower fluorides of sulphur. It is also particularly stable thermally[160] and kinetic investigations of its decomposition are noticeably lacking. On the other hand it has been used rather extensively as a third-body or so-called "inert gas" in kinetic systems. While its inertness to methyl radicals has been questioned[161], there appears to be no doubt that its activation energy for reaction with hydrogen atoms[162] exceeds 20 kcal. mole^{-1}.

This compound also possesses a comparatively large ionisation potential (15.3 eV)[163,164], and one of the largest known cross-sections for the capture of thermal electrons. The latter process has been studied in considerable detail by beam, swarm and microwave techniques[104,165–170]. The initial attachment gives rise to a vibrationally excited ion[169,170], *viz.*

$$e + SF_6 \rightleftharpoons SF_6^{-*}$$

When formed from thermal electrons at room temperature its lifetime with respect to auto-ionisation is about 25 μsec[170]. Consequently few of these ions could escape collisional stabilization

$$SF_6^{-*} + M \rightarrow SF_6^- + M$$

in a gas at normal pressure, and in most situations the capture reaction can be regarded as a two-body process, *viz.*

$$e + SF_6 \rightarrow SF_6^- \tag{35}$$

The cross-section for electron attachment shows an inverse dependence on electron velocity[170], and for this reason there has been a marked inconsistency in the cross-sections obtained by different methods. Mahan and Young[104] have reported a capture rate coefficient for thermal electrons of 2×10^{14} l.mole^{-1}.sec^{-1}. This was obtained by a microwave technique in the presence of helium as a moderating gas.

The comparatively high ionisation potential of sulphur hexafluoride and its inertness toward attack by thermal hydrogen atoms have lead to its use as a specific scavenger for electrons in several irradiated systems. This has already been illustrated in section 1.7.2. The ionisation processes in SF_6 have been studied by beam techniques[171], but to date there has been no investigation of its radiolysis *per se*. Such a study would be well worthwhile.

5.2 DISULPHUR DECAFLUORIDE

Disulphur decafluoride is thermally less stable than sulphur hexafluoride. Its tendency to react with glass and mercury precludes kinetic investigations in conventional pyrex apparatus, but Trost and McIntosh[172] have studied the thermal decomposition in a copper vessel fitted with a diaphragm manometer. Within the experimental error of ±3 % the stoichiometry of the reaction was

$$S_2F_{10} = SF_6+SF_4 \tag{68}$$

Over the temperature range 434–455 °K the rate was first order in the decafluoride up to about 70 % decomposition. Although the main part of the reaction was homogeneous, there was evidence of a surface-catalysed component. The importance of this increased in vessels packed with copper wire, and its rate was also dependent on pressure, attaining a maximum above 220 torr at 444 °K. By using surface to volume ratios which differed over a range of five, the contributions of the homogeneous and heterogeneous reactions were resolved. The first-order rate coefficient of the former was found to obey the Arrhenius expression

$$k_{68} = 4.98\times10^{18} \exp\left(-\frac{49{,}200}{RT}\right) \text{ sec}^{-1}$$

The activation energy of the heterogeneous reaction was estimated to be 39–40 kcal.mole^{-1}.

The authors suggested a Rice–Herzfeld mechanism of the form

$$S_2F_{10} \rightarrow 2\,SF_5 \tag{69}$$

$$SF_5+S_2F_{10} \rightarrow SF_6+S_2F_9 \tag{70}$$

$$S_2F_9 \rightarrow SF_4+SF_5 \tag{71}$$

$$S_2F_9+SF_5 \rightarrow SF_6+2\,SF_4 \tag{72}$$

The choice of reaction (72) as the termination reaction is dictated by the overall order of unity, if reaction (69) is in its first-order pressure region. This is probably true above 200 torr. With the usual approximations the mechanism leads to the rate expression

$$-\frac{d[S_2F_{10}]}{dt} = \left(\frac{k_{69}k_{70}k_{71}}{k_{72}}\right)^{\frac{1}{2}} [S_2F_{10}]$$

Since E_{72} will be small, $E_{69}+E_{70}+E_{71}$ must be about 98 kcal.mole^{-1}. The strength of the S–S bond[173, 174] should be close to 65 kcal.mole^{-1}. Thus the sum of E_{70} and E_{71} would have to be approximately 33 kcal.mole^{-1}, which does not appear unreasonable.

The homogeneous reaction was found to be catalysed by small percentages of nitric oxide and acetylene dichloride without any detectable change in the overall stoichiometry. This observation suggests the occurrence of additional initiation processes of the type

$$S_2F_{10}+NO \rightarrow S_2F_9+FNO$$

These are analogous to the reactions postulated for hydrocarbon systems by Laidler *et al.*[175,176]. With certain hydrocarbons there is almost no inhibition region, and the main effect of nitric oxide is to accelerate the rate of decomposition as in the present instance. Further work on this interesting aspect of the decomposition of S_2F_{10} is obviously desirable. In addition, attention should be given to the effects of inert gases on the orders of the individual reactions in the above mechanism, and on the rate of the heterogeneous reaction.

REFERENCES†

1 G. K. Rollefson and J. E. Booher, *J. Am. Chem. Soc.*, 53 (1931) 1728.
2 C. F. Goodeve and A. W. C. Taylor, *Proc. Roy. Soc. (London), Ser. A*, 154 (1936) 181.
3 J. Romand, *Ann. Phys. (Paris)*, 4 (1948) 527; *Compt. Rend.*, 227 (1948) 117.
4 R. S. Mulliken, *Phys. Rev.*, 51 (1937) 310.
5 E. Warburg, *Sitzber. Preuss. Akad. Wiss.*, 26 (1918) 300.
6 M. Bodenstein and F. Lieneweg, *Z. Physik. Chem. (Leipzig)*, 119 (1926) 123.
7 B. Lewis, *Proc. Natl. Acad. Sci. U.S.*, 13 (1927) 720; *J. Phys. Chem.*, 32 (1928) 270.
8 K. F. Bonhoeffer and L. Farkas, *Z. Physik. Chem. (Leipzig)*, 132 (1928) 235.
9 R. A. Ogg and R. R. Williams, *J. Chem. Phys.*, 11 (1943) 214.
10 R. A. Ogg and R. R. Williams, *J. Chem. Phys.*, 15 (1947) 691, 696.
11 R. A. Ogg and R. R. Williams, *J. Chem. Phys.*, 13 (1945) 586.
12 H. A. Schwarz, R. R. Williams and W. H. Hamill, *J. Am. Chem. Soc.*, 74 (1952) 6007.
13 J. H. Sullivan, *J. Chem. Phys.*, 30 (1959) 1292.
14 J. H. Sullivan, *J. Chem. Phys.*, 36 (1962) 1925.
15 J. L. Holmes and P. Rodgers, *Trans. Faraday Soc.*, 64 (1968) 2341.
16 R. M. Martin and J. E. Willard, *J. Chem. Phys.*, 40 (1964) 2999; 41 (1964) 3032.
17 R. J. Donovan, and D. Husain, *Trans. Faraday Soc.*, 62 (1966) 1050.
18 R. J. Donovan and D. Husain, *Trans. Faraday Soc.*, 62 (1966) 2643.
19 M. Bodenstein, *Chem. Ber.*, 26 (1893) 2603; *Z. Physik. Chem.*, 13 (1894) 56, 22 (1897) 1; 29 (1899) 295;
M. Bodenstein and W. Jost, *J. Am. Chem. Soc.*, 49 (1927) 1416;
D. Rittenberg and H. C. Urey, *J. Am. Chem. Soc.*, 56 (1934) 1885; *J. Chem. Phys.*, 2 (1934) 106;
K. H. Geib and A. Lendle, *Z. Physik. Chem.*, B32 (1936) 463;
J. C. L. Blagg and G. M. Murphy, *J. Chem. Phys.*, 4 (1936) 631;
A. H. Taylor and R. H. Crist, *J. Am. Chem. Soc.*, 63 (1941) 1377;
N. F. H. Bright and R. P. Hagerthy, *Trans. Faraday Soc.*, 43 (1947) 697.
20 S. W. Benson and R. Srinivasan, *J. Chem. Phys.*, 23 (1955) 200.

† Complete at the time the manuscript was submitted (September 1968).

21 J. H. Sullivan, *J. Chem. Phys.*, 39 (1963) 3001.
22 N. Semenov, *Problems in Chemical Kinetics*, Moscow, 1958, p. 407.
23 J. H. Sullivan, *J. Chem. Phys.*, 46 (1967) 73.
24 R. Engleman and N. R. Davidson, *J. Am. Chem. Soc.*, 82 (1960) 4770.
25 R. M. Noyes, *J. Chem. Phys.*, 47 (1967) 3097.
26 K. J. Laidler and J. N. Murrell, *Trans. Faraday Soc.*, 64 (1968) 371.
27 C. F. Goodeve and A. W. C. Taylor, *Proc. Roy. Soc. (London), Ser. A*, 152 (1935) 221.
28 W. C. Price, *Proc. Roy. Soc. (London), Ser. A*, 167 (1938) 216.
29 R. M. Martin and J. E. Willard, *J. Chem. Phys.*, 40 (1964) 3007.
30 R. S. Mulliken, *J. Chem. Phys.*, 8 (1940) 382.
31 M. Bodenstein and G. Jung, *Z. Physik. Chem. (Leipzig)*, 121 (1926) 127.
32 D. Britton and R. M. Cole, *J. Phys. Chem.*, 65 (1961) 1302.
33 M. Bodenstein and S. C. Lind, *Z. Physik. Chem. (Leipzig)*, 57 (1906) 168.
34 J. A. Christiansen, *Kgl. Danske Videnskab. Selskab., Mat.-Fys. Medd.*, 1 (1919) 14.
35 K. F. Herzfeld, *Z. Elektrochem.*, 25 (1919) 301; *Ann. Physik*, 59 (1919) 635.
36 M. Polanyi, *Z. Elektrochem.*, 26 (1920) 50.
37 A. Levy, *J. Phys. Chem.*, 62 (1958) 570.
38 M. Bodenstein and H. Lutkemeyer, *Z. Physik. Chem. (Leipzig)*, 114 (1924) 208; M. Bodenstein and G. Jung, *Z. Physik. Chem. (Leipzig)*, 121 (1926) 127.
39 F. Bach, K. F. Bonhoeffer and E. A. Moelwyn-Hughes, *Z. Physik. Chem.*, 27B (1934) 71.
40 G. B. Kistiakowsky and E. R. van Artsdalen, *J. Chem. Phys.*, 12 (1944) 469.
41 R. N. Pease, *Equilibrium and Kinetics of Gas Reactions*, Princeton, 1942, p. 121.
42 G. C. Fettis and J. H. Knox, *Progress in Reaction Kinetics*, G. Porter (Ed.), Vol. 2, Pergamon, London, 1964.
43 M. Bodenstein, *Trans. Faraday Soc.*, 27 (1931) 413.
44 P. G. Ashmore and J. Chanmugam, *Trans. Faraday Soc.*, 49 (1953) 254; P. G. Ashmore, *5th Intern. Symp. Combustion, Pittsburgh*, 1954, p. 700.
45 H. Steiner and E. K. Rideal, *Proc. Roy. Soc. (London), Ser. A*, 173 (1939) 503.
46 W. H. Rodebush and W. C. Klingelhoefer, *J. Am. Chem. Soc.*, 55 (1933) 130.
47 M. A. A. Clyne and D. H. Stedman, *Trans. Faraday Soc.*, 62 (1966) 2164.
48 W. M. Jones, *J. Chem. Phys.*, 19 (1951) 78.
49 J. Bigeleisen, F. S. Klein, R. E. Weston and M. Wolfsburg, *J. Chem. Phys.*, 30 (1959) 1340.
50 F. S. Klein and M. Wolfsberg, *J. Chem. Phys.*, 34 (1961) 1494.
51 E. S. Fishburne, *J. Chem. Phys.*, 45 (1966) 4053.
52 T. A. Jacobs, N. Cohen and R. R. Giedt, *J. Chem. Phys.*, 46 (1967) 1958.
53 M. Bodenstein, H. Jockusch and Shing Hou-Chong, *Z. Anorg. Allgem. Chem.*, 231 (1937) 24.
54 J. B. Levy and B. K. W. Copeland, *J. Phys. Chem.*, 67 (1963) 2156.
55 J. B. Levy and B. K. W. Copeland, *J. Phys. Chem.*, 69 (1965) 408.
56 T. A. Jacobs, R. R. Giedt and N. Cohen, *J. Chem. Phys.*, 43 (1965) 3688.
57 J. A. Blauer, *J. Phys. Chem.*, 72 (1968) 79.
58 R. N. Compton, R. H. Huebner, P. W. Reinhardt and L. G. Christophorou, *J. Chem. Phys.*, 48 (1968) 901.
59 S. A. Frances, *Ph.D. Thesis*, Ohio State University, 1947.
60 R. F. Barrow and J. C. Stamper, *Proc. Roy. Soc. (London), Ser. A*, 263 (1961) 259, 277.
61 J. K. Jacques and R. F. Barrow, *Proc. Phys. Soc.*, 73 (1959) 538.
62 D. C. Frost, C. A. McDowell and D. A. Vroom, *J. Chem. Phys.*, 46 (1967) 4255.
63 H. J. Lempka, T. R. Passmore and W. C. Price, *Proc. Roy. Soc. (London), Ser. A*, 304 (1968) 53.
64 D. C. Frost and C. A. McDowell, *Can. J. Chem.*, 36 (1958) 39.
65 A. O. Nier and E. E. Hanson, *Phys. Rev.*, 50 (1936) 722.
66 J. D. Morrison and A. J. C. Nicholson, *J. Chem. Phys.*, 20 (1952) 1021.
67 R. Thornburn, *Proc. Phys. Soc.*, 73 (1959) 122.
68 R. E. Fox, *J. Chem. Phys.*, 32 (1960) 385.
69 E. E. Hanson, *Phys. Rev.*, 51 (1937) 86.
70 R. E. Fox, *J. Chem. Phys.*, 26 (1957) 1281.

71 R. S. Davidow, R. A. Lee and D. A. Armstrong, *J. Chem. Phys.*, 45 (1966) 3364.
72 W. G. Johnston and J. R. Arnold, *J. Chem. Phys.*, 21 (1953) 1499.
73 F. H. Dorman and J. D. Morrison, *J. Chem. Phys.*, 35 (1961) 575.
74 *Manufacturing Chemists Association Research Project on Mass Spectral Data, Serial Number* 9, Carnegie Institute of Technology, Pittsburgh, Pa., 1959.
75 (a) R. E. Huffman, J. C. Larabee and Y. Tanaka, *J. Chem. Phys.*, 47 (1967) 856.
(b) C. E. Moore, *Atomic Energy Levels, N.B.S. (U.S.A.) Circ. Number* 467, 1–3, 1949, 1952 and 1958.
76 G. Herzberg, *Atomic Spectra and Atomic Structure*, Dover Publications, New York, 1944.
77 A. G. Harrison and J. C. J. Thynne, *Can. J. Chem.*, 45 (1967) 1321.
78 D. P. Stevenson and D. O. Schissler, *The Chemical and Biological Action of Radiations*, M. Haissinsky (Ed.), Vol. V, Academic Press, New York, 1961, p. 249.
79 R. S. Davidow and D. A. Armstrong, *Radiation Res.*, 28 (1966) 143.
80 R. M. Leblanc and J. A. Herman, *J. Chim. Phys.*, (1966) 1055.
81 T. S. Taylor, *Phil. Mag.*, 21 (1911) 571.
82 A. Henglein and G. A. Muccini, *Z. Naturforsch.*, 17a (1962) 452.
83 L. B. Loeb, *Proc. Natl. Acad. Sci. U.S.*, 12 (1926) 35.
84 A. M. Hogg, R. M. Haynes and P. Kebarle, *J. Am. Chem. Soc.*, 88 (1966) 28.
85 S. C. Lind, *Radiation Chemistry of Gases*, Reinhold, New York, 1961, Chap. 3.
86 J. Vandame, *Bull. Soc. Chim. Belges*, 41 (1932) 597.
87 J. D. Chen, *M. Sc. Thesis*, University of Calgary, 1967.
88 R. A. Lee, R. S. Davidow and D. A. Armstrong, *Can. J. Chem.*, 42 (1964) 1906.
89 R. A. Lee, *Ph.D. Thesis*, University of London, 1967.
90 R. Gillerot, *Bull. Soc. Chim. Belges*, 39 (1930) 503.
91 S. C. Lind and R. Livingston, *J. Am. Chem. Soc.*, 58 (1936) 612.
92 J. D. Chen and D. A. Armstrong, *J. Chem. Phys.*, 48 (1968) 2310.
93 E. G. Zubler, W. H. Hamill and R. R. Williams, *J. Chem. Phys.*, 23 (1955) 1263.
94 K. G. Brattain, *J. Phys. Chem.*, 42 (1938) 617.
95 P. Gunther and H. Leichter, *Z. Physik. Chem.*, B34 (1936) 443.
96 H. Eyring, J. O. Hirschfelder and H. S. Taylor, *J. Chem. Phys.*, 4 (1936) 570.
97 R. S. Davidow and D. A. Armstrong, *J. Chem. Phys.*, 48 (1968) 1235.
98 L. G. Christophorou, R. N. Compton and H. W. Dickson, *J. Chem. Phys.*, 48 (1968) 1949.
99 J. L. Magee and M. Burton, *J. Am. Chem. Soc.*, 73 (1951) 523.
100 R. A. Lee and D. A. Armstrong, *Intern. J. Appl. Radiation Isotopes*, 19 (1968) 585.
101 G. Herzberg, *Spectra of Diatomic Molecules*, Van Nostrand, New York, 1965.
102 R. S. Berry and C. W. Reimann, *J. Chem. Phys.*, 38 (1963) 1540.
103 E. W. McDaniel and M. R. C. McDowell, *Phys. Rev.*, 114 (1959) 1028.
104 B. H. Mahan and C. E. Young, *J. Chem. Phys.*, 44 (1966) 2192.
105 D. A. Armstrong and R. A. Back, *Can. J. Chem.*, 45 (1967) 3079.
106 G. R. A. Johnson and J. L. Redpath, *J. Phys. Chem.*, 72 (1968) 765.
107 R. C. Rumfeldt and D. A. Armstrong, *J. Phys. Chem.*, 68 (1964) 761.
108 C. J. Delbecque, W. Hayes and P. H. Yuster, *Phys. Rev.*, 121 (1961) 1043.
109 W. B. Person, *J. Chem. Phys.*, 38 (1963) 109.
110 D. C. Frost and C. A. McDowell, *Can. J. Chem.*, 38 (1960) 407.
111 L. Raff and H. A. Pohl, *Solvated Electron, Advances in Chemistry Series*, E. J. Hardt (Ed.), Vol. 50, Am. Chem. Soc., Washington, 1965, p. 173.
112 H. J. Bernstein, private communication, 1967.
113 D. H. McDaniel and R. E. Valee, *Inorg. Chem.*, 2 (1963) 996.
114 R. A. Lee, *Nature*, 216 (1967) 57.
115 (a) G. Glockner and R. E. Peck, *J. Chem. Phys.*, 4 (1936) 658.
(b) M. E. Peach and T. C. Waddington, *J. Chem. Soc.*, (1960) 2329; (1961) 1238.
116 E. J. Hart and R. L. Platzmann, *Mechanisms in Radiobiology*, M. Errera and A. Forseberg (Eds.), Academic Press, New York, 1961.
117 D. A. Armstrong, *Can. J. Chem.*, 40 (1962) 1385.
118 M. Walker and D. N. Eldred, *Ind. Eng. Chem.*, 17 (1925) 1074;
T. H. Volker, *Angew. Chem.*, 69 (1957) 728; 72 (1960) 379.

119 D. Hummel and O. Janssen, *Z. Physik. Chem. (Frankfurt)*, 31 (1962) 111.
120 K. D. J. Root, M. C. R. Symonds and B. C. Weatherley, *Mol. Phys.*, 11 (1966) 161.
121 S. C. Lind, D. C. Bardwell and J. H. Perry, *J. Am. Chem. Soc.*, 48 (1926) 1557.
122 S. C. Lind and D. C. Bardwell, *J. Am. Chem. Soc.*, 48 (1926) 1575.
123 P. A. Staats, H. W. Morgan and J. H. Goldstein, *J. Chem. Phys.*, 25 (1956) 582.
124 Saku Takahashi, *Mem. Defense Acad., Math., Phys., Chem., Eng., (Yokosuka, Japan)*, 3 (3) (1964) 161.
125 M. Bodenstein, S. Lenher and C. Wagner, *Z. Physik. Chem.*, B3 (1929) 459.
126 G. K. Rollefson and S. Lenher, *J. Am. Chem. Soc.*, 52 (1930) 500.
127 L. Fowler and J. J. Beaver, *J. Am. Chem. Soc.*, 75 (1953) 4186.
128 F. S. Dainton and W. G. Burns, *Trans. Faraday Soc.*, 48 (1952) 39.
129 M. Bodenstein and G. Plaut, *Z. Physik. Chem.*, 110 (1924) 399.
130 R. Thompson, *Trans. Faraday Soc.*, 37 (1941) 254.
131 P. Goldfinger, *J. Chem. Phys.*, 29 (1958) 456.
132 D. R. Stranks, *Trans. Faraday Soc.*, 51 (1955) 499, 514, 524.
133 J. N. Bradley and R. Tuffnell, *Proc. Roy. Soc. (London), Ser. A*, 280 (1964) 198.
134 M. E. Jacox and D. E. Milligan, *J. Chem. Phys.*, 43 (1965) 866.
135 S. Lenher and H. J. Schumacher, *Z. Physik. Chem.*, 135 (1928) 85.
136 H. J. Schumacher and P. Bergman, *Z. Physik. Chem.*, B13 (1931) 269.
137 O. Ruff and Shih-Chang Li, *Z. Anorg. Allgem. Chem.*, 242 (1939) 272.
138 Shih-Chang Li, *J. Chinese Chem. Soc.*, 11 (1944) 14.
139 L. H. Piette, F. A. Johnson, K. A. Booman and C. B. Colburn, *J. Chem. Phys.*, 35 (1961) 1481.
140 F. A. Johnson and C. B. Colburn, *J. Am. Chem. Soc.*, 83 (1961) 3043.
141 M. D. Harmony and R. J. Myers, *J. Chem. Phys.*, 37 (1962) 636.
142 J. T. Herron and V. H. Dibeler, *J. Res. Natl. Bur. Std.*, 65A (1961) 405–9.
143 A. Kennedy and C. B. Colburn, *J. Chem. Phys.*, 35 (1961) 1892.
144 F. A. Johnson, *Advan. Chem. Ser.*, 36 (1962) 123.
145 A. P. Modica and D. F. Hornig, *Techn. Rept. No.* 4, *O.N.R. Contract Nonr* 1858 (26), *N.R.* 357-275, Princeton University, Oct. 1963.
146 L. M. Brown and B. de B. Darwent, *J. Chem. Phys.*, 42 (1965) 2158.
147 H. S. Johnson, *Gas Phase Reaction Rate Theory*, Ronald Press, New York, 1966.
148 R. W. Diesen, *J. Chem. Phys.*, 41 (1964) 3256; 45 (1966) 759; *J. Phys. Chem.*, 72 (1968) 108.
149 A. P. Modica and D. F. Hornig, *J. Chem. Phys.*, 43 (1965) 2739; 45 (1966) 760.
150 C. L. Bumgardner and M. Lustig, *Inorg. Chem.*, 2 (1963) 662.
151 J. J. Comeford and D. E. Mann, *Spectrochim. Acta*, 21 (1965) 197.
152 J. G. A. Griffiths and R. G. W. Norrish, *Proc. Roy. Soc. (London), Ser. A*, 130 (1931) 591; 135 (1932) 69.
153 A. G. Briggs and R. G. W. Norrish, *Proc. Roy. Soc. (London), Ser. A*, 278 (1964) 27.
154 A. J. Apin, *Acta Physicochem. U.S.S.R.*, 12 (1940) 406.
155 P. G. Ashmore, *Nature*, 172 (1953) 449.
156 W. A. Noyes and W. F. Tuley, *J. Am. Chem. Soc.*, 47 (1925) 1336.
157 T. L. Cottrel, *The Strength of Chemical Bonds*, Butterworths, 2nd edn., London, 1958.
158 V. A. Shushunov and L. Z. Pavlova, *Zh. Neorgan. Khim.*, 2 (1957) 2272.
159 E. J. Bowen, *J. Chem. Soc.*, 123 (1923) 1199.
160 W. C. Schumb, J. G. Trump and G. L. Priest, *Ind. Eng. Chem.*, 41 (1949) 1348.
161 L. Batt and F. R. Cruickshank, *J. Phys. Chem.*, 70 (1966) 723.
162 C. P. Fenimore and G. W. Jones, *Combust. Flame*, 8 (1964) 231;
D. R. Safrany, private communication, 1966.
163 D. C. Frost, C. A. McDowell, J. S. Sandhu and D. A. Vroom, *J. Chem. Phys.*, 46 (1967) 2008.
164 K. Codling, *J. Chem. Phys.*, 44 (1966) 4401.
165 N. S. Buchel'nikova, *Soviet Phys. JETP*, 35 (1959) 783.
166 R. K. Asundi and J. D. Craggs, *Proc. Phys. Soc. (London)*, 83 (1964) 611.
167 J. B. Hasted and S. Beg, *Brit. J. Appl. Phys.*, 16 (1965) 74.
168 D. Rapp and D. D. Briglia, *J. Chem. Phys.*, 43 (1965) 1480.

169 D. Edeleson, J. E. Griffiths and K. B. McAfee, Jr., *J. Chem. Phys.*, 37 (1962) 917.
170 R. N. Compton, L. G. Christophorou, G. S. Hurst and P. W. Reinhardt, *J. Chem. Phys.*, 45 (1966) 4634.
171 R. E. Fox and R. K. Curran, *J. Chem. Phys.*, 34 (1961) 1595.
172 W. R. Trost and R. L. McIntosh, *Can. J. Chem.*, 29 (1951) 508.
173 S. Unner, *Acta Chem. Scand.*, 9 (1955) 837.
174 H. Mackle, *Tetrahedron*, 19 (1963) 1159.
175 B. W. Wojciechowski and K. J. Laidler, *Can. J. Chem.*, 38 (1960) 1027.
176 M. H. Back and K. J. Laidler, *Can. J. Chem.*, 44 (1966) 215.

Chapter 4

The Decomposition of Metal Alkyls, Aryls, Carbonyls and Nitrosyls

S. J. W. PRICE

1. Introduction

The literature cited in this article covers references listed in chemical abstracts to the end of 1961 and in current chemical papers after that. A degree of selection has been exercised in omitting some references that are now of limited value. Although metal nitrosyls are included in the scope of this chapter, no kinetic data on their decomposition is available and they will not be considered further. The data on metal carbonyls is limited and will be dealt with in the first section. The decomposition of metal alkyls and aryls has been extensively investigated. These compounds will be discussed in groups based on the position of the central metal atom in the periodic table and, when warranted, a further subdivision will be made based on the attached organic radicals.

2. Homogeneous decomposition of metal carbonyls

2.1 BORINE CARBONYL

Four studies of the thermal decomposition of borine carbonyl have been made[1-4]. In the absence of a radical scavenger, Burg[1] proposed a two-step mechanism

$$BH_3CO \rightleftharpoons BH_3 + CO \tag{1}$$

$$BH_3 + BH_3CO \rightleftharpoons B_2H_6 + CO \tag{2}$$

Subsequent work by Fu and Hill[2] indicated Burg's mechanism was inadequate, but recalculation of their results by Garabedian and Benson[5] confirmed the original mechanism. Thus, assuming reaction (−2) is negligible in the early stages of the reaction

$$2\,\frac{k_1k_2}{k_{-1}}\,t = \frac{a}{(a-x)} - \left(1 - \frac{k_2}{k_{-1}}\right) \ln \frac{a}{(a-x)}$$

References pp. 254–257

where a and $(a-x)$ are the concentrations of carbonyl initially and at time t, respectively. Using both Burg's[1] and Fu and Hill's[2] data, Garabedian and Benson[5] obtain

$$\frac{k_1 k_2}{k_{-1}} = 3.8\times 10^{13} \exp(-26{,}750/RT) \text{ sec}^{-1}$$

Combining this equation with standard entropy and enthalpy data, they estimate $A_1/A_{-1} = 2.63\times 10^5$ mole.l^{-1}, $A_2 = 1.45\times 10^8$ l.mole^{-1}.sec^{-1}, $A_{-2} = 2.09\times 10^8$ l.mole^{-1}.sec^{-1}, $E_{-2} = 17.7$ kcal.mole^{-1}. Assuming $A_{-1} = 10^8$–10^9 l.mole^{-1}.sec^{-1} gave $A_1 = 2.5\times 10^{13}$–2.5×10^{14} sec^{-1}. Based on the observed value $k_2/k_{-1} \leqslant 10^{-3}$ they calculate $E_2 - E_{-1} \geqslant 3.02$ kcal.mole^{-1} and hence $E_1 \leqslant 23.7$ kcal.mole^{-1}.

Felhner and Koski[3] tried to isolate reaction (1) by studying the decomposition in a fast flow system coupled to a mass spectrometer. Using the mean of Garabedian and Benson's estimate[5] of A_1, 8.0×10^{13} sec^{-1}, they obtain $E_1 = 23.1\pm 2$ kcal.mole^{-1}. Because of the experimental difficulties encountered, the significance of this result is questionable.

Grotewold *et al.*[4] determined k_1 by studying the decomposition of borine carbonyl in the presence of triethylamine. At constant total pressure the process may be described by reactions (1), (3) and (4)

$$BH_3CO \rightarrow BH_3 + CO \qquad (1)$$

$$BH_3 + Et_3N \rightarrow Et_3NBH_3 \qquad (3)$$

$$BH_3CO + Et_3N \rightarrow Et_3NBH_3 + CO \qquad (4)$$

The first-order experimental rate coefficient, $k = k_1 + k_4[Et_3N]$, was found to be independent of the triethylamine concentration indicating that the contribution of reaction (4) is negligible. A previous study using trimethylamine[6] showed that the analogue of reaction (4) was much faster than reaction (1).

The Lindemann–Hinshelwood treatment of values of k_1 determined over the pressure range 22–615 torr (7–19 torr borine carbonyl, remainder propane) at 0 °C gives a high-pressure limiting value for k_1, $k_\infty = 1.88\times 10^{-4}$ sec^{-1}, and $P_{\frac{1}{2}} = 90$ torr. At this pressure over the temperature range −6–32 °C $k_1 = 5.75\times 10^{10} \exp(-18{,}600/RT)$ sec^{-1}. Direct determination of the high-pressure activation energy, E_∞, was not attempted but a value of 19.6 ± 1 kcal.mole^{-1} was estimated by applying Slater's Theory[7] to the 90 torr results. An alternate procedure using a "selected" value of $A_1 = 2.5\times 10^{13}$ sec^{-1} gave $E = 21.4\pm 1.2$ kcal.mole^{-1}. If E_{-1} is zero E_∞ may be identified with D(B–C) in borine carbonyl. Using D(B–C) = 21.4 kcal.mole^{-1} and Burg's[1] equilibrium data places an upper limit on D(B–B) in diborane of 33.5 ± 2.4 kcal.mole^{-1}.

2.2 CHROMIUM, MOLYBDENUM AND TUNGSTEN CARBONYLS

The photolysis of $M(CO)_6$, M = Cr, Mo or W, in the presence of triethylamine[8] is adequately represented by reactions (1) and (2)

$$M(CO)_6 \rightarrow M(CO)_5+CO \quad (1)$$

$$M(CO)_5+Et_3N \rightarrow M(CO)_5NEt_3 \quad (2)$$

Using light of 3660 A the quantum yield of reaction (1), ϕ_1, was unity at zero time. At any other time some light is absorbed by $M(CO)_5NEt_3$ giving an apparent value of $\phi_1 < 1$. In the absence of Et_3N quantitative recombination of $M(CO)_5$ and CO occur when the irradiation is stopped[9].

Electron impact studies[10] of the appearance potentials and reaction cross-sections for the decomposition of $W(CO)_6$ have shown that this decomposition occurs by a cascade mechanism

$$\begin{aligned} W(CO)_6 &\rightarrow W(CO)_5{}^+ + CO \\ W(CO)_5 &\rightarrow W(CO)_4{}^+ + CO \\ &\ \vdots \\ W(CO)^+ &\rightarrow W^+ + CO \end{aligned}$$

although the direct decomposition of excited $W(CO)_6{}^+$ to $W(CO)_3{}^+$ was also observed. Similar experimental results[11, 12] for $Fe(CO)_5$, $Ni(CO)_4$, $Co_2(CO)_8$, $Mn_2(CO)_{10}$, $C_2H_5V(CO)_4$, $C_2H_5Co(CO)_2$, $C_6H_5Mo(CO)_2NO$, $(C_2H_5)_2Ni$ and $(C_2H_5)_2Fe$ indicate that these compounds may also decompose by a cascade mechanism. Negative ion mass spectra[13] of $Ni(CO)_4$, $Fe(CO)_5$, $Cr(CO)_6$, $Mo(CO)_6$ and $W(CO)_6$ also indicate cascade dissociation.

The decomposition of $Cr(CO)_6$ has also been studied using a shock tube[14]. The maximum in the chromium emission spectra occurred at 1000 μsec at 1000 °C and at 150 μsec at 1950 °C. The time required for cascade dissociation to Cr + 6 CO is approximately 1 μsec. The delay in the emission spectra was attributed to excitation of the chromium by vibrationally excited CO.

2.3 IRON AND NICKEL CARBONYLS

The decomposition of $Fe(CO)_5$ by photolysis[15] and ^{60}Co-γ radiolysis[16] have been reported. The mechanism of the photolysis is adequately represented by reactions (1) and (2)

$$Fe(CO)_5 \rightarrow Fe(CO)_4+CO \quad (1)$$

$$Fe(CO)_4+Fe(CO)_5 \rightarrow Fe_2(CO)_9 \quad (2)$$

References pp. 254–257

In the radiolysis the overall process shown as reaction (3)

$$3\,Fe(CO)_5 = Fe_3(CO)_{12}+3\,CO \qquad (3)$$

also occurs. The G values (yield per 100 eV) at 22 °C are $G_{CO} = 12.5$, $G_{Fe_2(CO)_9}=6.8$ and $G_{Fe_3(CO)_{12}} = 1.5$. These values are essentially independent of dose. The temperature dependence of G_{CO} is given by $G_{CO} = 15.1\exp(-104/RT)$. The observed value of G_{CO} at 22 °C is 1.2 units above the value required by the mechanism, $G_{CO} = G_{Fe_2(CO)_9}+3\,G_{Fe_3(CO)_{12}} = 11.3$. Barzynski and Hummel[16] attribute this to the formation of additional CO by reaction (4).

$$n\,Fe(CO)_5 \rightleftharpoons [Fe(CO)_m]_n+n\,(5-m)CO \qquad (4)$$

^{60}Co-γ irradiation of $Ni(CO)_4$[16] gave $G_{CO} < 0.1$ at -18 °C indicating almost complete recombination of any dissociation products. This confirmed the observation of Garrett and Thompson[15,17] that even though $Ni(CO)_4$ vapour gave continuous adsorption from 3950 A to shorter wavelengths, no overall decomposition was observed. Decomposition in *n*-hexane or carbon tetrachloride produced Ni and CO. The experimental results are consistent with the reactions

$$Ni(CO)_4 \rightleftharpoons Ni(CO)_3+CO \qquad (5)$$

$$Ni(CO)_3 = Ni+3\,CO \qquad (6)$$

Reaction (6) is a complex process which has been partially resolved by flash photolysis experiments[18]. At low conversion the rate of reaction was proportional to the square of the flash energy. This was taken as evidence that reaction (7)

$$Ni(CO)_3 \rightleftharpoons Ni(CO)_2+CO \qquad (7)$$

followed reaction (5). The mechanism by which $Ni(CO)_2$ is converted to Ni+2 CO is still not firmly established. Callear[18] proposed the overall process

$$2\,Ni(CO)_2 = Ni_2+4\,CO \qquad (8)$$

With[19] D(Ni–Ni) = 55 kcal.mole^{-1} and assuming the average Ni–CO bond dissociation energy in $Ni(CO)_2$ is the same (35.2 kcal.mole^{-1})[20] as in $Ni(CO)_4$, reaction (8) would be endothermic by approximately 85 kcal.mole^{-1}. It is therefore unlikely that this reaction occurs by a single homogeneous step. It might proceed by reaction (9)

$$2\,Ni(CO)_2 = Ni_2(CO)_2+2\,CO \qquad (9)$$

followed by the heterogeneous dissociation of $Ni_2(CO)_2$ to 2 Ni+2 CO. These steps are consistent with the observation of Callear[18] that adsorption spectra taken during and after the flash show only a minute concentration of Ni atoms.

The thermal decomposition of $Ni(CO)_4$ has been extensively investigated by Chan and McIntosh[21]. They assume a mechanism for the homogeneous portion of the reaction based on reactions (5) and (6). This gives

$$\left(\frac{dx}{dt}\right)_{\text{homo}} = \frac{k_5(p-x)}{1+b_2x}$$

where p is the initial pressure of $Ni(CO)_4$, x the amount decomposed (expressed in units of pressure), and $b_2 = 4\,k_{-5}/k_6$.

The heterogeneous reaction is assumed to depend on the surface area and fraction of the surface covered. Allowing for the competition between $Ni(CO)_4$ and CO for adsorption sites

$$\left(\frac{dx}{dt}\right)_{\text{hetero}} = \frac{kAb(p-x)}{1+b(p-x)+b_1x}$$

where A is the surface area, b the Langmuir constant for adsorption of $Ni(CO)_4$ and $b_1/4$ the Langmuir constant for adsorption of CO. The decomposition was studied from 35 °C to 80 °C over the pressure range 15–80 torr. Glass wool was used as a packing material when studying the heterogeneous reaction. The initial composite rate is given by

$$\frac{dx}{dt} = \frac{k\,Abp}{1+bp} + k_5p$$

Extrapolation of the nearly linear initial composite rate *vs.* A to zero surface gave values of k_5 at various temperatures and pressures. The homogeneous reaction was first order and the temperature dependence gave $k_5 = 4.16\times10^9$ $\exp(-19{,}100/\boldsymbol{R}T)$ $\sec^{-1}$. Using the same data but plotting $(dx/dt)_{\text{hetero}}$ *vs.* p^{-1} gave $k = 6.3\times10^4 \exp(-14{,}300/\boldsymbol{R}T)^{-1}$ torr. $\sec^{-1}.\text{cm}^{-2}$. The activation energy for the heterogeneous decomposition on a steel or nickel surface[21a] is 22 kcal. mole^{-1}.

The experimental work of Chan and McIntosh[21] seems self-consistent and reliable. However, the mechanism used as a basis for their calculations is over-simplified. They report values of $b_2 = 4\,k_{-5}/k_6$ which increase with decreasing temperature and pressure. This tendency is consistent with a mechanism based on reactions (5) and (7), followed by competition between reaction (9) and reactions (10) and (11). Reactions (12) and (13) would complete the dissociation process.

$$Ni(CO)_2 \rightarrow NiCO + CO \tag{10}$$

$$2\ NiCO \rightarrow Ni_2(CO)_2 \tag{11}$$

$$Ni_2(CO)_2 \rightarrow 2\ Ni + 2\ CO \tag{12}$$

$$Ni(CO) \rightarrow Ni + CO \tag{13}$$

Even with these possible modifications, the mechanism used by Chan and McIntosh[21] may be unsatisfactory. A_5 is abnormally low for a unimolecular decomposition. $E_5 \geqslant D[(CO)_3Ni\text{–}CO] = 19.1$ kcal.mole^{-1} also seems low in view of the fact that the mean Ni–CO bond dissociation energy[20] in $Ni(CO)_4$ is 35.2 kcal.mole^{-1}. It is reasonable to assume that $A_5 \geqslant A_7$. For reaction (5) to be rate-controlling the maximum value of E_7 would therefore be approximately 17 kcal.mole^{-1}. Assuming all back reactions have zero activation energy would then give a minimum value $D[(CO)Ni\text{–}CO] + D(Ni\text{–}CO) = 105$ kcal.mole^{-1}. This would eliminate the sequence of reactions (10)–(13) and would render reaction (9) unlikely because it would now be expected to be at least 50 kcal.mole^{-1} endothermic. Therefore, if the value of E_5 reported is correct, no logical sequence of reactions has been found that would yield the observed overall process $Ni(CO)_4 = Ni + 4\ CO$.

2.4 COBALT CARBONYLS

Heck has studied the reaction of triphenylphosphine[22–24] and trimethylolpropane phosphite[25] with the substituted cobalt carbonyls listed in Tables 1–4. The general mechanism for the reaction of the acyl cobalt carbonyls shown in Table 1 in the presence of triphenylphosphine is

$$RCOCo(CO)_4 \rightleftharpoons RCOCo(CO)_3 + CO \tag{1}$$

$$RCOCo(CO)_3 + P(C_6H_5)_3 \rightarrow RCOCo(CO)_3P(C_6H_5)_3 \tag{2}$$

Applying the steady-state approximation

$$-\frac{d[RCOCo(CO)_4]}{dt} = k_1[RCOCo(CO)_4]\frac{k_2[P(C_6H_5)_3]}{k_{-1}[CO] + k_2[P(C_6H_5)_3]}$$

Heck found that the reaction was first-order in $RCOCo(CO)_4$ and independent of $P(C_6H_5)_3$ and CO concentrations over at least two half-lives of the reaction. Neither substitution of trimethyl phosphite for $P(C_6H_5)_3$ nor carrying out the reaction in the dark had any effect on the observed rate. He concluded that under his experimental conditions $k_2[P(C_6H_5)_3] \gg k_{-1}[CO]$ so that the measured rate coefficient is k_1.

TABLE 1

THE DECOMPOSITION OF $RCOCo(CO)_4$ IN THE PRESENCE OF TRIPHENYL PHOSPHINE

R	*Solvent*	*Temperature* (°*C*)	*k* (*sec*$^{-1}$)
CH_3	Ether	0	1.01×10^{-3}
CH_3	Ether	25	2.55×10^{-2}
CH_3	Tetrahydrofuran	0	5.90×10^{-4}
CH_3	Methylenechloride	0	5.28×10^{-4}
CH_3	Toluene	0	6.46×10^{-4}
CH_3	Toluene	25	1.89×10^{-2}
$CH_3(CH_2)_4$	Ether	0	1.07×10^{-3}
$(CH_3)_2CH$	Ether	0	2.11×10^{-3}
$(CH_3)_3C$	Ether	0	$8.6\ \times10^{-2}$
$(CH_3)_2CHCH_2$	Ether	0	1.21×10^{-3}
$(CH_3CH_2)_2CH$	Ether	0	1.87×10^{-3}
$Cl(CH_2)_3$	Ether	0	7.65×10^{-4}
CH_3OCH_2	Ether	0	2.80×10^{-4}
CF_3CO	Ether	0	9.41×10^{-5}
$CH_3CH{=}CH$	Ether	0	6.18×10^{-2}
C_6H_5	Ether	0	3.44×10^{-2}
p-$CH_3OC_6H_4$	Ether	0	4.96×10^{-2}
m-$CH_3OC_6H_4$	Ether	0	1.81×10^{-2}
p-$NO_2C_6H_4$	Ether	0	1.65×10^{-2}
o-$CH_3C_6H_4$	Ether	0	2.82×10^{-2}
2,4,6-$(CH_3)_3C_6H_2$	Ether	0	1.98×10^{-4}
$C_6H_5CH_2$	Ether	0	1.32×10^{-3}

Concentration of cobalt compound 0.03–0.06 *M*.
Concentration of $P(C_6H_5)_3$, 0.063 *M* in either, 0.12–0.15 *M* in other solvents.

The compounds used allow the steric and electronic effects of the acyl ligand to be studied relatively independently of each other. From the data in Table 1, Heck[22] reached the following conclusions:

(*i*) Variation in electronic properties of the first six acyl groups listed in Table 1 are insignificant. The observed variation in rates correspond to variations in steric strain with tertiary group substitution on the acyl carbon having the major effect.

(*ii*) Electron-withdrawing substituents (Cl, CH_3O, CF_3) make dissociation more difficult by removing electrons from the metal so that loss of a pair of electrons with the dissociating CO ligand is less favourable.

(*iii*) The result for crotonoylcobalt tetracarbonyl was obtained using an equilibrium mixture of 44 % cyclic tricarbonyl and 56 % open chain tetracarbonyl. With excess triphenylphosphine both forms were converted to crotonoylcobalt tricarbonyl triphenylphosphine. Under the experimental conditions used, reaction (3) is much faster than reaction (−1).

$$CH_3CH{=}CHCOCo(CO)_4 \rightleftharpoons CH_3CH{=}CHCOCo(CO)_3 + CO \quad (1)$$

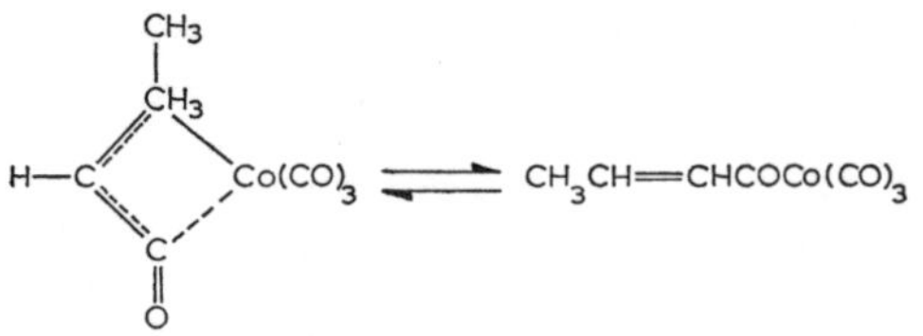

$$\underset{H-C}{\overset{CH_3}{\underset{|}{\overset{|}{C}}}}\underset{\overset{\|}{O}}{\overset{CH_3}{\underset{|}{C}}}Co(CO)_3 \rightleftharpoons CH_3CH=CHCOCo(CO)_3 \qquad (2)$$

$$CH_3CH=CHCOCo(CO)_3 + P(C_6H_5)_3 \longrightarrow CH_3CH=CHCOCo(CO)_3P(C_6H_5)_3 \quad (3)$$

Hence the presence of the cyclic tricarbonyl does not influence the rate of reaction of the tetracarbonyl and the observed rate coefficient is k_1. Instead of acting as an electron-withdrawing group, the α, β double bond accelerates the rate of dissociation. This is attributed to direct coupling of the R group to the cobalt atom to form a cyclic transition state.

$$RCOCo(CO)_4 \rightleftharpoons \left[\text{cyclic intermediate}\right] \longrightarrow RCOCo(CO)_3 + CO$$

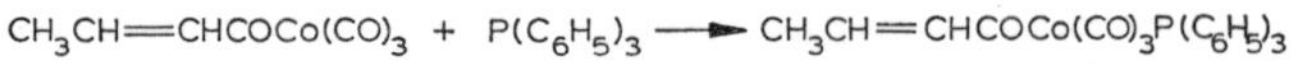

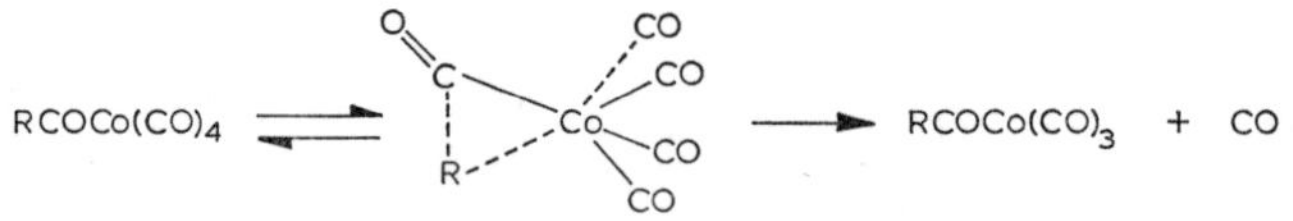

The propenyl group can then supply electrons to the cobalt atom by a resonance effect. This basic mechanism may also be applied to explain the marked accelerating effect of the aroyl group. Except in cases where ortho substitution sterically hinders formation of a cyclic transition state substitution on the aromatic ring has relatively little effect on the rate of decomposition.

The reaction of the π-complexes[23] shown in Table 2 may be represented by a mechanism similar to that proposed for the acylcobalt carbonyls.

$$\pi\text{-}RCo(CO)_3 \rightleftharpoons \pi\text{-}RCo(CO)_2 + CO \qquad (1)$$

$$\pi\text{-}RCo(CO)_2 + P(C_6H_5)_3 \rightarrow \pi\text{-}RCo(CO)_2P(C_6H_5)_3 \qquad (2)$$

With $[P(C_6H_5)_3] \geqslant 0.06\ M$ the CO evolution is first-order with respect to the cobalt complex and zero-order in phosphine, indicating that as in the case of the acyl carbonyls, the observed rate coefficient is k_1.

The reaction of binuclear cobalt complexes is somewhat different. For $Co_2(CO)_8$ the overall mechanism proposed by Heck[24] is

$$Co_2(CO)_8 + P(C_6H_5)_3 \rightarrow Co_2(CO)_8P(C_6H_5)_3 \qquad (1)$$

$$Co_2(CO)_8P(C_6H_5)_3 \rightarrow Co_2(CO)_7P(C_6H_5)_3 + CO \qquad (2)$$

$$Co_2(CO)_7P(C_6H_5)_3 + P(C_6H_5)_3 \rightarrow Co(CO)_3[P(C_6H_5)_3]_2^+Co(CO)_4^- \qquad (3)$$

Provided $[P(C_6H_5)_3] \geqslant 2\ [Co_2(CO)_8]$ the reaction is independent of phosphine concentration and first-order in cobalt octacarbonyl indicating that reaction (2) is the rate-controlling step so that the observed first-order rate coefficient is k_2.

TABLE 2

THE DECOMPOSITION OF ETHER SOLUTIONS OF π-ALLYLCOBALT TRI-CARBONYLS IN THE PRESENCE OF $P(C_6H_5)_3$

Compound	*Temperature* (°*C*)	*k* (*sec*$^{-1}$)
π-$C_3H_5Co(CO)_3$	0	3.35×10^{-4}
π-$C_3H_5Co(CO)_3$	25	1.14×10^{-2}
2-CH_3-π-$C_3H_4Co(CO)_3$	0	2.81×10^{-3}
2-Br-π-$C_3H_4Co(CO)_3$	0	4.10×10^{-3}
2-Cl-π-$C_3H_4Co(CO)_3$	0	6.47×10^{-3}
2-C_6H_5-π-$C_3H_4Co(CO)_3$	0	1.12×10^{-3}
1-CH_3-π-$C_3H_4Co(CO)_3$	0	1.82×10^{-3}
1-Cl-π-$C_3H_4Co(CO)_3$	0	9.36×10^{-5}
1-CH_3OCO-π-$C_3H_4Co(CO)_3$	0	7.85×10^{-4}

Compound 0.01–0.06 *M*, $[P(C_6H_5)_3] > [\text{Compound}]$

When one CO in the octacarbonyl is replaced by $C_4H_2O_2$ direct decomposition is again the rate-controlling process provided $[P(C_6H_5)_3] > 0.1$ *M*. The mechanism proposed by Heck[24] is

$$Co_2(CO)_7C_4H_2O_2 \rightleftharpoons Co_2(CO)_6C_4H_2O_2 + CO \quad (1)$$

$$CO_2(CO)_6C_4H_2O_2 + P(C_6H_5)_3 \rightarrow Co_2(CO)_6C_4H_2O_2[P(C_6H_5)_3] \quad (2)$$

$$Co_2(CO)_6C_4H_2O_2[P(C_6H_5)_3] \rightleftharpoons Co_2(CO)_5C_4H_2O_2[P(C_6H_5)_3] + CO \quad (3)$$

$$Co_2(CO)_5C_4H_2O_2[P(C_6H_5)_3] + P(C_6H_5)_3 \rightarrow Co_2(CO)_5C_4H_2O_2[P(C_6H_5)_3]_2 \quad (4)$$

Under the experimental conditions used, the overall rate coefficient was identified with k_1.

The mechanism of the reactions of 3-hexyne and acetylene-cobalt hexacarbonyl in toluene solution is similar to reactions (1) and (2) of the lactone decomposition. These reactions are followed by some disproportionation of the monophosphine

TABLE 3

THE DECOMPOSITION OF COBALT OCTACARBONYL AND SUBSTITUTED BINUCLEAR COBALT CARBONYL DERIVATIVES IN THE PRESENCE OF TRIPHENYL PHOSPHINE

Compound	*Solvent*	*Temperature* (°*C*)	*k* (*sec*$^{-1}$)
$Co_2(CO)_8$ [a]	Methylene chloride	−72	5.50×10^{-4}
$Co_2(CO)_7C_4H_2O_2$ [a]	Methylene chloride	0	7.79×10^{-4}
$Co_2(CO)_7C_4H_2O_2$ [a]	Methylene chloride	25	2.20×10^{-2}
$Co_2(CO)_6C_2H_5C_2C_2H_5$	Toluene	59.8	8.51×10^{-3}
$Co_2(CO)_6C_2H_2$	Toluene	25	5.87×10^{-5}
$Co_2(CO)_6C_2H_2$	Toluene	59.7	1.24×10^{-3}

Compound conc., 0.009–0.110 *M*.

[a] $[P(C_6H_5)_3] \geqslant$ twice the concentration of the compound.

to give a bisphosphine derivative. However, this does not affect the observed rate coefficients which are identified with the initial decomposition process.

The reaction of acylcobalt tetracarbonyls and π-allyl cobalt tricarbonyls in the presence of trimethylolpropane phosphite can be represented by the following mechanism[25].

$$RCo(CO)_n \rightleftharpoons RCo(CO)_{n-1}+CO \quad (1)$$

$$RCo(CO)_{n-1}+PO_3C_6H_{11} \rightleftharpoons RCo(CO)_{n-1}(PO_3C_6H_{11}) \quad (2)$$

$$RCo(CO)_{n-1}(PO_3C_6H_{11}) \rightleftharpoons RCo(CO)_{n-2}(PO_3C_6H_{11})+CO \quad (3)$$

$$RCo(CO)_{n-2}(PO_3C_6H_{11})+PO_3C_6H_{11} \rightleftharpoons RCo(CO)_{n-2}(PO_3C_6H_{11})_2 \quad (4)$$

Rate coefficients for reactions (1) and (3) are shown in Table 4. The values for reaction (1) are in good agreement with those found in the presence of triphenylphosphine (see Table 1 and 2).

Arrhenius parameters calculated from the data in Tables 1–4 are shown in Table 5. The pre-exponential factors are all within the range expected for unimolecular decompositions, with the exception of $Co_2(CO)_6C_2H_2$. The low value for its decomposition has been attributed to formation of a CO bridge in the transition state[24].

Basolo and Wojcicki[26] have reported that at 0 °C in toluene solution $Ni(CO)_4$ exchanges with radioactive CO by a first-order reaction with $k = 7.5\times10^{-4}$ sec^{-1}. Heck[24] reports that under the same conditions nickel carbonyl reacts with triphenyl phosphine with $k = 4.3\times10^{-4}$ sec^{-1}. The fact that both processes are first-order and have approximately the same rate coefficient was originally in-

TABLE 4

DECOMPOSITION OF DIGLYME SOLUTIONS OF VARIOUS COBALT CARBONYL COMPLEXES IN THE PRESENCE OF TRIMETHYLOLPROPANE PHOSPHITE

Compound	*Temperature (°C)*	*k (sec⁻¹)*
$CH_3COCo(CO)_4$	0	5.55×10^{-4}
	25	1.84×10^{-2}
$CH_3COCo(CO)_3(PO_3C_6H_{11})$	25	3.75×10^{-5}
	50.1	1.22×10^{-3}
	70.1	1.22×10^{-2}
π-$C_3H_5Co(CO)_3$	0	2.19×10^{-4}
	25	9.76×10^{-3}
π-$C_3H_5Co(CO)_2(PO_3C_6H_{11})$	50	3.49×10^{-4}
	74.9	6.51×10^{-3}
$HCo(CO)_3(PO_3C_6H_{11})$	−24.5	1.72×10^{-2}

[Compound] = 0.04–0.10 *M*, $[PO_3C_6H_{11}] \geqslant 2$ [compound]

TABLE 5

ARRHENIUS PARAMETERS FOR THE DECOMPOSITION OF COBALT CARBONYL COMPLEXES

Reaction	*Solvent*	*Temperature range (°C)*	*E (kcal.mole⁻¹)*	*A (sec⁻¹)*
$CH_3COCo(CO)_4 \rightarrow CH_3COCo(CO)_3+CO$	Ether	0–25	20.8	4.5×10^{13}
$CH_3COCo(CO)_4 \rightarrow CH_3COCo(CO)_3+CO$	Toluene	0–25	21.2	1.6×10^{14}
$CH_3COCo(CO)_4 \rightarrow CH_3COCo(CO)_3+CO$	Diglyme	0–25	20.5	1.4×10^{13}
π-$C_3H_5Co(CO)_3 \rightarrow \pi$-$C_3H_5Co(CO)_2+CO$	Ether	0–25	22.8	4.7×10^{14}
π-$C_3H_5Co(CO)_3 \rightarrow \pi$-$C_3H_5Co(CO)_2+CO$	Diglyme	0–25	23.7	2.2×10^{15}
$Co_2(CO)_7C_4H_2O_2 \rightarrow Co_2(CO)_6C_4H_2O_2+CO$	Methylene chloride	0–25	21.5	1.2×10^{14}
$Co_2(CO)_6C_2H_2 \rightarrow Co_2(CO)_5C_2H_2+CO$	Toluene	25–60	17.0	1.7×10^{8}
$CH_3COCo(CO)_3(PO_3C_6H_{11}) \rightarrow$ $\rightarrow CH_3COCo(CO)_2(PO_3C_6H_{11})+CO$	Diglyme	25–70	24.9	5.3×10^{13}
π-$C_3H_5Co(CO)_2(PO_3C_6H_{11}) \rightarrow$ $\rightarrow \pi$-$C_3H_5Co(CO)(PO_3C_6H_{11})+CO$	Diglyme	50–75	27.0	6.9×10^{14}

terpreted by Heck as evidence that both processes have reaction (1)

$$Ni(CO)_4 \rightarrow Ni(CO)_3+CO \tag{1}$$

as their rate-controlling step. A recent reinvestigation[27] has shown that the correspondence of the rate coefficients observed was fortuitous and that the enthalpies and entropies of activation for the two processes differ widely (Table 6). It is suggested that many processes which have been thought of as dissociation processes probably proceed by an alternate mechanism. The situation is further confused by the fact that the value of k_1 at 0 °C obtained by extrapolating results of the vapour-phase decomposition[21], $k_1 = 10^{-6}$ sec^{-1}, is approximately 400 times less than the value in toluene solution. However, it should be noted that the difference, within experimental error, is entirely due to differences in $\Delta S^\ddagger$ and that the value of the pre-exponential factor in toluene, 1.6×10^{13} sec^{-1}, is more reasonable for a unimolecular process than the value of 4.16×10^{9} sec^{-1} reported for the vapour-phase reaction.

TABLE 6

ENTHALPY AND ENTROPY OF ACTIVATION FOR $Ni(CO)_4+{}^{14}CO$ AND $Ni(CO)_4+P(C_6H_5)_3$

Solvent	^{14}Co *exchange*		$P(C_6H_5)_3$ *reaction*	
	$\Delta H^\ddagger$ *(kcal)*	$\Delta S^\ddagger$ *(e.u.)*	$\Delta H^\ddagger$ *(kcal)*	$\Delta S^\ddagger$ *(e.u.)*
Acetonitrile	10±1	−36±2	21±1	2±2
Toluene	12±1	−26±2	21±1	2±2
Heptane	10±1	−36±2	21±1	1±2

2.5 MANGANESE CARBONYLS

The decomposition of $Mn(CO)_5Cl$, $Mn(CO)_5Br$ and $Mn(CO)_5I$ in the presence of a wide variety of ligands [$P(C_6H_5)_3$, $As(C_6H_5)_3$, C_5H_5N, $P(OC_4H_9)_3$, etc.] proceeds by the slow loss of CO followed by fast coupling with the ligand[28], *viz.*

$$Mn(CO)_5X \rightarrow Mn(CO)_4X+CO \tag{1}$$

$$L+Mn(CO)_4X \rightarrow Mn(CO)_4XL \tag{2}$$

The rate of disappearance of $Mn(CO)_5X$ is independent of the nature of L but decreases with increasing dielectric constant (ε) of the solvent. For $Mn(CO)_5Br$ at 40 °C $k^{\varepsilon=2}_{\text{cyclohexane}} = 7.44\times10^{-4}$ sec^{-1}, $k^{\varepsilon=34}_{\text{nitrobenzene}} = 1.08\times10^{-4}$ sec^{-1}. For a given solvent, the rate decreases with changes in X in the order Cl > Br > I. The values of k_1 (chloroform solution) are

$$Mn(CO)_5Cl \qquad k_1 = 3.9\times10^{16} \exp(-27{,}500/RT) \text{ sec}^{-1}$$

$$Mn(CO)_5Br \qquad k_1 = 2.1\times10^{17} \exp(-29{,}800/RT) \text{ sec}^{-1}$$

$$Mn(CO)_5I \qquad k_1 = 5.2\times10^{17} \exp(-32{,}200/RT) \text{ sec}^{-1}$$

For $Mn(CO)_5Br$ in nitrobenzene, $k_1 = 3.6\times10^{7} \exp(-30{,}900/RT)$ sec^{-1}. The variation in k_1 with solvent and the unusually high pre-exponential factors reflect the decrease in solvation on formation of the activated complex.

A single value for the rate coefficient for loss of CO from $Mn_2(CO)_{10}$ has been obtained[29]. At 80 °C, $k = 3.9\times10^{-6}$ sec^{-1}. Solvent effects should play only a minor role in this decomposition. Therefore, an assumed value for the pre-exponential factor of 10^{14} sec^{-1} is reasonable. This gives an activation energy of 31.4 kcal.mole^{-1} which is in accord with the values for $Mn(CO)_5X$.

The work cited in sections 2.4 and 2.5 is representative of the S_N1 substitution reactions of metal carbonyls. However, a much more extensive and detailed account has recently been published covering similar reactions of vanadium, chromium, molybdenum, tungsten, rhenium, iron and nickel carbonyls in addition to those of manganese and cobalt[29a].

3. Homogeneous decomposition of metal alkyls and aryls

3.1 COPPER, SILVER, GOLD

Although many studies of the decomposition of copper, silver and gold alkyls and aryls have been reported[30–45], all but four of these[42–45] deal with suspensions or pure solids. Gilman and Woods[42] report that methyl gold in ether solution begins to decompose at $-40°$ to -35 °C giving a gold mirror and gaseous products containing 76.8 % C_2H_6 and 18.4 % CH_4. The thermal decomposition of methyl silver[42] and ethyl silver[44] in ethanol solution are both first-order processes. Methyl silver decomposes quantitatively to give silver and ethane, *viz.*

$$AgCH_3 \rightarrow Ag + CH_3 \quad (1)$$

$$2\,CH_3 \rightarrow C_2H_6 \quad (2)$$

Over the temperature range -60–0 °C $k_1 = 5.5 \times 10^5 \exp(-10{,}500/RT)$ sec^{-1}. Silver ethyl also undergoes decomposition by Ag–C bond rupture

$$AgC_2H_5 \rightarrow Ag + C_2H_5 \quad (3)$$

$$2\,C_2H_5 \rightarrow C_2H_4 + C_2H_6 \quad (4)$$

$$2\,C_2H_5 \rightarrow C_4H_{10} \quad (5)$$

$$C_2H_5 + C_2H_5OH \rightarrow C_2H_6 + C_2H_4OH \quad (6)$$

The product yield (at -23 °C, approximately 11 % C_2H_4, 78 % C_2H_6, 11 % C_4H_{10}) indicates that ethane is formed predominately by reaction (6). The total yield of gaseous products was about 60 % of the value expected based on reactions (3)–(6). Bawn and Johnson[44] attribute the apparent deficiency in gaseous products to formation of silver by reaction of silver ion or undissociated salt with the intermediate free radical formed in reaction (6). The rate coefficient for the decomposition process between -50 °C and -20 °C is $k_3 = 5.0 \times 10^8 \exp(-14{,}200/RT)$ sec^{-1}. The low pre-exponential factors for methyl silver and ethyl silver are probably a consequence of the ionic character of these compounds[43].

The competitive decomposition of phenyl silver and *p*-tolyl silver in pyridine solution[45] gives $(C_6H_5)_2$, $(C_6H_4CH_3)_2$ and $(C_6H_5C_6H_4CH_3)$ in yields consistent with simple metal–carbon bond rupture. Kinetic data is not available.

3.2 DIMETHYL ZINC

The pyrolysis of dimethyl zinc has been studied in a toluene carrier flow system[46], in sealed ampoules using both $Zn(CH_3)_2$ and $Zn(CD_3)_2$[47] and in an argon carrier system coupled to a mass spectrometer[48]. The decomposition in the toluene carrier system may be discussed in terms of the mechanism

$$Zn(CH_3)_2 \rightarrow ZnCH_3 + CH_3 \quad (1)$$

$$ZnCH_3 \rightarrow Zn + CH_3 \quad (2)$$

$$CH_3 + C_6H_5CH_3 \rightarrow CH_4 + C_6H_5CH_2 \quad (3)$$

References pp. 254–257

$$2\,CH_3 \rightarrow C_2H_6 \tag{4}$$

$$ZnCH_3 + C_6H_5CH_2 \rightarrow C_6H_5CH_2ZnCH_3 \tag{5}$$

$$2\,C_6H_5CH_2 \rightarrow \text{dibenzyl} \tag{6}$$

plus other minor reactions such as

$$CH_3 + C_2H_6 \rightarrow C_2H_5 + CH_4 \tag{7}$$

$$C_2H_5 \rightarrow C_2H_4 + H \tag{8}$$

$$H + C_6H_5CH_3 \rightarrow H_2 + C_6H_5CH_2 \tag{9}$$

$$CH_3 + C_6H_5CH_2 \rightarrow C_6H_5C_2H_5 \tag{10}$$

At 16 torr pressure using a contact time of 0.9 sec and a toluene to alkyl ratio of the order of 40 : 1, reaction (1) is 97 % complete at 700 °C. Under the same conditions, reaction (2) is only 7 % complete. Therefore, above approximately 730 °C all the dimethyl zinc is converted to methyl zinc in a very small fraction of the contact time. The rate of decomposition of methyl zinc is then readily determined. A correction must be made for methane formed by decomposition of the toluene carrier. Price and Trotman-Dickenson[46] based this correction on the equation given by Szwarc[49] which was roughly checked by a single experiment at 783 °C. Applying this correction $k_2 = 6.3\times10^6 \exp(-35{,}000/RT)\ \text{sec}^{-1}$. Price[50] has shown that Szwarc's equation is inadequate but in the temperature range used to study the $ZnCH_3$ decomposition, 730–827 °C, the error is not large. A correction based on this more recent work lowers E_2 to approximately 32 kcal.mole^{-1}.

Applying the usual steady-state treatment for consecutive first-order reactions k_1 at 16 torr pressure over the temperature range 597–701 °C is given by 1.8×10^{11} $\exp(-47{,}000/RT)\ \text{sec}^{-1}$. Within experimental error, reactions (1) and (2) were homogeneous processes. However, both k_1 and k_2 were functions of the total pressure in the system. This dependence is shown in Fig. 1. The methyl zinc decomposition is apparently in its second-order region. Therefore, assuming four effective oscillators and a mean temperature of 1050 °K, $E_\infty = E_{\text{obs.}} + \frac{1}{2}nRT = 32+4 = 36$ kcal.mole^{-1}. For the dimethyl zinc decomposition E_∞ has been estimated[51] at approximately 49.5 kcal.mole^{-1}. The E_∞ values should be a good approximation to the corresponding bond dissociation energies. From thermochemistry[52, 53], assuming[19] $\Delta H_f(CH_3) = 34.0$ kcal.mole^{-1}, $D(CH_3Zn\text{–}CH_3) + D(Zn\text{–}CH_3) = 85.9$ kcal.mole^{-1}. The close correspondence between this figure and the kinetic result, 85.5 kcal.mole^{-1}, indicates that when zinc methyl decomposes the zinc atom is released in its ground state.

Reaction (5) was not included in the mechanism originally suggested by Price and Trotman-Dickenson[46]. A more recent study of the thermal decomposition of dimethyl zinc in a benzene carrier system[54] gave $C_6H_5ZnCH_3$ as a major product. The activation energy of reaction (10) has been shown to be approx-

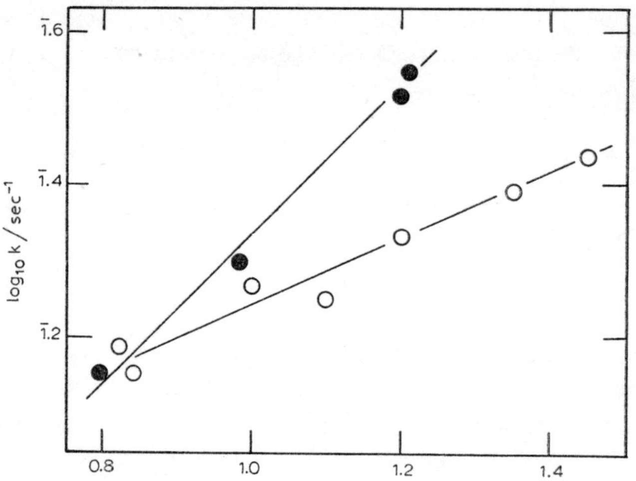

Fig. 1. Pressure dependence of the rate coefficients for the decomposition of dimethyl zinc in a toluene carrier system. ○, k_1(597 °C); ●, k_2(783 °C).

imately zero[55]. An analogous study of $CH_3+C_6H_5$ using dimethyl mercury and dimethyl cadmium as the source of methyl radicals[54] has shown that the reaction

$$C_6H_5+CH_3 \rightarrow C_6H_5CH_3 \tag{11}$$

also has approximately zero activation energy.

The basic assumption is made in these calculations that the benzyl or phenyl radicals formed are removed in the hot zone only by recombination with methyl radicals. When dimethyl zinc is used, the apparent energy of activation for reaction (11) is negative indicating a loss of phenyl radicals in the hot zone by an additional process. The only species present which is absent when the mercury and cadmium alkyls are used is $ZnCH_3$. The additional process by which phenyl radicals are lost is therefore probably

$$C_6H_5+ZnCH_3 \rightarrow C_6H_5ZnCH_3 \tag{12}$$

Reaction (5) may also occur in the reaction zone. However, the $C_6H_5CH_2$–Zn bond should be somewhat weaker than the C_6H_5–Zn bond so that the benzyl–methyl zinc recombination may occur in a cooler region. Some dimerization of $ZnCH_3$ probably also occurs.

The static system studies carried out by Lambert[47] utilized $Zn(CH_3)_2$, $Zn(CD_3)_2$, C_6H_{12} and C_6D_{12}. The decomposition of $Zn(CH_3)_2$ in the presence of C_6D_{12} was studied with $C_6D_{12}/Zn(CH_3)_2$ ratios of 0–17. At 348 °C using a ratio of 13, the gaseous products contained 97.5 % methane, 1.6 % ethane plus ethylene and 0.9 % propane. This ratio was used in subsequent work. Over the temperature range 297–370 °C, the ratio of CH_3D to CH_4 was approximately unity. Lambert proposed the following mechanism

$$Zn(CH_3)_2 \rightarrow CH_3 + ZnCH_3 \quad (1)$$

$$CH_3 + C_6D_{12} \rightarrow CH_3D + C_6D_{11} \quad (13)$$

$$ZnCH_3 \overset{H}{=} Zn + CH_4 \quad (14)$$

followed by dimerization or other stabilization of C_6D_{11} that does not involve CH_3. The processes leading to the formation of ethane, ethylene and propane were considered negligible. From the increase in the CH_4/CH_3D ratio as $C_6D_{12}/Zn(CH_3)_2$ decreases, a maximum value of $k_{15} = 6.0 \times 10^8 \exp(-15{,}000/\boldsymbol{R}T)$ l.mole^{-1}.sec^{-1} was estimated. (A_{15} was derived from transition state theory, assuming a linear activated complex.)

$$CH_3 + Zn(CH_3)_2 \rightarrow CH_4 + CH_3ZnCH_2 \quad (15)$$

This reaction was therefore negligible when $C_6D_{12}/Zn(CH_3)_2 = 13$. The rates of reactions (1) and (14) are strongly dependent on the nature of the surface. In an ampoule coated with "tar" from decomposition of a large quantity of dimethyl zinc, the rate of both reactions is only $\frac{1}{3}$–$\frac{1}{4}$ the rate observed in unconditioned vessels (tested at 348 °C with fraction $Zn(CH_3)_2$ reacted in conditioned vessel $= 0.035$). It has also been shown that in 90 min at 290 °C, the overall decomposition of dimethyl zinc in the absence of cyclohexane is 94 % complete if a zinc oxide surface is used, but only 4.5 % complete in a conditioned vessel. Decompositions carried out in conditioned vessels were assumed to be homogeneous.

The overall process shown as reaction (14) is a necessary consequence of the observed CH_3D/CH_4 ratio. To be consistent with a ratio of unity, this reaction must proceed without the liberation of free methyl radicals and must account quantitatively for the fate of the methyl zinc. The exact nature of reaction (14) is unknown but several important observations have been made. Decomposition of $Zn(CD_3)_2$ with C_6H_{12} in a vessel conditioned using $Zn(CH_3)_2$ produced the expected yield of CD_4 indicating that the additional hydrogen needed for reaction (14) does not come from the coating on the conditioned vessel. Since reaction (15) cannot compete successfully under the experimental conditions used, it is doubtful if the reaction

$$ZnCH_3 + Zn(CH_3)_2 = Zn + CH_4 + CH_3ZnCH_2 \quad (16)$$

can play any significant role. However, assuming $E_6 = 0$, Lambert has estimated that

$$2\,ZnCH_3 \rightarrow (ZnCH_3)_2 \tag{17}$$

has a rate which should be much greater than that of the reaction

$$ZnCH_3 \rightarrow Zn + CH_3 \tag{2}$$

It is therefore possible that the initial fate of $ZnCH_3$ in the static system work is dimerization. Under the conditions used diffusion to the surface can compete successfully with reaction (2) so formation of an intermediate dimer could be either a homogeneous or a heterogeneous process. By elimination, the two hydrogen atoms required to convert the dimer to $2\,Zn + 2\,CH_4$ must come from dimethyl zinc itself and must leave a product that does not undergo subsequent decomposition. It is possible that this occurs via a cyclic intermediate, with adsorbed dimethyl zinc leaving surface-adsorbed radicals which may undergo polymerization, viz.

$$\begin{array}{c}\text{-Zn} \quad \text{ZnZn} \quad \text{Zn-} \\ \text{CH}_3\ \text{CH}_3 \quad \text{CH}_3\ \text{CH}_3 \\ \text{H}\ \text{H} \quad \text{H}\ \text{H} \\ \text{CH}_2\text{ZnCH}_2 \quad \text{CH}_2\text{ZnCH}_2 \end{array} = 2n\,CH_4 + n\,Zn_2 + (CH_2ZnCH_2)_n \downarrow \\ 2n\,Zn$$

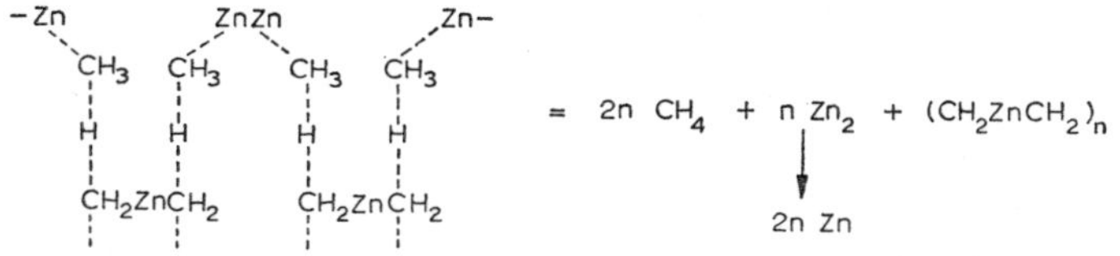

If the abstraction process involves only the first hydrogen from each methyl-group of the adsorbed dimethyl zinc, the true value of k_1 should be approximately that calculated from Lambert's expression, $k_1 = 1 \times 10^{11} \exp(-45,500/RT)$ sec^{-1}, at low % decomposition, but about 1.3 times greater at 75 % decomposition (50 % decomposition as calculated by Lambert). The results of Price and Trotman-Dickenson[46] and Lambert[47] are compared in Fig. 2. Increasing Lambert's value by 15 % and converting Price and Trotman-Dickenson values to infinite pressure (based on the observed pressure dependence of the pyrolysis of dimethyl mercury in a similar system[56]) gives

$$k_1 = 1 \times 10^{13} \exp(-50,700/RT)\ \text{sec}^{-1}$$

If all the hydrogens in the adsorbed dimethyl zinc are abstracted, then Lambert's value for k_1 is increased by a neglible amount at all % decompositions used, and the combined calculation with the estimated infinite pressure data of Price and Trotman-Dickenson gives

$$k_1 = 1.25 \times 10^{13} \exp(-51,200/RT)\ \text{sec}^{-1}$$

Nearly complete loss of hydrogen by the adsorbed alkyl would be consistent with the normal CD_4 yield from $Zn(CD_3)_2$ pyrolysis in a vessel coated by the de-

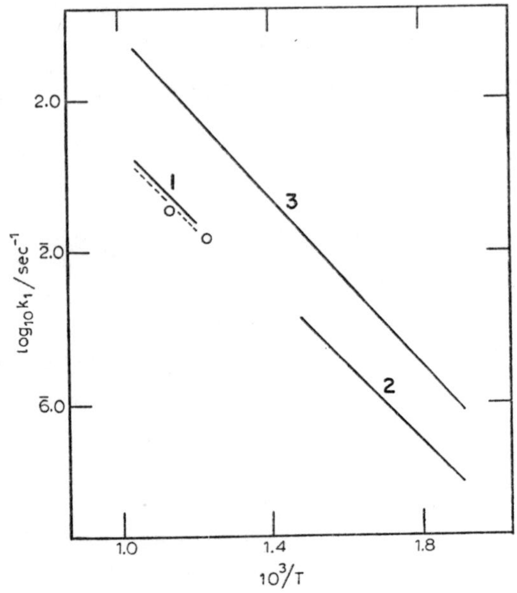

Fig. 2. Arrhenius plots for the decomposition of dimethyl zinc. 1, Price and Trotman-Dickenson: solid line, 16 torr toluene; dotted line, 5 torr toluene; 2, Lambert, approx. 760 torr (cyclohexane +7 % alkyl); 3, composite curve obtained from rate coefficients of Price and Trotman-Dickenson corrected to infinite pressure and from Lambert's values. The composite curve has been displaced upwards by 2.0 log units. Open circles: Topor, 22–24 torr (argon+2.4 torr dimethyl zinc).

composition of $Zn(CH_3)_2$.

Topor's results using argon as a carrier[48] are too limited to be considered in detail. The observed rate coefficient at 606 °C and 24 torr pressure for the removal of the first methyl radical from dimethyl zinc agrees fairly well with that observed by Price and Trotman-Dickenson at 4.6 torr pressure and 597 °C. Topor's mixture contained approximately 2.4 torr dimethyl zinc in argon. Considering the expected low efficiency of argon as an energy transfer agent, the agreement is quite reasonable.

Recent work[153] with $Zn(CH_3)_2$ and $Zn(CD_3)_2$ using the toluene carrier technique (surface effects negligible) confirms the mechanism previously proposed[46]. Negligible yields of CD_4 were observed. At infinite pressure

$$k_1 = 2.0 \times 10^{13} \exp(-54{,}000/RT) \text{ sec}^{-1}$$

obtained by extrapolation using $1/k$ *versus* $1/P^{0.6}$ plots which were approximately linear. At 100 torr (k_2 strongly pressure-dependent)

$$k_2 = 4.0 \times 10^{6} \exp(-31000/RT) \text{ sec}^{-1}$$

3.3 DIMETHYL CADMIUM

The decomposition of dimethyl cadmium in a toluene or benzene carrier system proceeds by the consecutive release of both methyl radicals[57, 58]. The suggested mechanism is

$$Cd(CH_3)_2 \rightarrow CdCH_3 + CH_3 \qquad (1)$$
$$CdCH_3 \rightarrow Cd + CH_3 \qquad (2)$$
$$CH_3 + ArH \rightarrow CH_4 + Ar \qquad (3)$$
$$2\,CH_3 \rightarrow C_2H_6 \qquad (4)$$
$$2\,Ar \rightarrow Ar_2 \qquad (5)$$
$$CH_3 + Ar \rightarrow ArCH_3 \qquad (6)$$
$$CH_3 + C_2H_6 \rightarrow C_2H_5 + CH_4 \qquad (7)$$
$$C_2H_5 \rightarrow C_2H_4 + H \qquad (8)$$
$$H + ArH \rightarrow H_2 + Ar \qquad (9)$$
$$C_2H_5 + CH_3 \rightarrow C_3H_8 \qquad (10)$$

where reaction (2) is much faster than reaction (1).

In the benzene carrier system[57] ($Ar = C_6H_5$) using 20–160 torr benzene pressure over the temperature range 475–527 °C, approximately 3–6 % of the methyl radicals are removed by reaction (6). An additional 1–5 % are found as ethylene and propane (less than 2 % under most conditions used). At all temperatures, k_1 is independent of pressure above approximately 8 torr. The Arrhenius equations for the decomposition at infinite pressure ($P > 8$ cm) and at 18 mm respectively are

$$k_1 = 2.5 \times 10^{13} \exp(-48{,}800/RT) \text{ sec}^{-1}$$

and

$$k_1 = 2.0 \times 10^{12} \exp(-45{,}850/RT) \text{ sec}^{-1}$$

In the toluene carrier system[58] reactions (6) and (10) were assumed to be negligible. Including the contribution of reaction (6), the rate coefficient at 18 torr pressure is

$$k_1 = 3.5 \times 10^{12} \exp(-47{,}600/RT) \text{ sec}^{-1}$$

The activation energy at the high pressure limit, E_∞, should be 3–5 kcal.mole^{-1} greater than $E_{18\text{ torr}}$. The value of 48.8 kcal.mole^{-1} from the benzene flow system work appears to be 2–4 kcal.mole^{-1} too low. Since an error of 4 kcal.mole^{-1} has been observed experimentally in the case of dimethyl mercury, (see Fig. 3 and discussion of the inhibited decomposition of dimethyl mercury), E_∞ for dimethyl cadmium may reasonably be assigned a value of 52.6 kcal.mole^{-1}. k_1 is then given approximately by $4.0 \times 10^{14} \exp(-52{,}600/RT)$ sec^{-1}.

Laurie and Long[59] studied the thermal decomposition of dimethyl cadmium in the absence of inhibitors. Rate coefficients were calculated on the basis of the undecomposed alkyl. A marked surface affect was noted. The homogeneous rate coefficient was obtained at 258 °C by studying the pyrolysis with various surface to volume ratios and extrapolating to zero surface. The mechanism proposed is

$$Cd(CH_3)_2 \rightarrow CdCH_3 + CH_3 \tag{1}$$

$$CdCH_3 \rightarrow Cd + CH_3 \tag{2}$$

$$2\,CH_3 \rightarrow C_2H_6 \tag{4}$$

$$CH_3 + C_2H_6 \rightarrow CH_4 + C_2H_5 \tag{7}$$

$$C_2H_5 + CH_3 \rightarrow C_3H_8 \tag{10}$$

$$CH_3 + C_3H_8 \rightarrow CH_4 + C_3H_7 \tag{11}$$

$$C_3H_7 \rightarrow CH_3 + C_2H_4 \tag{12}$$

$$\frac{n}{2}\,C_2H_4 \rightarrow (CH_2)_n \tag{13}$$

Based on this mechanism, the homogeneous rate coefficient at 258 °C and 158 torr is $k_1 = 1.43 \times 10^{-5}$ sec^{-1}. At the temperature and pressure used, this value should be pressure independent. Using the corresponding A factor from the flow system results, 3.0×10^{14} sec^{-1}, gives $E_1 = 47.0$ kcal.mole^{-1}.

A more significant method for testing the agreement of the flow and static system results is to use the high-pressure flow system value of k_1 at 477 °C, and the static system value at 258 °C. These values give $k_1 = 1.6 \times 10^9 \exp(-34{,}000/RT)$ sec^{-1}. There is obviously a serious discrepancy between the two methods. A possible source of error in the static system work is the neglect of reaction (14)

$$CH_3 + CH_3CdCH_3 \rightarrow CH_4 + CH_2CdCH_3 \tag{14}$$

and the subsequent chain process that might be initiated by this reaction. Neither the high-pressure flow system results nor the static system results are entirely satisfactory but the flow system work is probably the most significant. Using $E_1 = 52.6$ kcal.mole^{-1} as a measure of $D(CH_3Cd\text{–}CH_3)$ and taking the mean metal–carbon bond dissociation energy in this alkyl as 34.3 kcal.mole^{-1} based on Long's suggested value[60] recalculated using[19] $D(CH_3\text{–}H) = 104$ kcal.mole^{-1} gives $D(Cd\text{–}CH_3) = 16.0$ kcal.mole^{-1}. This value is consistent with the assumption that under all experimental conditions used $k_2 \gg k_1$.

3.4 MERCURY ALKYLS

3.4.1 Dimethyl mercury

The thermal decomposition of dimethyl mercury in the presence of radical scavengers has been thoroughly investigated[61-65]. The basic mechanism, the pre-exponential factor and the activation energy are all well established. There is still considerable doubt about the mechanism of the pyrolysis in the absence of chemically active additives. Consequently, the quantitative interpretation of rate data from such systems is of doubtful value. Systems using effective scavengers will be discussed first. The quantitative results from these systems will be used in assessing the data obtained in the absence of additives.

Cyclopentane[61, 62], propene[63] and toluene[64] are all effective radical scavengers under the conditions in which they have been used in studying the pyrolysis of dimethyl mercury. Russell and Bernstein[61] report that when cyclopentane is used in at least equimolar amounts, methane accounts for > 95 % of the carbon in the dimethyl mercury decomposed. The ratio C_2H_6/CH_4 ranged from 0.02 to 0.04 but approximately one-half of the ethane was derived from the cyclopentane carbon atoms. Less than 2 % of the methane arises from this source. The first-order rate coefficient based on the quantity of dimethyl mercury decomposed is a function of both alkyl/cyclopentane ratio and pressure. To obtain high-pressure rate coefficients for the fully inhibited reaction the fully inhibited value at each pressure was determined by extrapolating k *vs.* alkyl/cyclopentane to zero. These rate coefficients were then used to plot $1/k$ *vs.* $1/P$ from which k_∞, the rate coefficient for the fully inhibited reaction at infinite pressure, is determined by a short extrapolation to $1/P = 0$. Rate coefficients obtained with alkyl/cyclopentane = 0.05 are approximately 4 % above the extrapolated value. The fully inhibited decomposition occurs by the release of two methyl radicals followed by abstraction and recombination processes, *viz.*

$$Hg(CH_3)_2 \rightarrow HgCH_3 + CH_3 \quad (1)$$

$$HgCH_3 \rightarrow Hg + CH_3 \quad (2)$$

$$CH_3 + C_5H_{10} \rightarrow CH_4 + C_5H_9 \quad (3)$$

$$C_5H_9 = \text{products} \quad (4)$$

The overall process given by reaction (1) and reaction (2) is common to all mechanisms proposed for the decomposition of this alkyl. Whether this process occurs as written or by the simultaneous release of both methyl radicals is uncertain. Gowenlock *et al.*[66] tried to resolve this problem and the similar problem that arises with other mercury alkyls by determining $D(\text{RHg–R})$ [R = CH_3, C_2H_5, $(CH_3)_2CH$, $CH_3CH_2CH_2CH_2$] from appearance potential measurements.

Such studies may yield useful information, but only when more refined values of the ionization potential of HgR are available. For purposes of discussion, it will be assumed that the stepwise mechanism is correct.

For the fully inhibited decomposition Russell and Bernstein[61] give $k_1 = 5.0 \times 10^{15} \exp(-57{,}900/RT)$ sec^{-1}. This result has been criticized by Cattanach and Long[67] who point out that the presence of hydrogen from the decomposition of cyclopentane should initiate the chain process

$$CH_3 + H_2 \rightarrow CH_4 + H \quad (5)$$

$$H + Hg(CH_3)_2 = CH_4 + Hg + CH_3 \quad (6)$$

Weston and Seltzer[62], who have also studied the cyclopentane inhibited decom-

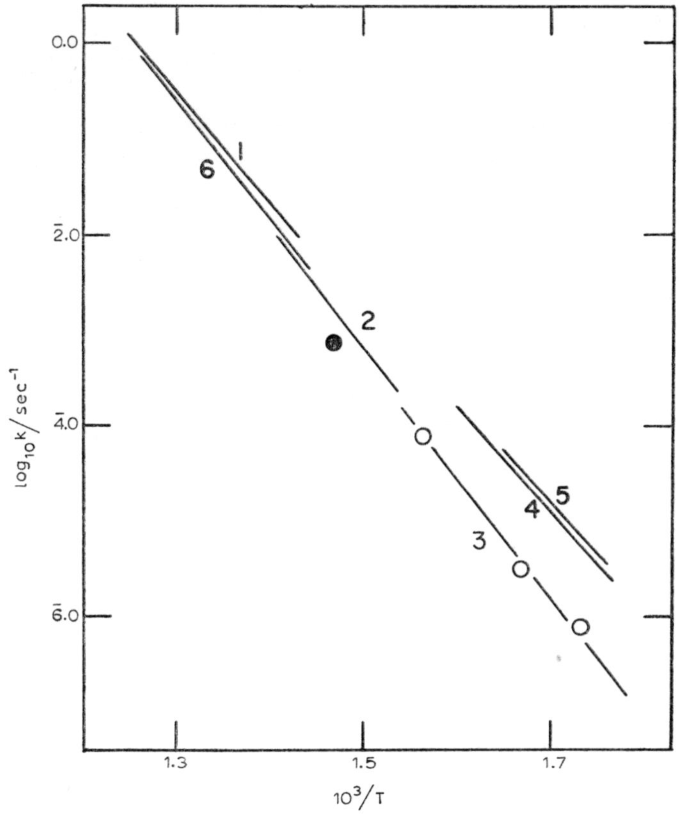

Fig. 3. Arrhenius plots for the decomposition of dimethyl mercury. All rate coefficients are at or near the high-pressure limit. If a radical scavenger has been used it is shown in brackets following the authors' names. 1, Krech and Price (benzene); 2, Kallend and Purnell (propene); 3, Russell and Bernstein (cyclopentane); 4, Russell and Bernstein; 5, Laurie and Long; 6, Kominar and Price (toluene); ○, Weston and Seltzer (cyclopentane); ●, point calculated from the steady-state equation of Kallend and Purnell.

position, reject this argument because $k_5 \approx k_3$ while $[C_5H_{10}] \gg [H_2]$. The rate coefficients obtained by these workers are in good agreement with those of Russell and Bernstein[61] (see Fig. 3).

The cyclopentane-inhibited studies were done over the temperature range 290–375 °C. Results at much higher temperatures, 420–519 °C, have been obtained using toluene carrier[64] and benzene carrier[65] flow systems. The toluene carrier work was done with toluene/alkyl ratios in the range 75–100. Contact time and contact time/pressure ratios were used that are in the range that should give plug flow and satisfy thermal equilibrium requirements[68, 69]. The first-order rate coefficient was independent of total pressure above approximately 10 cm. The value obtained is $k_1 = 5.5 \times 10^{15} \exp(-57{,}500/\boldsymbol{RT})$ sec^{-1}. Kallend and Purnell[63] have used added propene to study the high pressure fully inhibited decomposition over a temperature range which covers the gap between the toluene and cyclopentane experiments. They report $k_{\text{overall}} = 5.0 \times 10^{15} \exp(-57{,}600/\boldsymbol{RT})$ sec^{-1}. Under these conditions, k_{overall} should be effectively equal to k_1. Comparison of the cyclopentane, toluene- and propene-inhibited decompositions is shown in Fig. 3. The combined results give $k_1 = 1 \times 10^{16} \exp(-58{,}600/\boldsymbol{RT})$ sec^{-1}. The rate coefficients from the high-pressure benzene carrier studies[65] are in excellent agreement with this expression at the highest temperature used but at low decomposition ($\sim 5\,\%$) the observed values are about a factor of 1.8 greater than expected. Taking[19] $\Delta H_f(CH_3) = 34.0$ kcal.mole^{-1} in conjunction with[70] $\Delta H^0_{f298}(HgMe_2, g) = 22.4$ kcal.mole^{-1} gives $D(CH_3Hg–CH_3)+D(Hg–CH_3) = 60.2$ kcal.mole^{-1}. Systematic errors may influence the absolute value of k_1 in any set of experiments, particularly in flow system studies. The individual values of E_1 are therefore probably more reliable than the value obtained by combining the propene, cyclopentane and toluene results. The average value from these studies, $E_1 = 57.7$ kcal.mole^{-1}, should be a reasonable measure of $D(CH_3Hg–CH_3)$*. Hence, $D(Hg–CH_3) = 2.5$ kcal.mole^{-1} provided the recombination of Hg and CH_3 proceeds with zero energy of activation. The possible range of error is sufficiently large that $D(Hg–CH_3)$ may be zero. However, for seven other dimethyl and trimethyl metal alkyls discussed elsewhere $D(M–CH_3)$ has a finite value and there is no particular reason to expect dimethyl mercury to be a special case.

The gas-phase thermal decomposition of dimethyl mercury by itself and in the presence of inert gas has been extensively investigated[61, 63, 67, 71–79]. The

* To be consistent the value of E_1 should be corrected to constant pressure so that it represents ΔH for the process involved (flow system studies and static system work with excess inhibitor are essentially constant pressure experiments). Then $D < E < D+\boldsymbol{RT}$. In the present work a reasonable estimate gives $D = E-0.7 = 57.0$ kcal.mole^{-1}. Similarly, D_1+D_2 should be corrected to 0 °K, giving an estimated value of 59.0 kcal.mole^{-1}. This gives $D_2 = 2.0$ kcal.mole^{-1}. Such corrections are normally within the limits of experimental error, so that experimental values of E are associated directly with dissociation energies, and thermochemical data at 25 °C are used.

mechanisms proposed are generally complex and none of them can adequately explain all the observed facts. By far the most detailed mechanism is that proposed by Kallend and Purnell[63], *viz.*

$$CH_3HgCH_3 \rightarrow CH_3+HgCH_3 \quad (1)$$

$$HgCH_3 \rightarrow Hg+CH_3 \quad (2)$$

$$CH_3+CH_3HgCH_3 \rightarrow CH_4+CH_2HgCH_3 \quad (7)$$

$$CH_3+CH_2HgCH_3 \rightarrow [C_2H_5HgCH_3] = C_2H_5+CH_3+Hg \quad (8)$$

$$C_2H_5+CH_3HgCH_3 \rightarrow C_2H_6+CH_2HgCH_3 \quad (9)$$

$$C_2H_5+CH_2HgCH_3 \rightarrow [C_3H_7HgCH_3] = C_3H_7+CH_3+Hg \quad (10)$$

$$C_3H_7+CH_3HgCH_3 \rightarrow C_3H_8+CH_2HgCH_3 \quad (11)$$

$$2\,CH_3 \rightarrow C_2H_6 \quad (12)$$

$$CH_3+C_2H_5 \rightarrow C_3H_8 \quad (13)$$

$$2\,C_2H_5 \rightarrow C_4H_{10} \quad (14)$$

$$C_3H_7 \rightarrow C_2H_4+CH_3 \quad (15)$$

$$2\,CH_3HgCH_2 \rightarrow [CH_3HgC_2H_4HgCH_3] = $$
$$= CH_3+C_2H_4HgCH_3+H \quad (16)$$

$$C_2H_4HgCH_3 = CH_3+C_2H_4+Hg \quad (17)$$

$$CH_3+C_2H_4HgCH_3 \rightarrow [CH_3HgC_3H_7] = C_3H_7+CH_3+Hg \quad (18)$$

$$C_2H_4+CH_2HgCH_3 \rightarrow [C_3H_6HgCH_3] = C_3H_6+CH_3+Hg \quad (19)$$

$$C_2H_4+C_2H_4HgCH_3 \rightarrow [C_4H_8HgCH_3] = C_4H_8+CH_3+Hg \quad (20)$$

Before considering this mechanism and comparing it with other proposals an important fact should be noted. This work was carried out over the approximate temperature range 380–440 °C. Most of the special experiments carried out to test the mechanism were done above 400 °C. All other significant static system work has been done at much lower temperatures, 265–366 °C.

Reactions (1) and (2) are common to all proposed mechanisms. Either $k_2 \gg k_1$ or (1) and (2) occur as a single concerted step. Kallend and Purnell consider the former most likely. Activation energies ranging from 9 kcal.mole^{-1} to a maximum of 19 kcal.mole^{-1} have been estimated for reaction (7). It is usually assumed that E_7 is either 10 kcal.mole^{-1} or 15 kcal.mole^{-1}. The corresponding process for dimethyl zinc has a rate coefficient $k = 6\times10^{11}\exp(-15{,}000/\boldsymbol{R}T)$ sec^{-1} where the figure 15,000 kcal.mole^{-1} is considered a minimum value[47]. For comparative purposes it is assumed that this expression may also be used for k_7, but it must be recognized that this may give a lower limit on the contribution of this reaction. The reaction usually proposed as an alternate to reaction (7) is the reaction

$$CH_3+C_2H_6 \rightarrow CH_4+C_2H_5 \quad (21)$$

For this process[80], $k_{21} = 2\times10^{11} \exp(-10{,}400/RT)$ cm^3. mole^{-1}. sec^{-1}. Therefore, under the condition used by Kallend and Purnell the rate of reaction (7) should be at least four times that of reaction (21). Obviously, however, exclusion of reaction (21) from the mechanism may not be justified, particularly at the lower temperatures used in other static system work.

Reactions (8)–(11) were postulated in an attempt to provide a mechanism that is consistent with the effects observed when NO is added to the system. Addition of 2 torr of NO to 10 torr of dimethyl mercury at 437 °C changes the ratio $CH_4 : C_2H_6 : C_3H_8$ from 20 : 20 : 7 to approximately 40 : 6 : 0, but does not effect the overall rate of decomposition of dimethyl mercury. The doubling of the CH_4 yield was attributed to an increase by a factor of two in the methyl radical concentration. If this interpretation is correct, then since reaction (12) depends on the square of the methyl radical concentration an upper limit of 8 % of the C_2H_6 could be formed by this process under comparable conditions in the absence of NO. Reactions (8) and (9) which would be effectively inhibited by NO were, therefore, proposed as an alternate path for ethane formation in the uninhibited system. The failure of NO to effect the overall rate is then explained by the doubling of the rate of reaction (7) while reactions (9) and (11) are correspondingly decreased. Reactions (12)–(14) are considered minor processes but are included to cover possible chain termination steps. Reactions (15)–(20) account for certain minor products. The C_3H_6 shown in reaction (19) represents both propene and cyclopropane. Both of these minor products were greatly increased by addition of ethylene to the system.

The mechanism proposed by Kallend and Purnell explains many features of the dimethyl mercury pyrolysis but two difficulties arise. Their explanations are valid only if addition of NO does, in fact, increase the methyl radical concentration. The process by which this occurs has not been specified and none comes readily to mind. In fact, the equilibrium $CH_3+NO \rightleftharpoons CH_3NO$ might reasonably be expected to lower the methyl radical concentration. The second difficulty arises when high pressure limiting values of k_1 calculated from Kallend and Purnell's steady-state equation

$$\frac{-d[Hg(CH_3)_2]}{dt}\,\frac{1}{[Hg(CH_3)_2]} = 2\,k_1+B[Hg(CH_3)_2]^{\frac{1}{2}}$$

by extrapolation to zero concentration are compared with the values of k_1 shown in Fig. 3 for the fully inhibited decomposition. The values obtained by the extrapolation are apparently low by a factor of two. The significance of this discrepancy is not clear. It may be due to a combination of errors in the individual rate coefficients used in the steady-state expression and errors generated by using a simplified mechanism. However, it may also indicate that the proposed mechanism is not fully satisfactory. It is also significant that the k_∞ value for the overall decomposition at 440 °C is only about 10 % higher than the fully inhibited value, while at 385 °C it is still only about 50 % larger than the inhibited rate coefficient. This may imply that the chain carying steps are surface reactions, diffusion to the surface being more difficult at high pressures.

The discrepancies between the inhibited and uninhibited results are much greater in the temperature range 265–350 °C. Results of two sets of static system experiments[61, 72] using pure dimethyl mercury are shown in Fig. 3. Other work has been done in this region but these two studies cover the essential features. All rate coefficients plotted were calculated assuming first-order kinetics. Laurie and Long[72] report that at their highest temperature, 343 °C, the reaction has apparently become 1.5 order. However, in view of the major discrepancies between the inhibited and uninhibited experiments, this minor variation in behavior at 343 °C is not considered significant. The slight difference between Russell and Bernstein's[61] and Laurie and Long's[72] results may be a result of the different treatment used to prepare the surface of the reaction vessels for use. Laurie and Long used thorough nitric acid cleaning followed by a distilled water rinse. The vessels were then pumped while being baked out at 350 °C for one hour. A freshly cleaned vessel was used for each run. Russell and Bernstein used preliminary decomposition of a large amount of dimethyl mercury to condition their vessels. Although these procedures may lead to reproducible results, neither method definitely eliminates surface processes. Kallend and Purnell[79] report that a water rinse leads to a marked increase in the rate of decomposition. The treatment used by Laurie and Long must therefore be particularly suspect.

The mechanisms proposed by these two sets of workers represent two extreme views. Russell and Bernstein[61] propose a mechanism which involves short chains, *viz.*

$$Hg(CH_3)_2 \rightarrow HgCH_3 + CH_3 \quad (1)$$

$$HgCH_3 \rightarrow Hg + CH_3 \quad (2)$$

$$CH_3 + Hg(CH_3)_2 \rightarrow CH_2HgCH_3 + CH_4 \quad (7)$$

$$CH_3 + Hg(CH_3)_2 \rightarrow C_2H_6 + HgCH_2 \quad (22)$$

$$CH_3HgCH_2 \rightarrow CH_3 + Hg + CH_2 \quad (23)$$

$$2\,CH_3 \rightarrow C_2H_6 \quad (12)$$

Based on the loss of dimethyl mercury $k_{overall} = 5.0 \times 10^{13} \exp(-49{,}900/\boldsymbol{RT})$ sec^{-1}. Laurie and Long[72] propose an alternate scheme in which reactions (1) and (2) are the only processes involving dimethyl mercury. On this basis they calculate $k_1 = 1.9 \times 10^{14} \exp(-51{,}300/\boldsymbol{RT})\ sec^{-1}$. The methyl radicals released are assumed to undergo reactions (12) and (21), *viz.*

$$CH_3+C_2H_6 \rightarrow CH_4+C_2H_5 \tag{21}$$

$$2\,C_2H_5 \rightarrow C_2H_6+C_2H_4 \tag{24}$$

Reaction (24) produces C_2H_4 which is assumed to polymerize like the methylene produced in reaction (23). This is consistent with the observation that 40–50 % of the carbon from the decomposition is found in a solid product which does not contain mercury.

Neither of these mechanisms is complete. Only those reactions of interest for subsequent discussion have been included. Reaction (22) is included because it was originally proposed by Russell and Bernstein, but more recent work[81] indicates it is not a significant process. The chain processes proposed by Russell and Bernstein could explain why the uninhibited decomposition is 5–15 times faster than the fully inhibited pyrolysis. However, the work of Laurie and Long[72, 73] and the supporting work of Cattanach and Long[67] apparently rule out any chain process involving methyl radicals. The addition of up to 60 torr ethane to 40 torr $Hg(CH_3)_2$ (pressure as measured at 18 °C) has absolutely no effect on the rate of decomposition of the alkyl. Furthermore, when "hot" methyl radicals from the decomposition of azomethane are effectively thermalized by addition of sufficient nitrogen, they have no significant effect on the rate of decomposition. An increase in surface to volume ratio from 0.98 to 4.7 also had no effect on the rate.

If these facts reported by Cattanach and Long are accepted, it is impossible to explain the discrepancy between the inhibited and uninhibited studies. Some additional information may be obtained by considering Russell and Bernstein's results using cyclopentane as additive ($Q = Hg(CH_3)_2/C_5H_{10}$). At 303 °C $k_{Q=1} = 1.5\,k_{Q=0}$, but the yield of CH_4 accounted for at least 95 % of the $Hg(CH_3)_2$ decomposed with C_2H_6 accounting for another 2–4 %. These results indicate that reaction (23) is occurring to an extent not greater than 12 % of that of reaction (1) and (2), suggesting that $k_{Q=1} = 1.12\,k_{Q=0}$. Therefore, the differences between the fully inhibited and partly inhibited rate must be largely due to some process not covered by the postulated mechanism. It is apparent that complete reconciliation of all the observed facts is impossible. At this point it seems most likely that surface reactions either chain, nonchain or a mixture of both, play a significant role.

Flow system studies carried out in the pressure dependent region[58, 64, 65, 82, 83] are compared in Fig. 4. The toluene results at 16 torr were originally calculated assuming the contribution of the reaction

$$CH_3+C_6H_5CH_2 \rightarrow C_6H_5C_2H_5 \tag{25}$$

was negligible[58]. A correction based on more recent work[64] has been made. This

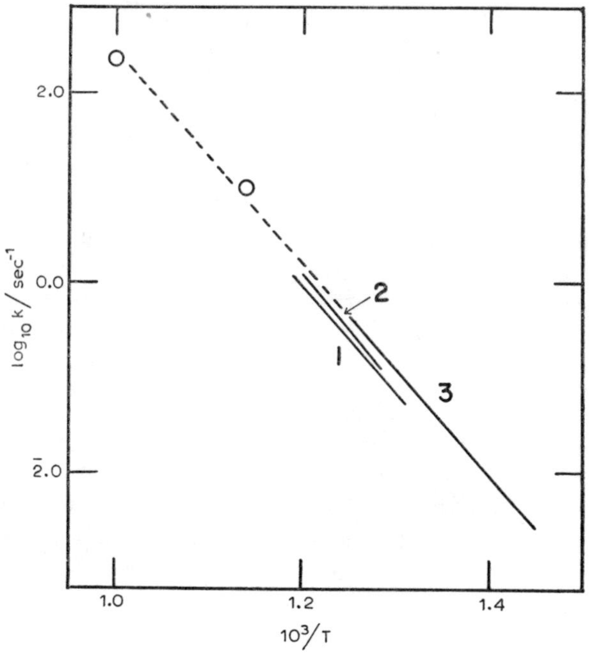

Fig. 4. Arrhenius plots for the pressure-dependent flow system decomposition of dimethyl mercury. 1, Gowenlock, Polanyi and Warhurst (7 torr CO_2+3 torr toluene), Kominar and Price (4.4 torr toluene); 2, Price and Trotman-Dickenson (16 torr toluene, rate coefficients corrected for methyl radicals found as ethylbenzene); 3, Krech and Price (16 torr benzene). ○, Lossing and Tickner (6–20 torr helium).

leads to $k_{16\text{torr}} = 6.3 \times 10^{13} \exp(-52{,}100/RT)\ \text{sec}^{-1}$. The curve for 4.4 torr was obtained[64] with toluene/alkyl ratios of approximately 10 : 1. Under these conditions 25–30 % of the methyl radicals released are removed by reaction (25). Allowing for pressure and energy transfer efficiency differences, the benzene carrier[65], toluene carrier[58, 64] and toluene plus CO_2 carrier[82] studies are in excellent agreement. The region of fall off, the decrease in the rate coefficient from the high-pressure limit and the observed decrease $E_\infty - E_{4.6\text{ torr}} = 5.5 \pm 2$ kcal.mole^{-1} are in reasonable agreement with the predictions of the theory of unimolecular reactions if 16–18 effective oscillators are assumed. Extrapolation of the flow system studies to 360 °C gives a value of k_1, which is close to that obtained by Russel and Bernstein under fully inhibited conditions at similar pressures[61]. Lossing and Tickner[83] used contact times of 0.0015–0.0010 sec with helium pressures of 6–20 torr. The partial pressure of dimethyl mercury was varied from 0.0028 to 0.0081 torr. These flow conditions are extreme and it is surprising that the rate coefficients obtained are in general agreement with those predicted by extrapolation of the other low-pressure flow system studies.

Photolysis of dimethyl mercury[84] proceeds by the overall process

$$Hg(CH_3)_2 h\nu = Hg + 2\,CH_3$$

The quantum yield for the decomposition of the liquid alkyl at 91 °C is 1.1 indicating the absence of any appreciable chain process[81]. This is consistent with the observation that CD_3CH_3 obtained in the photolysis of mixtures of $Hg(CD_3)_2$ and $Hg(CH_3)_2$ and of $Hg(CH_3)$ and CD_3COCD_3 arises through exchange of CH_3 and CD_3 *via* reactions such as (26)[81]

$$CD_3 + CH_3HgCH_3 \rightarrow CH_3 + CD_3HgCH_3 \qquad (26)$$

$$CD_3 + CD_3 \rightarrow C_2D_6 \qquad (12)$$

and not by methyl radical abstraction to form ethane as previously proposed[85]. Assuming $k_{12} = 4.8 \times 10^{13}$ l.mole^{-1}.sec^{-1} at approximately 80 torr over the temperature range 376–453 °K[86] gives[81] $k_{26} = 3.2 \times 10^{11} \exp(-12{,}600/RT)$ l.mole^{-1}.sec^{-1}.

3.4.2 Diethyl mercury

The pyrolysis of diethyl mercury has been studied using a nitrogen carrier flow system[87] both in the presence and absence of toluene. The experimental conditions used were: total pressure = 10±1 torr with 0.4 torr partial pressure of toluene, alkyl pressure $1\text{–}10 \times 10^{-2}$ torr, decomposition 10–75 % and contact time 0.1–0.3 sec. The presence of toluene had no effect on the rate coefficient, the observed ethane/ethylene ratio (~ 1) or the C_4/C_2 ratio (~ 4). These ratios were essentially independent of temperature.

An excellent material balance was obtained between the mercury collected and $C_4H_{10} + \frac{1}{2}(C_2H_6 + C_2H_4)$. The mechanism proposed for the pyrolysis both in the presence and absence of toluene therefore involves only four principal reactions

$$Hg(C_2H_5)_2 \rightarrow HgC_2H_5 + C_2H_5 \qquad (1)$$

$$HgC_2H_5 \rightarrow Hg + C_2H_5 \qquad (2)$$

$$2\,C_2H_5 \rightarrow C_4H_{10} \qquad (3)$$

$$2\,C_2H_5 \rightarrow C_2H_4 + C_2H_6 \qquad (4)$$

The failure to observe any appreciable temperature dependence of the C_4/C_2 ratio indicates $E_3 \approx E_4$. This is consistent with the difference $E_4 - E_3 = 0.8$ kcal.mole^{-1} reported by Ivin and Steacie[88]. The Arrhenius equation obtained over the temperature range 320–420 °C is $k = 1.26 \times 10^{14} \exp(-42{,}500/RT)$ sec^{-1}. If the observed activation energy is taken as a measure of D_1, $D_2 = 6.0 \pm 6$

kcal.mole^{-1} (Carson *et al.*[70]). It must be pointed out that although the experimental tests made do not indicate any problem, the flow conditions used may have caused significant errors. The expression for the homogeneous first-order rate coefficient obtained[89] with a static system over the temperature range 223–293 °C using pressure of 1–10 torr is $k = 2.7 \times 10^{14} \exp(-41{,}900/RT)$ sec^{-1}. This is generally good confirmation of the flow system result but at 370 °C an extrapolated rate coefficient of 0.73 sec^{-1} is obtained, almost twice that found for the nitrogen carrier system. A zero-order heterogeneous reaction with an activation energy of 22.1 kcal.mole^{-1} was also observed in the static system work. The distribution of products from both the homogeneous and heterogeneous reaction was consistent with the mechanism proposed for the flow system studies.

The mechanism for the photolysis of diethyl mercury is identical to that for the pyrolysis[88, 90]. The reaction of C_2H_5 with diethyl mercury, which apparently was of no importance under the conditions used in the pyrolysis studies, is also insignificant over the temperature range 25–75 °C. The photolysis produces ethyl radicals which initially have about 31 kcal.mole^{-1} excess energy so that the C_4/C_2 ratio is dependent on the surface/volume ratio and on the total pressure in the system. At 25 °C with essentially thermalized radicals (200 torr Ne added) $C_4/C_2 \approx 20$ while at 20 torr $C_4/C_2 \approx 3$.

In flash photolysis experiments at pressure < 0.1 torr reactions (1) and (2) presumably still occur, but the subsequent reactions involve vibrationally excited ethyl radicals[91], *viz.*

$$2\,C_2H_5^* \rightarrow C_2H_4 + 2\,CH_3 \quad (5)$$

$$2\,C_2H_5^* \rightarrow n\text{-}C_3H_7 + CH_3 \quad (6)$$

$$2\,CH_3 \rightarrow C_2H_6 \quad (7)$$

$$CH_3 + C_2H_5^* \rightarrow C_3H_8 \quad (8)$$

$$CH_3 + C_2H_5^* \rightarrow CH_4 + C_2H_4 \quad (9)$$

$$2\,C_3H_7 \rightarrow C_3H_8 + C_3H_6 \quad (10)$$

The relative importantce of these reactions can be judged from a typical product distribution reported by Thrush[91]: CH_4 15 %, C_2H_4 50 %, C_2H_6 10 %, C_3H_6 7 %, C_3H_8 10 %, C_4H_{10} 2 %.

More recently the flash photolysis of diethyl mercury has been re-investigated by Fischer and Mains[92]. At 1.54 torr and 24 °C the major products are butane (36 %), ethylene (32 %), ethane (22 %), propane (6 %) and hydrogen (4 %). Only traces of methane were detected. The addition of perfluorodimethylcyclobutane vapour did not alter the extent of photolysis, but the butane yield increased approximately 25 % while the yield of ethylene, ethane, hydrogen and propane all decreased. The change in product distribution occurred as the inert gas pressure was increased from zero to 4 torr. No further change was observed at pressures up to 30 torr. Under all conditions the ethane to butane ratio was higher than that expected if the products arose from disproportionation and recombination, respectively, of thermal ethyl radicals. However, the flash photolysis of mixture of mercury diethyl and mercury diethyl-d_{10} gave isotopic mixtures of ethanes, ethylenes and butanes that are consistent with the disproportionation and recombination of ethyl radicals. The isotopic distribution of the hydrogens indicated that these are formed from hydrogen atom reactions rather than molecular elimination processes.

The mechanism proposed by Fischer and Mains is

$$Hg(C_2H_5)_2 + h\nu = Hg + 2\,C_2H_5^* \qquad (11)$$

$$C_2H_5^* \rightarrow C_2H_4 + H \qquad (12)$$

$$C_2H_5^* + C_2H_5 \rightarrow C_2H_4 + 2\,CH_3 \qquad (5)$$

$$CH_3 + C_2H_5 \rightarrow C_2H_4 + CH_4 \qquad (9')$$

$$CH_3 + C_2H_5 \rightarrow C_3H_8 \qquad (8')$$

$$H + H + \text{wall} \rightarrow H_2 \qquad (13)$$

A material balance was observed that is consistent with the proposed mechanism within the limits of experimental error. The methane/propane ratio increases from 0.06 at 1 54 torr to 0.11 at 0.54 torr. Considerable uncertainty (approx. 50 %) must be attached to these ratios, but the trend is consistent with the higher yield of methane observed by Thrush[91] at pressure below 0.1 torr. Fischer and Mains[92] question the occurrence of reaction (6) as they could not detect any *n*-pentane in their reaction products. At the high ethyl radical concentrations obtained in flash photolysis this product would certainly be expected, if a significant concentration of thermal ethyl radicals were present. However, Thrush was unable to detect ethyl radicals spectroscopically under his experimental conditions. Therefore all reactions of ethyl in his system must involve $C_2H_5^*$ and the extent to which

$$C_2H_5^* + C_3H_7 \rightarrow C_5H_{12}$$

would be expected to occur is unknown.

3.4.3 *Divinyl mercury*

The mechanism proposed for the pyrolysis of divinyl mercury in a toluene carrier system[93] is

$$Hg(C_2H_3)_2 \rightarrow HgC_2H_3 + C_2H_3 \quad (1)$$

$$HgC_2H_3 \rightarrow Hg + C_2H_3 \quad (2)$$

$$C_2H_3 + C_6H_5CH_3 \rightarrow C_6H_5CH_2 + C_2H_4 \quad (3)$$

$$2\,C_2H_3 \rightarrow C_4H_6 \quad (4)$$

$$2\,C_2H_3 \rightarrow C_2H_4 + C_2H_2 \quad (5)$$

$$2\,C_6H_5CH_2 \rightarrow (C_6H_5CH_2)_2 \quad (6)$$

The rate coefficient calculated on the basis of the mercury released is first order at a fixed total pressure and is independent of surface-to-volume ratio. Below 590 °C concordant results are obtained based on the yield of gaseous products. Above this temperature $k_{Hg} > k_{gas}$ and at 696 °C approximately 22 % of the vinyl radicals form involatile products by reactions that are in part, at least, heterogeneous. The kinetics of the decomposition was studied over the temperature range 502–642 °C using contact times of 0.9–1.54 sec (packed vessel 2.5 sec) at a total pressure of 16 torr. The temperature dependence of the rate coefficient gives $k = 8.7 \times 10^{11} \exp(-48{,}300/RT)\ \text{sec}^{-1}$. The pre-exponential factor is of the same order of magnitude found for dimethyl cadmium under similar experimental conditions. This may indicate that the divinyl mercury pyrolysis was studied in its pressure-dependent region or that the mechanism is more complex than proposed. Considering these limitations, the observed activation energy of 48.3 kcal.mole^{-1} may place a lower limit on D_1. D_2 for dimethyl mercury and for diethyl mercury could be zero within the limits of experimental error, but in both cases seems more likely to have a small finite value. The thermodynamic data needed to calculate D_2 in divinyl mercury is not available. Representation of the overall decomposition, $Hg(C_2H_3)_2 = Hg + 2\,C_2H_3$, as a stepwise process is a reasonable postulate but is not based on any new experimental evidence.

The photolysis of divinyl mercury at 50 °C using light of 2200–2600 A occurs by a chain mechanism[94]

$$Hg(C_2H_3)_2 + h\nu = Hg + 2\,C_2H_3 \quad (7)$$

$$C_2H_3 + Hg(C_2H_3)_2 = C_4H_6 + Hg + C_2H_3 \quad (8)$$

$$C_2H_3 + Hg(C_2H_3)_2 \rightarrow C_2H_4 + C_2H_2HgC_2H_3 \quad (9)$$

$$C_2H_3 + C_2H_2HgC_2H_3 \rightarrow C_2H_4 + C_2HHgC_2H_3 \quad (10)$$

$$2\,C_2H_3 \rightarrow C_4H_6 \quad (4)$$

$$2\,C_2H_3 \rightarrow C_2H_2 + C_2H_4 \quad (5)$$

The rate of production of C_2H_2, C_2H_4 and C_4H_6 falls off as the reaction proceeds, indicating these products are consumed by secondary reactions. Addition of 400 torr of isobutane completely suppresses the formation of C_4H_6 and leads to a residual rate of photolysis about 1/25th of the rate calculated on the basis of the initial rate of formation of C_2H_2, C_2H_4 and C_4H_6 in the absence of inhibitor. This indicates a minimum chain length of approximately 25 which is consistent with the low acetylene yield and with the observation that both C_4H_6 and C_2H_4 are formed by processes that are first order in C_2H_3. The chain-carrying process, reaction (8), probably occurs in at least two steps

$$CH_2{=}CH\cdot + CH_2{=}CHHgCH{=}CH_2 \rightarrow [CH_2{=}CH{-}\underset{\displaystyle CH_2\cdot}{\underset{|}{C}}HHgCH{=}CH_2] \quad (8a)$$

$$\rightarrow CH_2{=}CH{-}CH{=}CH_2 + Hg + C_2H_3\cdot \quad (8b)$$

Addition of 400 torr of CO_2 decreased the overall rate of photolysis by only 25 % indicating that the vinyl radicals taking part in reaction (8) in the absence of added gas are largely thermal radicals. This is consistent with the proposed mechanism which indicates that most of the vinyl radicals are generated by reaction (8b) rather than reaction (7).

3.4.4 *Higher mercury alkyls*

Most of the significant kinetic work on the decomposition of these compounds has been carried out by Gowenlock and his co-workers[95-99]. The pyrolysis of di-*n*-propyl mercury[95], di-isopropyl mercury[96, 97] and di-*n*-butyl mercury[98] have been studied in a nitrogen carrier system, both in the presence and absence of NO. The decomposition of di-isopropyl mercury[99] and di-*n*-butyl mercury[99] have also been investigated using CO_2 and SF_6 as carrier gases. The decompositions all appear to be essentially first-order, homogeneous processes, and with minor exceptions have been studied at their high pressure limit. Warhurst[100] and Billinge and Gowenlock[99] have discussed reasons for assuming that these alkyls decompose into three fragments by the simultaneous rupture of both metal carbon bonds (with the exception of di-isopropyl mercury at low temperature). Further discussion of this point will be deferred until the experimental results for the individual compounds have been considered.

3.4.5 *Di-n-propyl mercury*[95]

The nitrogen carrier flow system pyrolysis of this compound was studied from 332–389 °C using total pressures of 4.4–9.5 torr and contact times of approximately 0.6 sec. The proposed mechanism is

$$Hg(CH_3CH_2CH_2)_2 = Hg + 2\,CH_3CH_2CH_2 \quad (1)$$

$$2\,CH_3CH_2CH_2 \rightarrow C_3H_6+C_3H_8 \quad (2)$$

$$2\,CH_3CH_2CH_2 \rightarrow C_6H_{14} \quad (3)$$

$$CH_3CH_2CH_2 \rightarrow CH_3+C_2H_4 \quad (4)$$

$$2\,CH_3 \rightarrow C_2H_6 \quad (5)$$

$$CH_3+CH_3CH_2CH_2 \rightarrow C_4H_{10} \quad (6)$$

The $C_2 : C_3 : C_4$ ratio varied from approximately 2 : 1 : 2 at low temperature to 2 : 1 : 1 at the high end of the temperature range. The C_2 fraction was 70–100 % C_2H_4 while the C_3 fraction contained 30–60 % C_3H_6. The mean rate coefficient is given by $k = 2.6\times10^{15}\exp(-46{,}500/RT)\ \text{sec}^{-1}$. In view of the wide divergence from the expected 1 : 1 ratio of C_3H_6 and C_3H_8 the significance of k is uncertain.

3.4.6 Di-isopropyl mercury

The flow system pyrolysis has been reported in three papers[96, 97, 99] although one of these[96] is a preliminary report of the final paper in this series[99]. The mechanism proposed[99] is

$$Hg(CH_3CHCH_3)_2 = Hg+2\,CH_3CHCH_3 \quad (1)$$

$$2\,CH_3CHCH_3 \rightarrow C_3H_6+C_3H_8 \quad (2)$$

$$2\,CH_3CHCH_3 \rightarrow (CH_3)_2CH\text{–}CH(CH_3)_2 \quad (3)$$

$$CH_3CHCH_3 \rightarrow C_2H_4+CH_3 \quad (4)$$

$$CH_3+CH_3CHCH_3 \rightarrow (CH_3)_3CH \quad (5)$$

$$CH_3+CH_3CHCH_3 \rightarrow C_3H_6+CH_4 \quad (6)$$

Over the temperature range studied, 170–300 °C, the yield ratios were $C_3H_6/C_3H_8 = 1.10\pm0.05$ and $C_3H_8/C_6H_{14} = 0.61\pm0.05$, and were independent of surface-to-volume ratio. Above 240 °C rate coefficients calculated on the basis of the mercury produced were in good agreement with those obtained from analysis of the hydrocarbon products. The expression for the rate coefficient obtained using nitrogen carrier at 7.4 torr[97] $k = 1.6\times10^{16}\exp(-40{,}400/RT)\ \text{sec}^{-1}$ is in excellent agreement with that obtained using CO_2, CO_2+NO and SF_6 at 5–12 torr pressure[96, 99], $k = 2.5\times10^{16}\exp(-40{,}700/RT)\ \text{sec}^{-1}$. Rate coefficients below 230 °C were calculated on the basis of the yield of hydrocarbon products. These experiments were carried out using nitrogen carrier at 8–15 torr pressure. As this lower temperature region is approached, a marked curvature is observed in the Arrhenius plot, and over the temperature range 170–230 °C the rate coefficient is best represented by $k = 1\times10^{11} \exp(-27{,}000/RT)\ \text{sec}^{-1}$. Billinge and Gowenlock[99] associate the activation energy in this low temperature region with D_1, the bond dissociation energy of the first isopropyl–mercury bond. Based on the high-temperature activation energy and on thermochemical studies[101], $D_1+D_2 = 40.4$ kcal.mole^{-1}. D_2 would then be 13.4 kcal.mole^{-1}. If the upper error limit placed on the rate coefficient by Billinge and Gowenlock[99] is accepted, then $k = 1\times10^{13} \exp(-32{,}000/RT)\ \text{sec}^{-1}$ and $D_2 = 8.4$ kcal.mole^{-1}. These results seem more reasonable, but there is little justification for taking them in preference to the observed values.

It is possible the reaction in the low-temperature region is a chain process similar to that observed in the photolysis of divinyl mercury, *viz.*

$$CH_3CHCH_3+Hg(CH_3CHCH_3)_2 = C_3H_8+Hg+C_3H_6+CH_3CHCH_3 \quad (7)$$

in competition with

$$CH_3CHCH_3+Hg(CH_3CHCH_3)_2 = C_6H_{14}+Hg+CH_3CHCH_3 \quad (8)$$

Direct tests for chain processes in the low-temperature region do not appear to have been made. The invariance of the ratio $C_3H_8/C_6H_{14} = 0.61\pm0.05$ over the temperature range used would require $E_7 \approx E_8$ and $k_7/k_8 = 0.61$. This is a rather stringent requirement and although it does not rule out the participation of a chain process it does make it less plausible. However, Razuvaev *et al.*[102] report that the decomposition of liquid di-isopropyl mercury at temperatures up to 150 °C does occur by a mechanism involving reaction (7). The rate coefficients that can be calculated from their results at 140 °C and 150 °C are in reasonable agreement with the value predicted by extrapolation the of data of Billinge and Gowenlock[99]. Razuvaev *et al.*[102] also report that at 140 °C with a twenty-fold excess of mono-deuterocyclohexane approximately 30 % of the isopropyl radicals form propane by hydrogen abstraction from the cycloalkane. It is therefore concluded that in the absence of solvents the high yield of propane observed is a result of similar abstraction process from the hydrocarbons produced by the pyrolysis. Similar abstraction processes in the flow system decomposition are unlikely because of the very low concentration of such products.

3.4.7 Di-n-butyl mercury

The two investigations of the flow system pyrolysis of this compound[98, 99] were carried out under conditions similar to those used in the study of the di-isopropyl mercury decomposition[99]. Over the temperature range 350–495 °C the rate coefficient based on mercury production is $k = 6.3\times10^{15} \exp(-47{,}800/RT)$

References pp. 254–257

sec^{-1}. Error limits of ± 2 $kcal.mole^{-1}$ were proposed to allow for the scatter observed. The products may be explained by the mechanism

$$Hg(CH_2CH_2CH_2CH_3)_2 = Hg + 2\,CH_3CH_2CH_2CH_2 \quad (1)$$

$$2\,CH_3CH_2CH_2CH_2 \rightarrow C_8H_{18} \quad (2)$$

$$2\,CH_3CH_2CH_2CH_2 \rightarrow C_4H_{10} + C_4H_8 \quad (3)$$

$$CH_3CH_2CH_2CH_2 \rightarrow C_2H_4 + C_2H_5 \quad (4)$$

$$CH_3CH_2CH_2CH_2 \rightarrow CH_3CH = CH_2 + CH_3 \quad (5)$$

plus the normal reactions of C_2H_5 and CH_3. The observed activation energy was associated with $D_1 + D_2$. No thermochemical value of this sum is available.

3.4.8 Alkyl–mercury bond dissociation

The available kinetic and thermochemical data are summarized in Table 7. Based on the approximate equality of E and $D_1 + D_2$ and on the magnitude of the frequency factor, Billinge and Gowenlock[98] would place dimethyl mercury, di-*n*-propyl mercury, di-isopropyl mercury (above 230 °C) and (on the basis of the frequency factor only, since thermochemical data are not available) di-*n*-butyl mercury in class II (simultaneous rupture into mercury and two alkyl radicals). If the high frequency factors are simply due to a general softening of the vibrations in the activated state, then in the case of di-isopropyl mercury $D_2 = 0$, while for dimethyl and di-*n*-propyl mercury D_2 is small but finite (2–3 $kcal.mole^{-1}$). However, within the limits of experimental error all of these alkyls for which thermochemical data are available may have $E = D_1 + D_2$, and thus all may belong to class II. At the same time it must be noted that some metal alkyls which are

TABLE 7

ARRHENIUS PARAMETERS AND THERMOCHEMICAL VALUES OF D_1+D_2 FOR MERCURY ALKYLS

Alkyl	*A* (sec^{-1})	*E* ($kcal.mole^{-1}$)	D_1+D_2 ($kcal.mole^{-1}$)
$Hg(CH_3)_2$	5.2×10^{15}	57.7	60.2
$Hg(C_2H_5)_2$	2.0×10^{14}	42.2	48.5
$Hg(C_2H_3)_2$	5.5×10^{13} [a]	55.3 [a]	
$Hg(CH_2CH_2CH_3)_2$	2.6×10^{15}	46.5	49.4
$Hg(CH_3CHCH_3)_2$	2.5×10^{16} [b]	40.7 [b]	40.4
$Hg(CH_2CH_2CH_2CH_3)_2$	6.3×10^{15}	47.8	

[a] Experimental values 8.7×10^{11} sec^{-1}, 48,300 $kcal.mole^{-1}$, assumed to be in pressure-dependent region. Estimated corrections based on experience with dimethyl mercury.
[b] Above 240 °C. Below 230 °C, $A = 1 \times 10^{11}$ sec^{-1}, $E = 27.0$ $kcal.mole^{-1}$.

known to decompose by a class I mechanism (initial decomposition into two fragments by loss of a single alkyl radical) have high frequency factors: $Ga(CH_3)_3$, $A = 3.2 \times 10^{15}$ sec^{-1}; $In(CH_3)_3$, $A = 5.0 \times 10^{15}$ sec^{-1}; $Sb(CH_3)_3$, $A = 3.1 \times 10^{15}$ sec^{-1}. It is the author's opinion that at this point it is impossible to distinguish between the possibilities of class II and class I mechanisms for the thermal decomposition of mercury alkyls.

3.5 MERCURY ARYLS

The pyrolysis of diphenyl mercury[87], phenylmercuric chloride[87], phenylmercuric bromide[87] and phenylmercuric iodide[103] has been studied using the nitrogen-plus-toluene flow procedures[87] described for diethyl mercury. As previously noted, the flow conditions are not optimum, and may have caused significant errors.

3.5.1 Diphenyl mercury[87]

The hydrocarbon products of this pyrolysis were not determined but the mechanism is probably very simple, *viz.*

$$Hg(C_6H_5)_2 = Hg + 2\,C_6H_5 \qquad (1)$$

$$2\,C_6H_5 \rightarrow (C_6H_5)_2 \qquad (2)$$

$$C_6H_5 + C_6H_5CH_3 \rightarrow C_6H_6 + C_6H_5CH_2 \qquad (3)$$

$$2\,C_6H_5CH_2 \rightarrow (C_6H_5CH_2)_2 \qquad (4)$$

3.5.2 Phenyl mercuric chloride and bromide[87]

The mechanism proposed for these compounds is

$$C_6H_5HgX \rightarrow C_6H_5 + HgX \qquad (1)$$

$$C_6H_5 + C_6H_5CH_3 \rightarrow C_6H_6 + C_6H_5CH_2 \qquad (2)$$

$$2\,C_6H_5 \rightarrow (C_6H_5)_2 \qquad (3)$$

$$HgX + C_6H_5CH_3 = Hg + HX + C_6H_5CH_2 \qquad (4)$$

$$HgX + HgX \rightarrow Hg + HgX_2 \xrightarrow[\text{exit tube}]{\text{in cold}} Hg_2X_2 \qquad (5)$$

$$HgX + C_6H_5 \rightarrow Hg + C_6H_5X \qquad (6)$$

Reaction (4) was proposed by Carter *et al.*[87] on the assumption that the overall process was exothermic and therefore had a reasonably low activation energy (5–10 kcal.mole^{-1} for X = Cl, 10–15 kcal.mole^{-1} for X = Br). These calculations were based on $D(C_6H_5CH_2{-}H) = 77.5$ kcal.mole^{-1} (Szwarc[104]). However, using the more recent value[50, 105] of 85 kcal.mole^{-1}, both processes are endothermic. It is more likely that reaction (4) occurs in two steps, *viz.*

$$HgX \rightarrow Hg + X \quad (4a)$$

$$X + C_6H_5CH_3 \rightarrow HX + C_6H_5CH_2 \quad (4b)$$

3.5.3 Phenyl mercuric iodide[103]

On the basis of results obtained with other mercury aryls the flow system decomposition of this compound in the presence of toluene was assumed to be first-order and homogeneous. The extent of decomposition for the 9 runs carried out was determined by isolation of the undecomposed aryl. Detailed analysis of other products was not attempted.

3.5.4 Aryl–mercury bond dissociation

The Arrhenius parameters and the thermochemical sum of the phenyl–carbon and phenyl–halogen bond dissociation energies are shown in Table 8. The extent of the diphenyl mercury decomposition was determined from the weight of mercury produced. It is the present author's opinion that in calculating the Arrhenius parameters for this compound Carter *et al.*[87] gave too great a statistical

TABLE 8

ARRHENIUS PARAMETERS AND THERMOCHEMICAL VALUES OF D_1+D_2 FOR MERCURY ARYLS

Aryl	*A* (*sec*$^{-1}$)	*E* (*kcal.mole*$^{-1}$)	D_1+D_2 (*kcal.mole*$^{-1}$) [a]
$Hg(C_6H_5)_2$	1.0×10^{14} [b]	59 ± 4 [b]	60.3 ± 3 [c]
C_6H_5HgCl	1.0×10^{13}	59 ± 3	85.8 ± 3
C_6H_5HgBr	2.0×10^{14}	63 ± 2	79.9 ± 3
C_6H_5HgI	5.0×10^{15} [d]	63 ± 2 [d]	67.2 ± 3 [d]

[a] Based on $D(C_6H_5{-}H) = 102$ kcal.mole^{-1}.
[b] Estimate based on re-evaluation of published data[87]. Values originally proposed $A = 1\times10^{16}$ sec^{-1}, $E = 68\pm4$ kcal.mole^{-1}.
[c] Refs. 106, 107.
[d] Ref. 103, all other data from ref. 87.

weight to the lowest rate coefficient. If this point, which presumably is the least precise, is omitted, the line of best fit is quite different and the approximate parameters shown in Table 8 are obtained. The activation energy calculated in this manner is more in accord with the thermochemical value of D_1+D_2[106, 107]. Phenyl mercuric chloride and phenyl mercuric bromide are obviously examples of class I decomposition. The transfer of activity to diphenyl mercury during thermal decomposition in radioactive benzene[108] indicates that the C_6H_5Hg radical is stable at 170 °C but data at higher temperatures is not available. Under the flow system conditions used, diphenyl mercury and phenyl mercuric iodide decompositions may be either class I or class II. Discussion of this point would follow the arguments already presented in the discussion of alkyl–mercury bond rupture and again no definite conclusion could be reached.

The photolysis of 2,2-diphenyl mercury produces diphenylene[109]

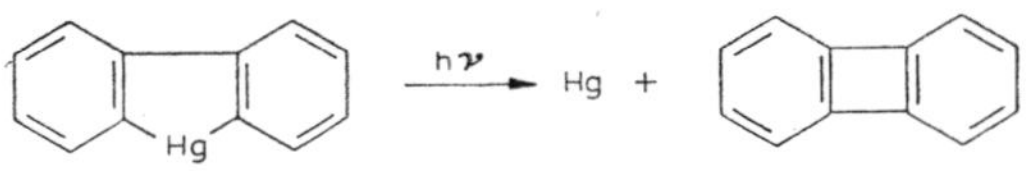

No kinetic data are available but numerous references to related pyrolysis and photolysis studies are given. The product distribution from the decomposition of dibenzyl mercury under a variety of conditions has also been reported (ref. 110 lists the pertinent papers).

The primary step in the photolysis of phenyl mercuric cyanide and cyclohexyl mercuric cyanide in methanol at room temperature is mercury–phenyl or mercury–cyclohexyl bond rupture[111]. The subsequent steps in the mechanism are complex, but do not involve the parent substance. Reactions involving HgCN must be postulated to explain the observed product, so that this radical must have a finite existence.

3.6 BORON ALKYLS

3.6.1 Trimethyl boron

The thermal decomposition and photolysis of this alkyl have been studied by Buchanan and Creutzberg[112]. The pyrolysis mechanism is not fully understood. The overall process is first-order and is unaffected by an 8.5-fold increase in surface-to-volume ratio. Based on measurements of pressure increase, the reaction exhibits an induction period ranging from 2–3 minutes at 513 °C to 40 minutes at 466 °C. Short chains are apparently involved. A polymer initially of empirical formula $(BCH_2)_n$ but slowly losing hydrogen to form $(BCH)_n$ is deposited on the surface.

The mechanism probably involves the reactions

$$B(CH_3)_3 \rightarrow B(CH_3)_2 + CH_3 \quad (1)$$

$$CH_3 + B(CH_3)_3 \rightarrow CH_2B(CH_3)_2 + CH_4 \quad (2)$$

$$CH_2B(CH_3)_2 \rightarrow CH_2BCH_3 + CH_3 \quad (3)$$

followed by polymerization steps which eventually yield hydrogen and $(BCH)_n$. Reaction (4)

$$B(CH_3)_3 \rightarrow CH_2B(CH_3)_2 + H \quad (4)$$

was also proposed by Buchanan and Creutzberg[112] on the assumption that $D[(CH_3)_2B–CH_3]$ was greater than 100 kcal.mole^{-1} and therefore $E_1 \geqslant E_4$. The best value[113–115] for the mean bond dissociation energy in this alkyl is 87 kcal.mole^{-1}. The dissociation energy for the first metal–carbon bond in trimethyl gallium is very close to the mean bond dissociation energy[116] and there is some indication that this may also be true for trimethyl aluminium[117]. $D[(CH_3)_2B–CH_3]$ is therefore probably less than 90 kcal.mole^{-1} and the participation of reaction (4) seems less likely. The overall rate coefficient for $B(CH_3)_3$ decomposition is given by

$$k = 1.2 \times 10^{12} \exp(-56{,}000/RT) \text{ sec}^{-1}$$

and the activation energies for methane and hydrogen production are 76 and 75 kcal.mole^{-1} respectively. With the limited knowledge of the mechanism these numerical values cannot be associated with particular kinetic process.

The photolysis of trimethyl boron at temperatures up to 300 °C may occur by reactions (1)–(3) of the pyrolysis process[112]. Subsequent steps which lead to hydrogen formation in the pyrolysis system are much less important [$CH_4 : H_2 \approx 2:1$ under pyrolysis conditions, $\approx 9:1$ in photolysis system].

TABLE 9

KINETIC PARAMETERS FOR THE ISOMERIZATION OF TRIALKYLBORANES

Reaction	*Temperature* (°*C*)	*A* (*sec*$^{-1}$)	*E* (*kcal.mole*$^{-1}$)
1. (*tert*.-Bu)B(iso-Bu)$_2$ → (iso-Bu)$_3$B	120–150	1.1×10^{16}	39
2. (iso-Pr)$_3$B → (iso-Pr)$_2$B(*n*-Pr)	121–135	7.0×10^{11}	29
3. (iso-Pr$_2$B(*n*-Pr) → (iso-Pr)B(*n*-Pr)$_2$	128–135	4.7×10^{11}	29
4. (iso-Pr)B(*n*-Pr)$_2$ → (*n*-Pr)$_3$B	128–135	2.3×10^{11}	29
5. (*sec*.-Bu)$_3$B → (*sec*.-Bu)$_2$B(*n*-Bu)		(5.7×10^{12}) [a]	(32) [a]
6. (*sec*.-Bu)$_2$B(*n*-Bu) → (*sec*.-Bu)B(*n*-Bu)$_2$		(3.8×10^{12}) [a]	(32) [a]
7. (*sec*.-Bu)B(*n*-Bu)$_2$ → (*n*-Bu)$_3$B	125–150	1.9×10^{12}	32

[a] Based on reaction (7), overall analysis for alkyl groups (alkaline oxidation and alcohol ratio determination) and ratio $k_5 : k_6 : k_7 = 3:2:1$ (assumed because $k_2 : k_3 : k_4 = 3:2:1$).

The kinetics of the decomposition of higher boron alkyls have not been investigated but in general these compounds undergo olefin elimination rather than metal–carbon bond rupture.

3.6.2 tert.-*Butyldiisobutyl, triisopropyl and tri*-sec.-*butyl boranes*

The thermal isomerization of *tert.*-butyl-diisobutylborane to triisobutylborane and the stepwise isomerization of triisopropyl and tri-*sec.*-butyl boranes to the corresponding straight-chain isomers have been studied over a range of temperature, both neat and using either the end product trialkylborane or diglyme as solvent[117a]. In all cases the reactions were first order and showed no solvent effect.

The results are shown in Table 9. Based on absolute rate theory Rossi *et al.*[117a] give $\Delta S^{\ddagger}$ for reaction (1) as 14.2±2.6 e.u. The mechanism proposed is

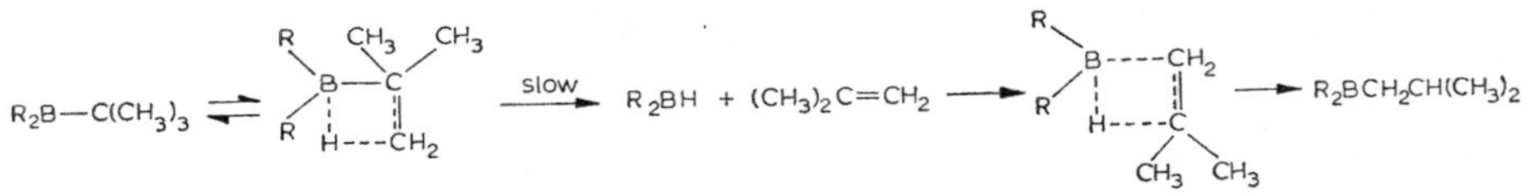

(1)

so that the rate-controlling step is olefin elimination.

The large negative entropies of activation for reactions (2)–(4)($\Delta S^{\ddagger}$ = −8 to −11 e.u.) and the reversible nature of these reactions is attributed to isomerization *via* a π-complex, *viz.*

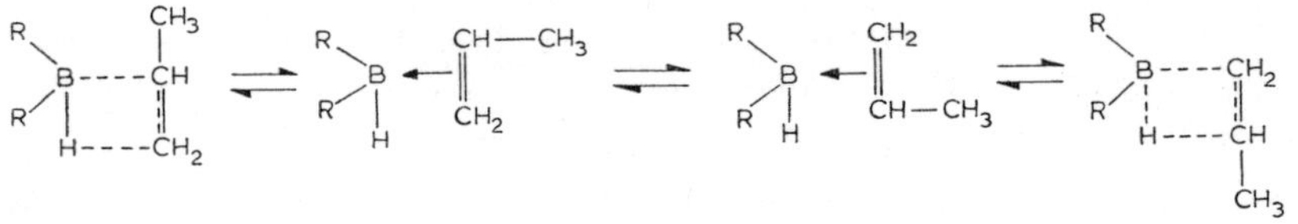
(2)

Of reactions (5)–(7) only reaction (7) has been directly investigated. The entropy of activation for this reaction, −3.0 e.u., probably indicates that the mechanism for the isomerization of *sec.*-butyl is closely related to that of iso-propyl.

3.7 ALUMINIUM ALKYLS

3.7.1 Trimethyl aluminium

The thermal decomposition of trimethyl aluminium has been investigated by Yeddanapalli and Shubert[117] and by Ouimet[118]. In both investigations the decomposition was found to occur by a predominantly homogeneous chain reaction and methane was the major gaseous product (CH_4 > 91 %; above 400 °C and 100 torr pressure $CH_4 \approx 99$ %).

The proposed initiation and propagation reactions are

$$Al(CH_3)_3 \rightarrow Al(CH_3)_2 + CH_3 \quad (1)$$

$$CH_3 + Al(CH_3)_3 \rightarrow CH_4 + (CH_3)_2AlCH_2 \quad (2)$$

$$(CH_3)_2AlCH_2 \rightarrow CH_3AlCH_2 + CH_3 \quad (3)$$

From thermochemistry[19, 119] the mean metal–carbon bond dissociation energy in trimethyl aluminium is 66.0 ± 2.0 kcal.mole^{-1}. By analogy with trimethyl gallium[120] $D[(CH_3)_2Al–CH_3]$ should be approximately equal to this value. Although the details differ slightly it was suggested in both investigations that $Al(CH_3)_2$ yields $Al + 2\,CH_3$. The stage $Al(CH_3)_2 \rightarrow AlCH_3 + CH_3$ is quite likely, but the $AlCH_3$ formed should be thermally stable and probable ends up in the observed polymer. Yeddanapalli and Shubert give, as the termination step, reaction (4)

$$2\,CH_3 \rightarrow C_2H_6 \quad (4)$$

Steady-state treatment then predicts the observed $\frac{3}{2}$-order kinetics and leads to

$$E_{\text{overall}} = E_2 + \tfrac{1}{2}(E_1 - E_4)$$

E_2 should be approximately 10–15 kcal.mole^{-1}, $E_4 = 0$ and the observed overall activation energy is 45 kcal.mole^{-1} ($A = 2.56 \times 10^{14}$ l.$^{\frac{1}{2}}$mole$^{-\frac{1}{2}}$.sec^{-1}). The value of $E_1 = 65$ kcal.mole^{-1} obtained from this equation seems quite reasonable. Ouimet[118] reports that above 400 °C and 100 torr pressure the decomposition is first order and that below this temperature the order is less than one and irreproducible induction periods are observed. However, he also found that at 405 °C and 100 torr pressure the reaction was 39 % heterogeneous. Yeddanapalli and Shubert using pressures of 20–170 torr over the temperature range 298–334 °C report only 6 % heterogeneous reaction. The rate coefficient reported by Ouimet is $1.02 \times 10^5 \exp(-20{,}000/RT)$ sec^{-1}. In view of the apparent heterogeneous contribution the significance of this is uncertain.

3.7.2 Triethyl aluminium

The thermal decomposition of this alkyl was studied in a static system over the temperature range 162–192.4 °C[121]. The surface of the reaction vessel was unintentionally activated before each run by rinsing with distilled water before baking out. Under these conditions the decomposition occurs by a chain mechanism with initiation and termination at the walls. The overall rate coefficient for the decomposition is $1.58 \times 10^8 \exp(-29{,}000/RT)$ sec^{-1}.

3.8 TRIMETHYL GALLIUM, INDIUM AND THALLIUM

Studies of the pyrolysis of these three alkyls may conveniently be discussed in a combined section. The decompositions were carried out in a conventional toluene carrier flow system using contact times of 1–2 sec[120, 122, 123]. The conditions used satisfy both plug flow and thermal equilibrium requirements[68, 69]. Toluene to alkyl ratios greater than 50 in the trimethyl gallium system and greater than 200 in the trimethyl indium and thallium studies were required to obtain first-order dependence in terms of the alkyl concentration. Under these conditions methane and ethane are produced by the reactions

$$CH_3 + C_6H_5CH_3 \rightarrow CH_4 + C_6H_5CH_2 \tag{1}$$

$$2\,CH_3 \rightarrow C_2H_6 \tag{2}$$

Some 3–4 % of the CH_4 is formed by abstraction of ring hydrogen atoms, but this is of no consequence in the present discussion. Values of $k_1/k_2^{\frac{1}{2}}$ can be calculated using the equation

$$\frac{k_1}{k_2^{\frac{1}{2}}} = \frac{[\text{moles CH}_4]}{[\text{moles C}_2\text{H}_6]^{\frac{1}{2}}} \times \frac{1}{V^{\frac{1}{2}} t^{\frac{1}{2}}\,[\text{toluene}]}$$

where V = volume of the reaction zone and t = length of the run. At a fixed total pressure (reaction (2) is strongly pressure dependent under the conditions used in most flow system work) a linear plot of $k_1/k_2^{\frac{1}{2}}$ *vs.* $10^3/T$ is obtained if an alkyl which, over the entire temperature range studied, releases a fixed number of methyl radicals per alkyl molecule undergoing decomposition is used (see, for example, ref. 124). However, in studying the thermal decomposition of dimethyl zinc a sharp break in the Arrhenius plot of $k_1/k_2^{\frac{1}{2}}$ was observed[46]. Similar patterns have now been found when either trimethyl gallium or trimethyl indium is used as the source of methyl radicals[120, 122]. The three curves are shown in Fig. 5.

The observed breaks may be explained if it is assumed that the following general mechanism applies

$$M(CH_3)_n = M(CH_3)_{n-y} + y\,CH_3 \tag{3}$$

$$M(CH_3)_{n-y} = M(CH_3)_{n-y-z} + z\,CH_3 \tag{4}$$

where reaction (4) occurs at an appreciable rate only under conditions such that reaction (3) is complete in a very small fraction of the contact time. One effect of such a mechanism is that when reaction (4) is occurring at an appreciable rate, the bulk of the methyl radicals are released at an average temperature below that of the reaction zone and throughout a volume much smaller than V, the normal

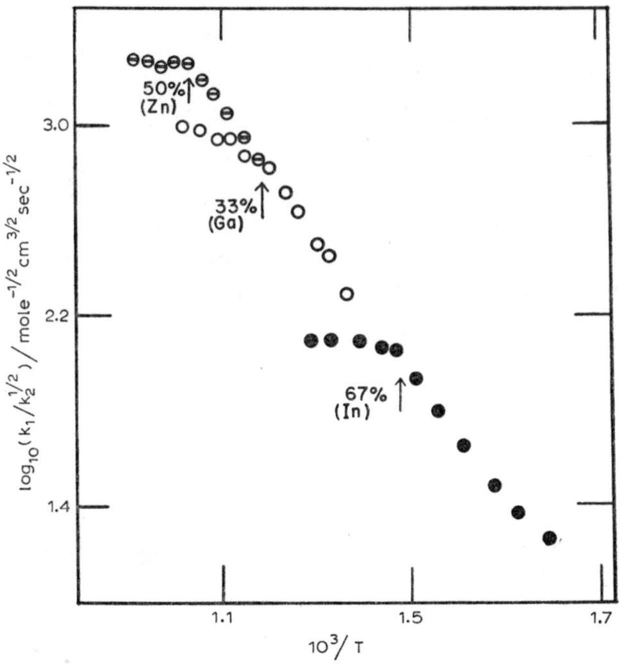

Fig. 5. The Arrhenius plot of $k_1/k_2^{\frac{1}{2}}$. ⊖, dimethyl zinc; ○, trimethyl gallium; ●, trimethyl indium. % figures are in terms of the theoretical yield of methyl radicals that would be obtained if all methyl–metal bonds were broken. All points are corrected to 13 torr pressure (toluene+0.3–2.5 % alkyl).

reaction volume. The lower values of $k_1/k_2^{\frac{1}{2}}$ arising from this are partially offset by the radicals released by reaction (4) giving the net results shown in Fig. 5.

In the case of dimethyl zinc the break in the Arrhenius plot occurs at 50 % theoretical yield of methyl radicals indicating that below 973 °K only the first bond is broken. For trimethyl gallium the break occurs at 33 % indicating that below 820 °K only the first bond is broken. However, the maximum methyl radical yield obtained in any experiment was 67 %. This would indicate that under obtainable experimental conditions the gallium methyl radical did not dissociate. Analysis for gallium in the outlet traps and in the reaction zone confirmed the stability of gallium methyl and indicated that it polymerized in the reaction zone[120]. For trimethyl indium the break in the Arrhenius plot of $k_1/k_2^{\frac{1}{2}}$ occurs at 67 % theoretical yield of methyl radicals. It therefore appears that between 550 °K and 670 °K the second methyl radical is released immediately after the first. Subsequent metal and methyl radical analysis showed that methyl indium retained its identity and polymerized in the reaction zone[122]. Above 680 °K appreciable decomposition of methyl indium occurred and above 780 °K the decomposition of methyl indium is complete in approximately one second.

When trimethyl thallium is used[123] the Arrhenius plot of $k_1/k_2^{\frac{1}{2}}$ is linear over

TABLE 10

MECHANISMS AND ARRHENIUS PARAMETERS FOR THE THERMAL DECOMPOSITIONS OF GROUP III METAL ALKYLS

Temperature range (°K)	*Reaction*	*Log A (sec⁻¹)*	*E (kcal.mole⁻¹)*	
686–820	$Ga(CH_3)_3 \rightarrow Ga(CH_3)_2+CH_3$	15.5	59.5	(5)
820–983	$Ga(CH_3)_2 \rightarrow Ga(CH_3)+CH_3$	7.9	35.4	(6)
	$n[Ga(CH_3)] = [Ga(CH_3)]_n$		[a]	(7)
550–670	$In(CH_3)_3 \rightarrow In(CH_3)_2+CH_3$	15.7	47.2	(8)
	$In(CH_3)_2 \rightarrow In(CH_3)+CH_3$		20.2 [b]	(9)
	$n[In(CH_3)] = [In(CH_3)]_n$			(10)
670–781	$In(CH_3) \rightarrow In+CH_3$	10.9	38.7	(11)
	$n[In(CH_3)] = [In(CH_3)]_n$			(10)
470–591	$Tl(CH_3)_3 \rightarrow Tl(CH_3)_2+CH_3$	10.8	27.4	(12)
	$Tl(CH_3)_2 \rightarrow Tl(CH_3)+CH_3$			(13)
	$Tl(CH_3) \rightarrow Tl+CH_3$			(14)

[a] Combining thermochemical results[19,60] with E_5 and $D_6 = E_6+4$ (estimated pressure cor rection), $D(Ga\text{–}CH_3) = 76.5$ kcal.mole^{-1}.
[b] From E_8, $D_{11} = E_{11}+2$ (estimated pressure correction) and thermochemical results[19,125].

the entire range of decomposition, indicating the simple consecutive release of three methyl radicals. Analysis of the material deposited in the reaction zone and in the outlet traps confirm the release of all three methyl radicals with the deposition of metallic thallium in the reaction zone. The mechanisms for the therma-decomposition of these Group III methyl metallic alkyls show an interesting progression (Table 10).

Reactions (5), (6) and (12) are straightforward. Reaction (7) is a natural consequence of the thermal stability of the $GaCH_3$ radical and the availability of two bonding electrons. Reaction (10) is observed under certain conditions for similar reasons. Reaction (11) occurs because the $In\text{–}CH_3$ bond is much weaker than the $Ga\text{–}CH_3$ bond and at the considerably lower temperatures unimolecular pressure effects play a smaller role (*i.e.* $ZnCH_3$, 1000–1100 °K, $\log A = 6.8$; $InCH_3$, 670–781 °K, $\log A = 10.9$). Reactions (12), (13) and (14) cannot be discussed in detail since only $D[(CH_3)_2Tl\text{–}CH_3]$ is known. For reaction (6) to occur there must be the simultaneous return of an electron from the $4P$ to the $4S$ level; the low A factor found may be attributed to this requirement. The A factor for the trimethyl thallium decomposition is abnormally low for a unimolecular process at its high-pressure limit. Extensive surface effects were observed and even under the best conditions achieved the overall reaction was estimated to be 23 % heterogeneous at 480 °C and 13 % heterogeneous at 560 °C. The tests made may not have been adequate from a quantitative standpoint so that the reported Arrhenius parameters may be unsatisfactory.

References pp. 254–257

3.9 SILICON ALKYLS

3.9.1 Tetramethyl silicon

Over the temperature range 659–717 °C and above 100 torr the thermal decomposition of this alkyl is first-order and homogeneous[126]. The overall rate coefficient is $1.6 \times 10^{15} \exp(-78{,}800/RT)\ \text{sec}^{-1}$. The detailed mechanism is undoubtedly complex. The main gaseous products at high conversion are H_2 and CH_4. The following steps have been proposed[127]

$$Si(CH_3)_4 \rightarrow Si(CH_3)_3 + CH_3 \tag{1}$$

$$2\,CH_3 = C_2H_4 + 2\,H \tag{2}$$

$$2\,H \xrightarrow[\text{or M}]{\text{wall}} H_2 \tag{3}$$

$$C_2H_4 = CH_4 + C \tag{4}$$

Rather than reactions (2) and (3), steps (5)–(8) seem more likely

$$2\,CH_3 \rightarrow C_2H_6 \tag{5}$$

$$CH_3 + C_2H_6 \rightarrow CH_4 + C_2H_5 \tag{6}$$

$$C_2H_5 \rightarrow C_2H_4 + H \tag{7}$$

$$H + C_2H_6 \rightarrow C_2H_5 + H_2 \tag{8}$$

The fate of the $Si(CH_3)_3$ radical was not specified. It can probably lose one methyl radical quite readily giving $Si(CH_3)_2$ which may then polymerize.

Cyclic trimers and tetramers have been reported by Fritz and Raabe[128]. These were presumed to arise from intermediates formed in the dissociation of tetramethyl silicon by C–H bond rupture. This seems unlikely, but the same precursors could be formed by hydrogen abstraction from the parent compound. The available evidence indicates that the decomposition does not involve long chains, but free radical attack on the alkyl by the reactions

$$CH_3 + Si(CH_3)_4 \rightarrow (CH_3)_3SiCH_2 + CH_4 \tag{9}$$

$$H + Si(CH_3)_4 \rightarrow (CH_3)_3SiCH_2 + H_2 \tag{10}$$

cannot be ruled out.

Sathyamurthy *et al.*[129] found an order closer to $\frac{3}{2}$ and suggested that the decomposition occurs by a chain process involving reactions (9) and

$$(CH_3)_3SiCH_2 = Si + C_2H_5 + 2\,CH_3 \tag{11}$$

However, (11) is approximately 180 kcal.mole^{-1} endothermic. If it occurs it must be a stepwise process. Partial stepwise decomposition may take place but complete dissociation to liberate Si atoms would be inconsistent† with the observation that the carbon content of the solid product under all decomposition conditions is approximately half that of the parent alkyl (low decomposition – grey solid, 25.3 % C; high decomposition – black solid, 28.3 % C; $Si(CH_3)_4$ = 54 % C)[130].

It is unlikely that the observed rate coefficient can be related to k_1. Conner *et al.*[131] report that the decomposition of hexamethyldisilane in a toluene carrier system occurs by Si–Si bond rupture giving trimethylsilane as the major product. The observed activation energy, 58±4 kcal.mole^{-1}, should be equal to $D[(CH_3)_3Si–Si(CH_3)_3]$. A later result[131a], 49±6 kcal.mole^{-1}, was obtained using pressures and flow rates which would be expected to produce a low result. Both flow system results have been criticized by Davidson and Stephenson[131b], who give $k_{12} = 3.16 \times 10^{13} \exp(-67{,}300/RT)$ sec^{-1}.

$$(CH_3)_3SiSi(CH_3)_3 \rightarrow 2\,(CH_3)_3Si \quad (12)$$

They estimate an uncertainty of ±1.0 log units in the pre-exponential factor and ±2.2 kcal.mole^{-1} in the activation energy.

Combining $D[(CH_3)_3Si–Si(CH_3)_3]$ = 58.0 kcal.mole^{-1} with electron impact data[131, 132] gives $D[(CH_3)_3Si–CH_3]$ = 69±5 kcal.mole^{-1}. This is close to the mean Si–CH_3 bond dissociation energy obtained from thermochemistry[19, 133], $\bar{E}$ = 67.6 kcal.mole^{-1}. If $D[(CH_3)_3Si–Si(CH_3)_3]$ = 67.3 kcal.mole^{-1} the corresponding value of $D[(CH_3)_3Si–CH_3]$ = 73.7±4 kcal.mole^{-1}. Neither determination of $D[(CH_3)_3Si–CH_3]$ is completely satisfactory. Until additional results are available a tentative value of 72 kcal.mole^{-1} is suggested as a weighed mean which puts slightly more reliance on the work of Davidson and Stephenson[131b] than on the early work of Connor *et al.*[131] and which disregards the later work of this group[131a].

3.9.2 *Tetraethyl and tetrapropyl silicon*[127]

The reaction schemes that can be proposed for these alkyls are basically analogous to those discussed for the tetramethyl compound. The initiation step should be Si–C bond rupture followed by various reactions of ethyl and propyl radicals, free radical attack on the parent alkyl and various polymerization processes. Significant chain reactions involving the alkyls are apparently homogeneous processes and lead to first-order kinetics. The rate coefficients for the

† Provided the solid is not a mixture of free silicon and polyethylene.

ethyl and propyl compounds respectively are

$$k = 5.0 \times 10^{10} \exp(-50,500/RT) \text{ sec}^{-1} \quad (530\text{–}600 \text{ °C})$$

and

$$k = 6.3 \times 10^{9} \exp(-46,000/RT) \text{ sec}^{-1} \quad (520\text{–}570 \text{ °C})$$

$D[(C_2H_5)_3Si\text{–}C_2H_5]^{131} = 65\pm5$ kcal.mole^{-1}. It seems unlikely that either rate coefficient can be associated with any elementary process.

3.9.3 Polyfluoroalkyl silicon compounds[134, 135]

A summary of the kinetic parameters is given in Table 11. With the exception of 3,3,3-trifluoropropyltrifluorosilane the decompositions are first-order homogeneous processes. The 2,2-difluoro compounds decompose by a four-centre transition state:

$$CHF_2CH_2SiX_3 \longrightarrow \begin{bmatrix} CHF\text{===}CH_2 \\ | \quad\quad | \\ F\text{-----}SiX_3 \end{bmatrix} \longrightarrow CHFCH_2 + SiFX_3$$

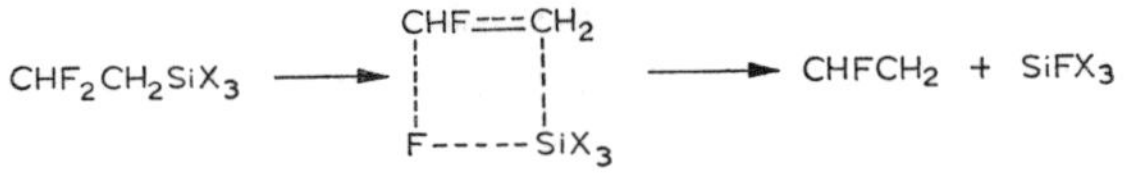

1,1,2,2-tetrafluoroethyltrifluorosilane and the corresponding trimethyl silane undergo α-elimination *via* a three-centre transition state

$$CHF_2CF_2SiX_3 \longrightarrow CHF_2C\overset{F}{\underset{SiX_3}{\diagup\!\!\!\diagdown}}F \longrightarrow SiFX_3 + CHF_2CF$$

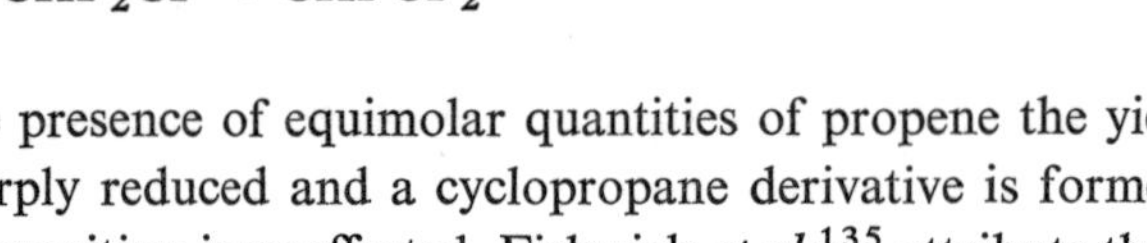

followed by

$$CHF_2CF \rightarrow CHFCF_2$$

In the presence of equimolar quantities of propene the yield of trifluoroethylene is sharply reduced and a cyclopropane derivative is formed. The overall rate of decomposition is unaffected. Fishwick *et al.*[135] attribute the much lower reactivity of 1,1,2,2-tetrafluoroethyltrimethylsilane (*cf.* the trifluorosilane derivative) to electron release by methyl to silicon which makes the silicon less electropositive and less receptive to attack by fluorine.

The decomposition of 3,3,3-trifluoropropyltrifluorosilane yields a complex mixture of products, including SiF_4, C_2H_4, CHF_3, CF_3CHCH_2, CH_2CHSiF_3, CH_3SiF_3 and H_2. The reaction is affected by surface conditions. The rate is reduced by inhibitors indicating a radical chain mechanism.

TABLE 11

KINETIC PARAMETERS FOR THE DECOMPOSITION OF POLYFLUOROALKYLSILICON COMPOUNDS

Compound	*Temperature* (°*C*)	*Order*	*A* (*sec*$^{-1}$)	*E* (*kcal.mole*$^{-1}$)
$CHF_2CH_2SiF_3$	150–220	1	2.0×10^{12}	32.7
$CHF_2CH_2SiCH_3F_2$	190–240	1	1.0×10^{13}	36.0
$CHF_2CH_2Si(O(CH_2)_3CH_3)_3$ [a]	230–270	1	3.2×10^{8}	30.2
$CHF_2CF_2SiF_3$	130–210	1	1.3×10^{11}	28.7
$CHF_2CF_2Si(CH_3)_3$	310–370	1	4.0×10^{13}	46.9
$CF_3CH_2CH_2SiF_3$	550–640	1.5	2.0×10^{16} [b]	74.0

[a] In solution in MS 550 silicone oil; all other decompositions are in the gas phase.
[b] $l^{\frac{1}{2}}.mole^{-\frac{1}{2}}.sec^{-1}$.

3.9.4 *Hexamethyl disilane*

The pyrolysis of this compound in a toluene carrier system gave $D[(CH_3)_3Si–Si(CH_3)_3] = 58\pm4$ kcal.mole^{-1} [131]. Results in a molecular flow system[131] at 10^{-5}–10^{-3} torr over the temperature range 670–750 °C gave an activation energy of 49 ± 6 kcal.mole^{-1} consistent with the expected fall off in activation energy of a unimolecular process in its pressure dependent region. In the absence of toluene[136] the initial Si–Si decomposition is followed by

$$(CH_3)Si+(CH_3)_3SiSi(CH_3)_3 \rightarrow (CH_3)_3SiH+(CH_3)_3SiSi(CH_3)_2CH_2 \quad (1)$$

$$(CH_3)_3SiSi(CH_3)_2CH_2 \rightarrow (CH_3)_3SiCH_2Si(CH_3)_2 \quad (2)$$

$$(CH_3)_3SiCH_2Si(CH_3)_2+(CH_3)_3SiSi(CH_3)_3 \rightarrow (CH_3)_3SiCH_2Si(CH_3)_2H + (CH_3)_3SiSi(CH_3)_2CH_2 \quad (3)$$

At 600 °C the chain length is approximately 4.

As discussed in section 3.9.1., more recent work[131b] gives $D[(CH_3)_3Si–Si(CH_3)_3] = 67.3$ kcal.mole.$^{-1}$

3.10 TETRAMETHYL GERMANIUM[137]

The thermal decomposition is first-order and the rate coefficient is given by $k = 1.7\times10^{14}\exp(-51{,}000/RT)$ sec^{-1}. The mean metal–carbon bond dissociation energy in this alkyl is 58.0 kcal.mole^{-1}. In view of the normal frequency factor, it might seem reasonable to relate the observed activation energy to

$D[(C_2H_5)_3Ge–C_2H_5]$. However, the product distribution indicates complexities in the mechanism that make this unlikely.

3.11 TETRAMETHYL TIN AND DIMETHYL TIN DICHLORIDE

Waring and Horton[138] found the pyrolysis of tetramethyl tin was approximately first-order. They give $k = 8.3 \times 10^{21} \exp(-82{,}400/\boldsymbol{R}T)$ sec^{-1}. Sathyamurthy *et al.*[129] reinvestigated the decomposition and report $\frac{3}{2}$ order and $E = 75.9$ kcal.mole^{-1}. They suggest the mechanism

$$Sn(CH_3)_4 \rightarrow Sn(CH_3)_3 + CH_3 \quad (1)$$

$$CH_3 + Sn(CH_3)_4 \rightarrow CH_4 + (CH_3)_3SnCH_2 \quad (2)$$

$$(CH_3)_3SnCH_2 \rightarrow CH_3 + (CH_3)_2SnCH_2 \quad (3)$$

$$2\,CH_3 \rightarrow C_2H_6 \quad (4)$$

This predicts the observed order, but requires a value of $E_1 = 121$ kcal.mole^{-1}. Neither investigation is satisfactory.

Recent work[156] using the toluene carrier technique (surface effects negligible) shows that the breaking of the first methyl–tin bond is rate-controlling in a process which produces four methyl groups for each molecule reacting. Above the high-pressure limit

$$k_1 = 4.7 \times 10^{15} \exp(-64{,}500/\boldsymbol{R}T)\ \text{sec}^{-1}$$

The pyrolysis of dimethyltindichloride has been studied in a toluene carrier flow system[139]. The decomposition is first-order and homogeneous. The proposed mechanism is

$$(CH_3)_2SnCl \rightarrow CH_3SnCl_2 + CH_3 \quad (5)$$

$$CH_3SnCl_2 \rightarrow CH_3 + SnCl_2 \quad (6)$$

followed by the usual reactions of methyl radicals in the presence of a large excess of toluene. Over the temperature range 554–688 °C, $k = 3.3 \times 10^{13} \exp(-56{,}100/\boldsymbol{R}T)$ sec^{-1}. The observed activation energy should be related to D_5. However, the rate of decomposition was slightly dependent on total pressure so that $D_5 \geqslant 56.1$ kcal.mole^{-1}. This gives [19, 140] $D_6 \leqslant 34.0$ kcal.mole^{-1}, a value consistent with the proposed mechanism where it was assumed that $k_6 \gg k_5$.

3.12 LEAD ALKYLS

3.12.1 Tetramethyl lead

Eltenton[141] studied the thermal decomposition of a very dilute stream of tetramethyl lead vapour in He (total pressure = 0.4 torr) in a fast flow system (contact time 0.1–0.001 sec) over the temperature range 400–700 °C. The decomposition was essentially complete at 600 °C. A small portion of the effluent from the reaction zone passed directly into the ionization chamber of a mass spectrometer. The reaction was followed by observing the methyl radical concentration. The rate-controlling step observed under these conditions is probably the loss of the first CH_3 group by the reaction

$$Pb(CH_3)_4 \rightarrow Pb(CH_3)_3 + CH_3 \quad (1)$$

The apparent first-order rate coefficient is $1.5 \times 10^{10} \exp(-28{,}200/\boldsymbol{R}T)$ sec^{-1}. This expression has undoubtedly been obtained for a pressure-dependent region. If, as an extreme case, it is assumed that the unimolecular process occurred in the second-order region and if approximately one half of the classical degree of vibrational freedom are active, an upper limit of $k_1 = 1.5 \times 10^{15} \exp(-46{,}000/\boldsymbol{R}T)$ sec^{-1} is obtained. The mean Pb–CH_3 bond dissociation energy in tetramethyl lead[19, 142] is 37.6 kcal.mole^{-1}. D_1 should therefore be about 40 kcal.mole^{-1}.

Recent work[155] using the toluene carrier technique (surface effects negligible) shows that release of the first methyl group is the rate-controlling step. Approximately four methyl radicals are released for each molecule undergoing reaction but the final step, the decomposition of $PbCH_3$, may not occur by simple bond fission. Above the high-pressure limit

$$k_1 = 3.0 \times 10^{14} \exp(-48{,}600/\boldsymbol{R}T) \text{ sec}^{-1}$$

3.12.2 Tetraethyl lead

The rate of decomposition of liquid tetraethyl lead and of solutions of tetraethyl lead in inert solvents, is independent of surface-to-volume ratio but is catalyzed by the finely divided lead formed during the reaction[143–146]. The proposed mechanism is

$$Pb(C_2H_5)_4 \rightarrow Pb(C_2H_5)_3 + C_2H_5 \quad (1)$$

$$Pb(C_2H_5)_3 + Pb(C_2H_5)_4 \rightarrow Pb_2(C_2H_5)_6 + C_2H_5 \quad (2)$$

$$Pb(C_2H_5)_6 \rightarrow Pb(C_2H_5)_4 + Pb(C_2H_5)_2 \quad (3)$$

$$Pb(C_2H_5)_2 = Pb + 2\,C_2H_5 \quad (4)$$

followed by the usual recombination and disproportionation reactions of ethyl radicals. Subsequent discussion will be based on this mechanism, but until it is shown that the processes leading to the very high ethane: butane and ethylene: butane ratios, 19 and 15 respectively, observed in the decomposition of hexaethyldiplumbane do not effect the rate of decomposition, the rate equations obtained and all other quantitative conclusions must be considered tentative.

The autocatalytic effect of lead on the rate of reactions (3) and (4) was demonstrated[146] by comparing the pyrolysis of mixtures of tetramethyl lead and hexaethyldiplumbane and of hexaethyldiplumbane and diethyl lead in the presence and absence of lead formed by reaction (4).

Up to 30 % conversion the rate of decomposition determined on the basis of the percent metallic lead precipitated is zero-order. The rate coefficient for the decomposition in solution is $2.43 \times 10^{12} \exp(-35{,}200/RT)$ mole.sec^{-1}. If reaction (2) is fast compared with reaction (1), $k_1 = 1.22 \times 10^{12} \exp(-35{,}200/RT)$ mole.sec^{-1}. The rate of decomposition of hexaethyldiplumbane was also studied. Based on spectroscopic observation of the concentration of the reactants, $k_3 = 5.89 \times 10^{10} \exp(-28{,}500/RT)$ mole.sec^{-1}. On the basis of formation of metallic lead, the rate coefficient was $2.44 \times 10^{10} \exp(-28{,}000/RT)$ mole.sec^{-1}. Therefore, under the conditions used $k_3 \gg k_1$ and reaction (3) is followed rapidly by reaction (4).

A detailed investigation of the gas-phase decomposition of tetraethyl lead has been carried out in a static system over the temperature range 233–267 °C[147]. The overall rate of decomposition determined in a variety of ways is in agreement with the earlier manometric determinations of Leermakers[148]. Over the pressure range 2–15 torr the decomposition is first order and essentially homogeneous up to at least 80 % conversion. However, at high conversions secondary reactions of ethylene and various free radicals complicate kinetic analysis, so Pratt and Purnell[147] concentrated on the initial stages of reaction. The gaseous products and their relative initial rates of formation at 11 torr and 252 °C are n-C_4H_{10} (7.3), C_2H_4(9.8), C_2H_6(3.2) and H_2(3.3). The rate coefficient calculated on the basis of these products was virtually identical with that calculated from the rate of removal of tetraethyl lead, indicating no other hydrocarbon products were formed and that each alkyl molecule undergoing reaction eventually produced metallic lead. The order and the activation energy (kcal.mole^{-1}) for formation of the products are: n-C_4H_{10} (0.8, 37), C_2H_4(1.2, 44), C_2H_6(1.0, 29), $\frac{1}{2}(C_2H_4-C_2H_6) = H_2(1.5, 56)$. The rate coefficient for the overall decomposition of tetraethyl lead was $1.6 \times 10^{12} \exp(-37{,}000/RT)$ sec^{-1}. A mechanism consistent with these results is

$$Pb(C_2H_5)_4 \rightarrow Pb(C_2H_5)_3 + C_2H_5 \quad (1)$$

$$C_2H_5 + Pb(C_2H_5)_4 \rightarrow (C_2H_5)_3PbC_2H_4 + C_2H_6 \quad (2)$$

$$C_2H_5 + Pb(C_2H_5)_n[0 \leqslant n \leqslant 3] \rightarrow C_2H_4 + (C_2H_5)_nPbH \quad (3)$$

$$(C_2H_5)_nPbH \rightarrow Pb(C_2H_5)_n + H \quad (4)$$

$$H + Pb(C_2H_5)_n \rightarrow H_2 + (C_2H_5)_3PbC_2H_2 \quad (5)$$

$$(C_2H_5)_3PbC_2H_2 = Pb + 2\,C_2H_5 + C_2H_4 \quad (6)$$

$$Pb(C_2H_5)_3 = Pb + 3\,C_2H_5 \quad (7)$$

$$2\,C_2H_5 \rightarrow C_4H_{10} \quad (8)$$

$$2\,C_2H_5 \rightarrow C_2H_6 + C_2H_4 \quad (9)$$

At 252 °C based on $k_9/k_8 = 0.15$ reaction (9) accounts for only 34 % of the ethane and 11 % of the ethylene. Reactions (6) and (7) are required to explain the concordance of results based on gas analysis and with those based on tetramethyl lead analysis. All observed orders and activation energies are consistent with this mechanism. If reaction (1) is the rate-controlling step in the initiation, the rate of this reaction can be calculated from

$$0.65\,\frac{\partial[C_4H_{10}]}{\partial t} - 0.25\,\frac{\partial[C_2H_4 + C_2H_6]}{\partial t}$$

From this expression $k_1 = 3.5 \times 10^{11} \exp(-37{,}000/\boldsymbol{R}T)$ sec^{-1}. Hence, $k_{\text{overall}}/k_1 = 4.6$ at all temperatures. With[142] $\Delta H_f[Pb(C_2H_5)_4, g] = 25.8$ kcal.mole^{-1}, the mean bond dissociation energy in tetraethyl lead is 31.0 kcal.mole^{-1}. Association of the observed activation energy with $D[(C_2H_5)_3Pb–C_2H_5]$ seems reasonable but the pre-exponential factor appears rather low. (The value originally reported by Pratt and Purnell[147] was 4×10^{12} sec^{-1}. The present value was calculated from the reported initial rates of product formation at 252.2 °C and $E = 37.0$ kcal.mole^{-1}).

3.13 PHOSPHORUS, ARSENIC, ANTIMONY AND BISMUTH

3.13.1 Tributyl phosphate[149]

The thermal decomposition of liquid tributyl phosphate at 178–240 °C may be followed by the rate of formation of dibutylphosphoric acid. Below 3 % decomposition the reaction is first-order. At higher acid concentrations acid catalysis is observed. The main reaction seems to be dealkylation

$$(C_4H_9O)_3PO = C_4H_8 + (C_4H_9O)_2POOH$$

but small amounts of butanol, dibutyl ether and tetrabutylpyrophosphate were found indicating some simultaneous dealkoxylation. The rate coefficient based on dibutylphosphoric acid formation is $k = 1 \times 10^{12} \exp(-40{,}000/\boldsymbol{R}T)$ sec^{-1}.

3.13.2 Trimethyl arsenic[150]

The pyrolysis in a static system at 50–200 torr from 410–450 °C is first order based on pressure measurements of $t_{\frac{1}{4}}$ and $t_{\frac{1}{2}}$), but at 450 °C the apparent rate coefficient increases 24 % over this pressure range. A partial mechanism may be written

$$As(CH_3)_3 \rightarrow As(CH_3)_2 + CH_3 \quad (1)$$

$$CH_3 + As(CH_3)_3 \rightarrow (CH_3)_2AsCH_2 + CH_4 \quad (2)$$

$$2\,CH_3 \rightarrow C_2H_6 \quad (3)$$

$$CH_3 + C_2H_6 \rightarrow CH_4 + C_2H_5 \quad (4)$$

$$C_2H_5 + CH_3 \rightarrow C_3H_8 \quad (5)$$

$$C_2H_5 \rightarrow C_2H_4 + H \quad (6)$$

$$H + As(CH_3)_3 \rightarrow H_2 + (CH_3)_2AsCH_2 \quad (7)$$

The volatile reaction products contain about 90 % CH_4 and account for about 70 % of the carbon expected if each alkyl molecule undergoes complete dissociation. $As(CH_3)_2$ and $(CH_3)_2AsCH_2$ must both undergo further decomposition. The extent and nature of this is uncertain. $As(CH_3)_2$ may give CH_3 and $AsCH_3$ (this could polymerize to $(AsCH_3)_n$) or $As + 2\,CH_3$. As long as neither process is rate-controlling it will not affect the overall rate of decomposition. However, the related decomposition of $(CH_3)_2AsCH_2$ must initiate a chain process although if its rate of decomposition is similar to that of the parent alkyl the chain length will be very short. The rate coefficient for the overall process is $k = 5.9 \times 10^{12}$ $\exp(-54{,}600/RT)$ sec^{-1}. The observed activation energy was assumed to be a measure of $D[(CH_3)_2As\text{–}CH_3]$ but this interpretation is uncertain.

Over 764–858°K the pyrolysis in the presence of toluene carrier (surface effects negligible) shows little pressure dependence when $P \geqslant 13$ torr[154]. At 21–25 torr

$$k_1 = 6.6 \times 10^{15} \exp(-62{,}800/RT)\text{sec}^{-1}$$

3.13.3 Perfluorotrimethyl arsenic[150]

The pyrolysis was studied using pressure of 50–200 torr over 350–410 °C in silicon and platinum vessels. The reaction in the platinum vessel was complex and will not be discussed. A probable mechanism for that in the silicon vessel is

$$As(CF_3)_3 \rightarrow As(CF_3)_2 + CF_3 \quad (1)$$

$$CF_3 + As(CF_3)_3 \rightarrow As(CF_3)_2 + C_2F_6 \quad (1a)$$

$$2\,CF_3 \rightarrow C_2F_6 \quad (2)$$

$$CF_3+SiO_2 = CO_2+SiF_3 \quad (3)$$

$$SiF_3+C_2F_6 \rightarrow SiF_4+C_2F_5 \quad (4)$$

$$C_2F_5+CF_3 \rightarrow C_3F_8 \quad (5)$$

$$2\,C_2F_5 \rightarrow C_4F_{10} \quad (6)$$

$As(CF_3)_2$ probably undergoes further decomposition but the extent of this is uncertain. However, from the distribution of the main products at 370 °C [C_2F_6 (61.4 %), SiF_4+CO_2 (29.4 %), $C_3F_8+C_4F_{10}$ (7.0 %)] and the observed pressure increase of somewhat over 50 % it appears likely that eventually quantitative decomposition occurs yielding primarily elementary arsenic. Reactions (3) and (4) are respectively at least 80 kcal.mole^{-1} and 15 kcal.mole^{-1} exothermic and therefore seem quite likely. Within experimental error they are consistent with the observed $(CO_2+SiF_4)/(C_3F_8+C_4F_{10})$ ratio. The overall rate coefficient is $2.6\times10^{15}\exp(-57{,}400/RT)$ sec^{-1}. The significance of the observed activation energy will be determined by the relative importance of reaction (1a) and the ensuing chain reaction that could be set up by the reaction

$$As(CF_3)_2 \rightarrow AsCF_3+CF_3 \quad (7)$$

With the available data no estimate of the importance of these reactions can be made.

Very limited data[150] on the decomposition of perfluorotriethyl arsenic at 280 °C indicates a mechanism similar to that for perfluorotrimethyl arsenic. If the reaction is first order and has the same pre-exponential factor, 2.6×10^{15} sec^{-1}, the overall activation energy is 48 kcal.mole^{-1}.

3.13.4 Trimethyl antimony[139]

The pyrolysis was studied in a toluene carrier flow system over the temperature range 475–603 °C. Most runs were carried out at 16–17 torr with a contact time of 1–2 sec. The ratio % decomposition (gas analysis)/% decomposition (antimony recovered from reaction zone) varied from 0.91 at 475 °C to 0.75 at 603 °C. Apparent first-order rate coefficients based on both metal and gas analysis increased with decreasing alkyl concentration ($\log k/\log[Sb(CH_3)_3] = 0.28$ at all temperatures). Corrected for this effect, $k_{24\,torr}/k_{6\,torr} = 1.3$, indicating a small unimolecular pressure effect.

A possible mechanism is

$$Sb(CH_3)_3 \rightarrow Sb(CH_3)_2+CH_3 \quad (1)$$

References pp. 254–257

$$Sb(CH_3)_2 \rightarrow Sb(CH_3)+CH_3 \quad (2)$$

$$SbCH_3 \rightarrow Sb+CH_3 \quad (3)$$

$$n\,(SbCH_3) = (SbCH_3)_n \quad (4)$$

$$2\,Sb(CH_3)_2 \rightarrow Sb(CH_3)_3+SbCH_3 \quad (5)$$

followed by the usual reaction of methyl radicals in the presence of a large excess of toluene. The relative extent of reactions (3) and (4) is unimportant as both processes deposit antimony in a region of the reaction system where it will be recovered for analysis. In the limit of zero alkyl concentration the observed rate coefficient should be k_1. Accurate extrapolation to zero is not possible with the data given, but the nearly constant log $k$log[(Sb(CH$_3$)$_3$] slope at various temperatures coupled with the fact that on the basis of rough extrapolation k at zero concentration is only about 25–30 % greater than k at the lowest alkyl concentration used (2×10^{-9} mole.ml^{-1}) makes it reasonable to associate the activation energy at 7.6×10^{-9} mole.ml^{-1}, 57 kcal.mole^{-1}, with $D[CH_3)_2Sb–CH_3]$. Applying an approximate correction to the A factor based on the zero concentration extrapolation gives $k_1 = 3.1\times10^{15}\exp(-57{,}000/RT)$ sec^{-1}.

3.13.5 *Trimethyl bismuth*[139]

The thermal decomposition in a toluene carrier flow system above 10 torr over the temperature range 367–409 °C is a first-order homogeneous process independent of total pressure. The experimental results are consistent with the simple consecutive release of three methyl radicals with the release of the first as the rate-controlling step. The observed rate coefficient is $k = 1.05\times10^{14}\exp(-44{,}030/RT)$ sec^{-1} with a statistical uncertainty in the action energy of only ±25 cal.mole^{-1}. From the heat of formation of trimethyl bismuth[151] the mean Bi–CH$_3$ bond dissociation energy may be calculated as 35.4 kcal.mole^{-1}. Associating the observed activation energy with $D[(CH_3)_2Bi–CH_3]$, $D[CH_3Bi–CH_3]+D(Bi–CH_3) = 62$ kcal.mole^{-1}. This is consistent with the proposed mechanism if $D(CH_3Bi–CH_3) \approx 2\,D(Bi–CH_3)$.

3.14 PERIODIC FUNCTION IN THE DECOMPOSITION OF METHYL METALLIC ALKYLS

The mean bond dissociation energies ($\bar{E}$) given in Table 12 are based on thermochemical data at 25 °C[19]. Unless previously discussed, the heat of formation of the metal alkyl used is that given by Long[60]. The higher values of $\bar{E}$ and D_2 for dimethyl mercury are obtained when Long's recommended value for the heat of formation is used. The lower value is that previously estimated (see p. 219). Activation energies, corrected to infinite pressure, have been associated directly with bond dissociation energies. The thermochemical data should be corrected to 0 °K and the limits $D \leqslant E \leqslant D+RT$ should be applied to the corrected experimental activation energies. However, the necessary heat capacity and vibrational data are not available. Reasonable estimates would place both corrections at about 1 kcal.mole^{-1}, well within the limits of experimental error.

It would be desirable to be able to use data such as that given in Table 12 to predict D_1 values for other methyl metallic alkyls and to set a pattern for ethyl and possibly higher alkyls. These dissociation energies should be approximately equal to the kinetic activation energy for the first stage of dissociation in a non-chain decomposition or to the activation energy of the initiation step in a chain decomposition.

Mortimer[152] and Long[60] have each attempted to set up a method of calculation relating $\bar{E}$ and D_1. At present insufficient auxiliary information is available to

TABLE 12

AVERAGE BOND ENERGIES, $\bar{E}$, AND BOND DISSOCIATION ENERGIES, D, (kcal.mole^{-1}) IN METHYL METALLIC ALKYLS

	$Al(CH_3)_3$ $\bar{E}$ = 66 D_1 = 65 (est.)	$Si(CH_3)_4$ $\bar{E}$ = 68 D_1 = 72	$P(CH_3)_3$ $\bar{E}$ = 67
$Zn(CH_3)_2$ $\bar{E}$ = 43 D_1 = 51 (54) D_2 = 36 (kinetic) D_2 = 35 (thermo) 522–582 °C [c]	$Ga(CH_3)_3$ $\bar{E}$ = 59 D_1 = 60 D_2 = 39 D_3 = 77 (thermo) 515–560 °C [c]	$Ge(CH_3)_4$ $\bar{E}$ = 64	$As(CH_3)_3$ $\bar{E}$ = 58 D_1=62.8
$Cd(CH_3)_2$ $\bar{E}$ = 34 D_1 = 53 D_2 = 16 (thermo) 471–527 °C [c]	$In(CH_3)_3$ $\bar{E}$ = 37 D_1 = 47 D_2 = 20 (thermo) D_3 = 41 340–390 °C [c]	$(CH_3)_2SnCl_2$‡ $\bar{E}$ = 45 [a] D_1 = 56 D_2 = 34 (thermo) $\bar{E}$ = 55 [b] 570–637 °C [c]	$Sb(CH_3)_3$ $\bar{E}$ = 51 D_1 = 57 496–545 °C [c]
$Hg(CH_3)_2$ $\bar{E}$ = 30–31 D_1 = 58 D_2 = 2–4 487–527 °C [c]	$Tl(CH_3)_3$ $\bar{E}$ – D_1 = 27 232–287 °C [c]	$Pb(CH_3)_4$ $\bar{E}$ = 38 D_1=48.6	$Bi(CH_3)_3$ $\bar{E}$ = 35 D_1 = 44 367–415 °C [c]

(est.) Estimated as outlined in text.
(thermo) Appropriate multiple of $\bar{E}$ minus the sum of the other methyl–carbon bond dissociation energies.
[a] For the two methyl–tin bonds.
[b] Tetramethyl tin.
[c] Approximate temperature range over which the high-pressure rate coefficient for release of the first methyl group rises from 0.1 to 1.0 sec^{-1}.
‡ $D[CH_3-Sn(CH_3)_3]$=64.5

usefully employ either method. The general discussion in both papers is excellent but much of the numerical data used is now out of date (*cf.* Table 12). The only regularity that now appears is a general increase in thermal stability as the period increases. Little regularity exists from left to right. Local patterns do emerge that may be useful. From these the following values may be estimated: $D_1[P(CH_3)_3] = 66–68$ kcal.mole^{-1}, $D_1[Ge(CH_3)_4] = 67–69$ kcal.mole^{-1}, $D_1[As(CH_3)_3]^{\ddagger} = 60–62$ kcal.mole^{-1}. Until more data is available for fully inhibited, pressure independent decompositions a detailed analysis is not warranted.

References

1 A. Burg, *J. Am. Chem. Soc.*, 74 (1952) 3482.
2 Y. C. Fu and G. R. Hill, *J. Am. Chem. Soc.*, 84 (1962) 353.
3 T. P. Felhner and W. S. Koski, *J. Am. Chem. Soc.*, 87 (1965) 409.
4 J. Grotewold, E. A. Lissi and A. E. Villa, *J. Chem. Soc.* (A), (1966) 1038.
5 H. E. Garabedian and S. W. Benson, *J. Am. Chem. Soc.*, 86 (1964) 176.
6 J. Grotewold, E. A. Lissi and A. E. Villa, *J. Chem. Soc.* (A), (1966) 1034.
7 N. B. Slater, *Theory of Unimolecular Reactions*, Methuen, London, 1959, p. 186.
8 W. Strohmeier and D. v. Hobe, *Chem. Ber.*, 94 (1961) 761.
9 W. Strohmeier and K. Geriach, *Chem. Ber.*, 94 (1961) 398.
10 B. Cantone, F. Grasso and S. Pignataro, *J. Chem. Phys.*, 44 (1966) 3115.
11 R. E. Winters and R. W. Kiser, *Inorg. Chem.*, 3 (1965) 699.
12 R. E. Winters and R. W. Kiser, *Inorg. Chem.*, 4 (1965) 157.
13 R. E. Winters and R. W. Kiser, *J. Chem. Phys.*, 44 (1966) 1964.
14 W. S. Watt and S. H. Bauer, *J. Chem. Phys.*, 44 (1966) 2206.
15 W. H. Thompson and A. P. Garrett, *J. Chem. Soc.*, (1934) 524.
16 H. Barzynski and D. Hummell, *Z. Physik. Chem. (Frankfurt)*, 38 (1963) 103.
17 A. P. Garrett and H. W. Thompson, *J. Chem. Soc.*, (1934) 1817.
18 A. B. Callear, *Proc. Roy. Soc. (London), Ser. A*, 265 (1961) 71.
19 *Handbook of Chemistry and Physics*, 47th edn., The Chemical Rubber Co., Cleveland, Ohio, 1966–67.
20 F. A. Cotton, A. K. Fischer and G. W. Wilkinson, *J. Am. Chem. Soc.*, 81 (1959) 800.
21 R. K. Chan and R. McIntosh, *Can. J. Chem.*, 40 (1962) 845.
21aH. E. Carlton and J. H. Oxley, *Am. Inst. Chem. Engs. J.*, 13 (1967) 86.
22 R. F. Heck, *J. Am. Chem. Soc.*, 85 (1963) 651.
23 R. F. Heck, *J. Am. Chem. Soc.*, 85 (1963) 655.
24 R. F. Heck, *J. Am. Chem. Soc.*, 85 (1963) 657.
25 R. F. Heck, *J. Am. Chem. Soc.*, 87 (1965) 2572.
26 F. Basolo and A. Wojcicki, *J. Am. Chem. Soc.*, 83 (1961) 520.
27 L. R. Kangas, R. F. Heck, P. M. Henry, S. Breitschaft, E. M. Thorsteinson and F. Basolo, *J. Am. Chem. Soc.*, 88 (1966) 2334.
28 R. J. Angelici and F. Basolo, *J. Am. Chem. Soc.*, 84 (1962) 2495.
29 C. H. Bamford and R. Denyer, *Trans. Faraday Soc.*, 62 (1966) 1567.
29aR. J. Angelici, *Organometal. Chem. Rev.*, 3 (1968) 173.
30 H. Gilman, R. G. Jones and L. A. Woods, *J. Org. Chem.*, 17 (1952) 1630.
31 C. E. H. Bawn and R. Johnson, *J. Chem. Soc.*, (1960) 4162.
32 G. Semerano and L. Riccoboni, *Chem. Ber.*, 74 (1941) 1089, 1297.
33 C. E. H. Bawn and F. J. Whitby, *J. Chem. Soc.*, (1960) 3926.

‡ It is interesting to note that the recent experimental value is in good agreement with this estimate.

34 G. M. Whitesides and C. P. Casey, *J. Am. Chem. Soc.*, 88 (1966) 4541.
35 G. Costa and G. DeAlti, *Atti Accad. Nazl. Lincei, Rend., Classe Sci. Fis., Mat. Nat.*, 28 (1960) 845.
36 G. Costa, G. DeAlti, L. Stefani and G. Boscarato, *Ann. Chim. (Rome)*, 52 (1962) 289.
37 F. Glockling, *J. Chem. Soc.*, (1955) 716.
38 G. Costa and A. Camus, *Gazz. Chim. Ital.*, 86 (1956) 77.
39 G. Semerano and L. Riccoboni, *Z. Physik. Chem.*, 189 (1941) 203.
40 G. Costa, A. M. Camus and E. Pauluzzi, *Gazz. Chim. Ital.*, 86 (1956) 997.
41 H. Gillman and L. A. Woods, *J. Am. Chem. Soc.*, 65 (1943) 435.
42 H. Gilman and L. A. Woods, *J. Am. Chem. Soc.*, 70 (1948) 550.
43 C. E. H. Bawn and F. J. Whitby, *Discussions Faraday Soc.*, 2 (1947) 228.
44 C. E. H. Bawn and R. Johnson, *J. Chem. Soc.*, (1960) 3923.
45 H. Hashimoto and T. Nakano, *J. Org. Chem.*, 31 (1966) 891.
46 S. J. W. Price and A. F. Trotman-Dickenson, *Trans. Faraday Soc.*, 53 (1957) 1208.
47 I. Lambert, *J. Chim. Phys.*, 62 (1965) 516.
48 D. C. Topor, *J. Chim. Phys.*, 63 (1966) 347.
49 M. Szwarc, *J. Chem. Phys.*, 16 (1948) 128.
50 S. J. Price, *Can. J. Chem.*, 40 (1962) 1310.
51 S. J. Price, *Ph.D. Thesis*, Edinburgh, 1958, p. 7.
52 L. H. Long and R. G. W. Norish, *Phil. Trans. Roy. Soc. London, Ser. A*, 241 (1949) 587.
53 A. S. Carson, K. Hartley and H. A. Skinner, *Trans. Faraday Soc.*, 45 (1949) 1159.
54 M. Krech and S. J. Price, unpublished results.
55 R. J. Kominar, M. G. Jacko and S. J. Price, *Can. J. Chem.*, 45 (1967) 575.
56 M. Krech and S. J. Price, *Can. J. Chem.*, 41 (1963) 224.
57 M. Krech and S. J. Price, *Can. J. Chem.*, 43 (1965) 1929.
58 S. J. W. Price and A. F. Trotman-Dickenson, *Trans. Faraday Soc.*, 53 (1957) 939.
59 C. M. Laurie and L. H. Long, *Trans. Faraday Soc.*, 53 (1957) 1431.
60 L. H. Long, *Pure Appl. Chem.*, 2 (1961) 61.
61 M. R. Russell and R. B. Bernstein, *J. Chem. Phys.*, 30 (1959) 607.
62 R. E. Weston and S. Seltzer, *J. Chem. Phys.*, 66 (1962) 2192.
63 A. S. Kallend and J. H. Purnell, *Trans. Faraday Soc.*, 60 (1964) 103.
64 R. J. Kominar and S. J. Price, *Can. J. Chem.*, 47 (1969) 991.
65 M. Krech and S. J. Price, *Can. J. Chem.*, 41 (1963) 224.
66 B. G. Gowenlock, R. M. Haynes and J. R. Majer, *Trans. Faraday Soc.*, 58 (1962) 1905.
67 J. Cattanach and L. H. Long, *Trans. Faraday Soc.*, 56 (1960) 1286.
68 J. J. Batten, *Australian J. Appl. Sci.*, 12 (1961) 11.
69 M. F. R. Mulcahy and M. R. Pethard, *Australian J. Chem.*, 16 (1961) 527.
70 A. S. Carson, E. S. Carson and B. Wilmshurst, *Nature*, 170 (1952) 320.
71 L. M. Yeddanapalli, R. Srinivasan and V. J. Paul, *J. Sci. Ind. Res. (India) B*, 13 (1954) 232.
72 C. M. Laurie and L. H. Long, *Trans. Faraday Soc.*, 51 (1955) 665.
73 L. H. Long, *Trans. Faraday Soc.*, 51 (1955) 673.
74 J. P. Cunningham and H. S. Taylor, *J. Chem. Phys.*, 6 (1938) 359.
75 R. Srinivasan, *J. Chem. Phys.*, 28 (1958) 895.
76 K. B. Yerrick and M. E. Russell, *J. Phys. Chem.*, 68 (1964) 3752.
77 R. Ganesan, *J. Sci. Ind. Res. (India) B*, 20 (1961) 228.
78 R. Ganesan, *Z. Physik. Chem. (Frankfurt)*, 31 (1962) 328.
79 A. S. Kallend and J. H. Purnell, *Trans. Faraday Soc.*, 60 (1964) 93.
80 A. F. Trotman-Dickenson, A. F. Birchard and E. W. R. Steacie, *J. Chem. Phys.*, 19 (1951) 163.
81 R. E. Rebbert and P. Ausloos, *J. Am. Chem. Soc.*, 85 (1963) 3086.
82 B. G. Gowenlock, J. C. Polanyi and E. Warhurst, *Proc. Roy. Soc. (London), Ser. A*, 218 (1953) 269.
83 F. P. Lossing and A. W. Tickner, *J. Chem. Phys.*, 20 (1952) 907.
84 R. E. Rebbert and E. W. R. Steacie, *Can. J. Chem.*, 31 (1953) 631.
85 H. G. Oswin, R. Rebbert and E. W. R. Steacie, *Can. J. Chem.*, 33 (1955) 472.

86 P. KEBARLE, *J. Phys. Chem.*, 67 (1963) 351.
87 H. V. CARTER, E. I. CHAPPELL AND E. WARHURST, *J. Chem. Soc.*, (1956) 106.
88 K. J. IVIN AND E. W. R. STEACIE, *Proc. Roy. Soc. (London), Ser. A*, 208 (1951) 25.
89 W. PASTFIELD, *Dissertation Abstr.*, 15 (1955) 1325.
90 J. N. BRADLEY, H. W. MELVILLE AND J. C. ROBB, *Proc. Roy. Soc. (London), Ser. A*, 236 (1956) 318.
91 B. A. THRUSH, *Proc. Roy. Soc. (London), Ser. A*, 243 (1958) 555.
92 L. C. FISCHER AND G. J. MAINS, *J. Phys. Chem.*, 68 (1964) 2522.
93 A. F. TROTMAN-DICKENSON AND G. J. O. VERBEKE, *J. Chem. Soc.*, (1961) 2580.
94 A. G. SHERWOOD AND H. E. GUNNING, *J. Phys. Chem.*, 69 (1965) 2323.
95 H. T. J. CHILTON AND B. G. GOWENLOCK, *Trans. Faraday Soc.*, 50 (1954) 824.
96 B. H. M. BILLINGE AND B. G. GOWENLOCK, *Proc. Chem. Soc.*, (1962) 24.
97 H. T. J. CHILTON AND B. G. GOWENLOCK, *Trans. Faraday Soc.*, 49 (1953) 1451.
98 H. T. J. CHILTON AND B. G. GOWENLOCK, *J. Chem. Soc.*, (1954) 3174.
99 B. H. M. BILLINGE AND B. G. GOWENLOCK, *Trans. Faraday Soc.*, 59 (1963) 690.
100 E. WARHURST, *Trans. Faraday Soc.*, 54 (1958) 1769.
101 C. T. MORTIMER, H. O. PRITCHARD AND H. A. SKINNER, *Trans. Faraday Soc.*, 48 (1952) 220.
102 G. A. RAZUVAEV, O. N. DRUZHKOV, S. F. ZHIL'TSOV AND G. G. PETUKHOV, *Zh. Obshch. Khim.*, 35 (1965) 174.
103 M. COWPERTHWAITE AND E. WARHURST, *J. Chem. Soc.*, (1958) 2429.
104 M. SZWARC, *J. Chem. Phys.*, 16 (1948) 128.
105 G. L. ESTEBAN, J. A. KERR AND A. F. TROTMAN-DICKENSON, *J. Chem. Soc.*, (1963) 3873.
106 C. L. CHERNICK, H. A. SKINNER AND I. WADSO, *Trans. Faraday Soc.*, 52 (1956) 1088.
107 A. S. CARSON, D. R. STRANKS AND B. R. WILMSHURST, *Proc. Roy. Soc. (London), Ser. A*, 244 (1958) 72.
108 I. A. KORSCHUNOV AND A. A. ORLOVA, *Zh. Obshch. Khim.*, 28 (1958) 45.
109 G. J. FONKEN, *Chem. Ind. (London)*, 21 (1961) 716.
110 K. C. BASS AND P. NABABSING, *J. Chem. Soc.*, (1966) 1184.
111 K. YOSHIDA AND S. TSUTSUMI, *Tetrahedron Letters*, (1966) 281.
112 A. S. BUCHANAN AND F. CREUTZBERG, *Australian J. Chem.*, 15 (1962) 744.
113 R. M. VARUSHCHENKO AND G. L. GAL'CHENKO, *Zh. Fiz. Khim.*, 38 (1964) 1474.
114 W. H. JOHNSON, J. V. KILDAY AND E. J. PROSEN, *J. Res. Natl. Bur. St., A*, 65 (1961) 215.
115 G. VERHAEGEN AND J. DROWART, *J. Chem. Phys.*, 37 (1962) 1367.
116 M. G. JACKO AND S. J. W. PRICE, *Can. J. Chem.*, 41 (1963) 1560.
117 L. M. YEDDANAPALLI AND C. C. SHUBERT, *J. Chem. Phys.*, 14 (1946) 1.
117a F. M. ROSSI, P. A. MCCUSKER AND G. F. HENNION, *J. Org. Chem.*, 32 (1967) 450.
118 A. J. OUIMET, *Dissertation Abstr.*, 23 (1962) 1532.
119 C. T. MORTIMER AND P. W. SELLERS, *J. Chem. Soc.*, (1963) 1978.
120 M. G. JACKO AND S. J. W. PRICE, *Can. J. Chem.*, 41 (1963) 1560.
121 W. L. SMITH, *Dissertation Abstr.*, 24 (1964) 3105.
122 M. G. JACKO AND S. J. W. PRICE, *Can. J. Chem.*, 42 (1964) 1198.
123 M. G. JACKO AND S. J. W. PRICE, *Can. J. Chem.*, 43 (1965) 1961.
124 S. J. W. PRICE AND A. F. TROTMAN-DICKENSON, *J. Chem. Soc.*, (1958) 851.
125 W. D. CLARK AND S. J. W. PRICE, unpublished results.
126 D. F. HELM AND E. MACK, *J. Am. Chem. Soc.*, 59 (1937) 60.
127 C. E. WARING, *Trans. Faraday Soc.*, 36 (1940) 1142.
128 G. FRITZ AND R. RAABE, *Z. Anorg. Allgem. Chem.*, 286 (1956) 149.
129 T. V. SATHYAMURTHY, S. SWAMINATHAN AND L. M. YEDDANAPALLI, *J. Indian Chem. Soc.*, 27 (1950) 509.
130 D. A. PETROV, G. T. DANILOVA-DOBRYAKOVA AND V. F. TROKHOVA, *Zh. Abshch. Khim.*, 30 (1960) 235.
131 J. A. CONNOR, G. FINNEY, G. J. LEIGH, R. N. HASZELDINE, P. T. ROBINSON, R. D. SEDGWICK AND R. F. SIMMONS, *Chem. Commun.*, (1966) 178.
131a J. A. CONNOR, R. N. HASZELDINE, G. J. LEIGH AND R. D. SEDGWICK, *J. Chem. Soc. (A)*, (1967) 768.
131b I. M. T. DAVIDSON AND I. L. STEPHENSON, *J. Chem. Soc. (A)*, (1968) 282.

132 I. M. T. Davidson and I. L. Stephenson, *Chem. Commun.*, (1966) 746.
133 V. I. Tel'noi and I. B. Rabinovich, *Zh. fiz. Khim.*, 39 (1965) 2076.
134 R. N. Haszeldine, P. J. Robinson and R. F. Simmons, *J. Chem. Soc.*, (1964) 1890.
135 G. Fishwick, R. N. Haszeldine, C. Parkinson, P. J. Robinson and R. F. Simmons, *Chem. Commun.*, (1965) 382.
136 H. Sakurai, R. Koh, A. Hosomi and M. Kumada, *Bull. Chem. Soc. Japan*, 39 (1966) 2050.
137 R. L. Geddes and E. Mack, *J. Am. Chem. Soc.*, 52 (1930) 4372.
138 C. E. Waring and W. S. Horton, *J. Am. Chem. Soc.*, 67 (1945) 540.
139 S. J. W. Price and A. F. Trotman-Dickenson, *Trans. Faraday Soc.*, 44 (1958) 1630.
140 G. A. Nash, H. A. Skinner and W. F. Stack, *Trans. Faraday Soc.*, 61 (1965) 640.
141 G. C. Eltenton, *J. Chem. Phys.*, 15 (1947) 455.
142 W. D. Good, D. W. Scott, I. L. Lacina and J. P. McCullough, *J. Phys. Chem.*, 63 (1959) 1139.
143 G. A. Razuvaev, N. S. Vyazankin and N. N. Vyshinskii, *Zh. Obshch. Khim.*, 29 (1959) 3662.
144 G. A. Razuvaev, N. S. Vyazankin and N. H. Vyshinskii, *Zh. Obshch. Khim.*, 30 (1960) 967.
145 G. A. Razuvaev, N. S. Vyazankin and D. A. Shchepetkova, *Zh. Obshch. Khim.*, 30 (1960) 2498.
146 G. A. Razuvaev, N. S. Vyazankin and N. N. Vyshinskii, *Zh. Obshch. Khim.*, 30 (1960) 4099.
147 G. L. Pratt and J. H. Purnell, *Trans. Faraday Soc.*, 60 (1964) 519.
148 J. A. Leermakers, *J. Am. Chem. Soc.*, 55 (1933) 4508.
149 C. E. Higgins and W. H. Baldwin, *J. Org. Chem.*, 26 (1961) 846.
150 P. B. Ayscough and H. J. Emeleus, *J. Chem. Soc.*, (1954) 3381.
151 L. H. Long and J. F. Sackman, *Trans. Faraday Soc.*, 50 (1954) 1177.
152 C. T. Mortimer, *J. Chem. Educ.*, 35 (1958) 381.
153 A. N. Dunlop and S. J. W. Price, *Can. J. Chem.*, 48 (1970) 3205.
154 S. J. W. Price and J. P. Richard, *Can. J. Chem.*, 48 (1970) 3209.
155 S. J. W. Price and N. J. Webster, to be published.
156 R. P. Johnson and S. J. W. Price, to be published.

Index

A

absorption spectrum, of ClO_2, 126
—, of CO_2, 54–57
—, of COS, 62
—, of CS_2, 59
—, of HBr, 150
—, of HI, 144, 145
—, of N_2O, 70
—, of NO_2, 88, 90
—, of N_2O_4, 88
—, of SO_2, 115, 116
—, of SO_3, 117
acetic acid, effect on radiolysis of HCN, 175
acetylene, from photolysis of $Hg(C_2H_3)_2$, 228, 229
acetylene dichloride, effect on decomposition of S_2F_{10}, 191
activation energy, see also enthalpy of activation
—, of $BH_3+B_2H_6$, 38
—, of B_2H_6+CO, 198
—, of CD_3+HgMe_2, 225
—, of $Cl+H_2$(HD, HT), 153
—, of Cl_2+H, 154
—, of ClO+ClO, 123, 128
—, of $CO+C_2O$, 48
—, of CO formation in $Fe(CO)_5$ radiolysis, 200
—, of $CO+O_2$, CO_2+O, 54
—, of COS+S, 61
—, of decomposition of AgEt, AgMe, 209
—, — $AlEt_3$, $AlMe_3$, 238
—, — $As(CF_3)_3$, $As(C_2F_5)_3$, $AsMe_3$, 250, 251
—, — B_2H_6, 38
—, — B_4H_{10}, $B_{10}H_{14}$, 40, 41
—, — BH_3CO, 198
—, — $BiMe_3$, 252
—, — BMe_3, 236
—, — $(BuO)_3PO$, 249
—, — $CdMe_2$, 215, 216
—, — Cl_2O, 122
—, — Cl_2O_7, 130
—, — cobalt carbonyls, 206
—, — CO, 51
—, — CO_2, 53, 54
—, — C_3O_2, 48
—, — $COCl_2$, 177
—, — COS, 61
—, — CS_2, 58
—, — fluoroalkylsilanes, 245
—, — F_2O, F_2O_2, 119, 120
—, — GeH_4, 34, 35
—, — $GeMe_4$, 245
—, — HCl, 154
—, — $HClO_4$, 131
—, — $HgBu_2$, $HgPr_2$, 230–232
—, — $Hg(C_2H_3)_2$, 228, 232
—, — $HgEt_2$, 225, 226, 232
—, — $HgMe_2$, 218, 219, 222, 224, 232
—, — $HgPh_2$(PhHgX), 234
—, — HNO_3, 101–103
—, — $Mn(CO)_5X$, 208
—, — NCl_3, 188
—, — NF_2, 183, 184
—, — N_2H_4, 20–22, 24
—, — $Ni(CO)_4$, 201
—, — N_2O, 67, 70
—, — N_2O, catalysed by Br_2, Cl_2, I_2, 129
—, — N_2O_4, 87
—, — N_2O_5, 95, 98, 99
—, — O_3, 106
—, — $PbEt_4$, $PbMe_4$, Pb_2Et_6, 247–249
—, — $SbMe_3$, 252
—, — S_2F_{10}, 190
—, — SiH_4, 28, 30
—, — Si_2H_6, Si_3H_8, 32, 33
—, — $SiMe_4$, 242
—, — Si_2Me_6, 243, 245
—, — $SnMe_4(Me_2SnCl_2)$, 246
—, — SO_3, 111, 112
—, — ZnMe, 210, 214
—, of $D_2+I(I_2)$, 148
—, of dissociation of HF, 155
—, — N_2F_4, 180, 182
—, — NO, 76
—, —$ZnMe_2$, 210, 213, 214
—, of Et+Et, 225
—, of $F+H_2$, H+HF, 155
—, of $H+Br_2$(HBr), 151
—, of $HCl^-_{(c)}+Cl_2$(HCl), 169
—, of H+HCl, 153
—, of $H+HI(I_2)$, H_2+I, $H_2+2I(I_2)$, HI+HI, $I+H_2$(HI), 146, 148, 149
—, of $H+HN_3$, 16
—, of $H+N_2H_4$, 25, 26
—, of $H+SF_6$, 189
—, of isomerisation of BR_3, 236
—, of $Me+AlMe_3$, 238
—, of $Me+C_2H_6$, 221
—, of $Me+HgMe_2$, 220
—, of $Me+Ph(PhCH_2)$, 210, 211

B

D

F

G

H

I

K

L

O

P

Q

R

sulphuric acid, effect on HCN radiolysis, 175
—, treatment of vessel for H_2O_2 pyrolysis, 7, 9
sulphur monoxide, in photolysis of SO_2, SO_3, 166, 117
—, in pyrolysis of COS, 62
—, — SO_2, 111–114
sulphur pentafluoride radical, in S_2F_{10} decomposition, 190
sulphur trioxide, decomposition of, 117
—, in decomposition of SO_2, 111–116
surface, effect on Cl_2+CO, 177
—, effect on decomposition of AlR_3, 238
—, — BMe_3, 235
—, — $CdMe_2$, 216
—, — $CF_3C_2H_4SiCF_3$, 244
—, — ClO_2, 125
—, — $COBr_2$, 178
—, — HgR_2, 222, 223, 226, 228
—, — $InMe_3$, 241
—, — $Ni(CO)_4$, 201
—, — S_2F_{10}, 190
—, — $ZnMe_2$, 212, 213
—, effect on photolysis of Cl_2/NCl_3, 186
—, — CO, 50
—, — CO_2, 55, 56
—, effect on pyrolysis of $As(CF_3)_3$, $AsMe_3$, 250, 251
—, — boron hydrides, 36, 40
—, — COS, 61
—, — F_2O, 118
—, — GeH_4, Ge_2H_6, 34–36
—, — HN_3, 26
—, — HNO_3, 101, 102
—, — H_2O_2, 6, 7, 9
—, — H_2S, 11
—, — Me_2SnCl_2, 246
—, — N_2H_4, 18, 23, 24
—, — NO, 76
—, — N_2O, 66–68
—, — N_2O_5, 95
—, — O_3, 104, 105
—, — $PbEt_4$, 247
—, — silanes, 28, 31, 33
—, — SO_3, 117

T

tetraborane, pyrolysis of, 40, 41
tetraethyllead, decomposition of, 247–249
1, 1, 2, 2-tetrafluoroethylsilanes, decomposition of, 244, 245
tetramethyllead, pyrolysis of, 247
tetramethyl tin, pyrolysis of, 246
tetrasilane, from pyrolysis of Si_2H_6 and Si_3H_8, 33
thionyl chloride, flash photolysis of, 113
third body, efficiency of in H+OH, 6
—, — I+I, 149
—, — S+S, 60
threshold energy, of $e+Br_2$, Cl_2, 168
time-of-flight mass spectrometer, and pyrolysis of N_2F_4, 182
—, — N_2H_4, 22
toluene, decomposition of cobalt carbonyls in, 203, 205, 207
—, $Ni(CO)_4$+CO, PPh_3 in, 206, 207
—, pyrolysis of $AsMe_3$ in, 250
—, — $BiMe_3$ in, 252
—, — $CdMe_2$ in, 215
—, — $GaMe_3$, $InMe_3$, $TlMe_3$ in, 239–241
—, — $Hg(C_2H_3)_2$ in, 227, 228
—, — $HgEt_2$, $HgMe_2$ in, 217–219, 223–225
—, — $HgPh_2$, PhHgX in, 233
—, — N_2H_4 in, 23
—, — $PbMe_4$ in, 247
—, — $SbMe_3$ in, 251
—, — Si_2Me_6 in, 243, 245
—, — $SnMe_4$, Me_2SnCl_2 in, 246
—, — $ZnMe_2$ in, 209–211, 214
transition state, for decomposition of $Co_2(CO)_6C_2H_2$, 206
—, — fluoroalkylsilanes, 244
—, — MeCH=CHCOCo$(CO)_4$, 204
—, for H+HCl, 153
—, for HI+HI, 150
—, for Me+$ZnMe_2$, 212
—, for NO_2+NO_2, 86
tributylphosphate, decomposition of, 249
trichlorotrifluoroethane, decomposition of N_2O_5 in, 99
triethylaluminium, decomposition of, 238
triethylamine, effect on decomposition of BH_3CO, 198
—, effect on photolysis of carbonyls, 199
trifluoroethylene, from decomposition of $CHF_2CF_2SiX_3$, 244
trifluoromethyl radicals, in decomposition of $As(CF_3)_3$, 250, 251
3,3,3-trifluoropropyltrifluorosilane, pyrolysis of, 244, 245
trimethylaluminium, decomposition of, 237, 238
trimethylamine, effect on BH_3CO decomposition, 198
trimethylantimony, pyrolysis of, 251, 252
trimethylarsenic, pyrolysis of, 250
trimethylbismuth, pyrolysis of, 252
trimethylboron, decomposition of, 235, 236
trimethylgallium(indium, thallium), decomposition of, 239–241
trimethylolpropane phosphate, effect on decomposition of cobalt carbonyls, 202, 206